The End

Joel Wainwright is a professor of geography at Ohio State and the author of *Geopiracy* and *Decolonizing Development*, which won the Blaut Award. He is also coauthor, with Geoff Mann, of *Climate Leviathan: A Political Theory of Our Planetary Future,* which won the Sussex Prize.

The End

Marx, Darwin, and the Natural History of the Climate Crisis

Joel Wainwright

VERSO

London • New York

First published by Verso 2025
© Joel Wainwright 2025

All rights reserved

The manufacturer's authorized representative in the EU for product safety (GPSR) is LOGOS EUROPE, 9 rue Nicolas Poussin, 17000, La Rochelle, France
contact@logoseurope.eu

The moral rights of the author have been asserted

1 3 5 7 9 10 8 6 4 2

Verso
UK: 6 Meard Street, London W1F 0EG
US: 207 East 32nd Street, New York, NY 10016
versobooks.com

Verso is the imprint of New Left Books

ISBN-13: 978-1-80429-941-8
ISBN-13: 978-1-80429-943-2 (US EBK)
ISBN-13: 978-1-80429-942-5 (UK EBK)

British Library Cataloguing in Publication Data
A catalogue record for this book is available from the British Library

Library of Congress Cataloging-in-Publication Data
A catalog record for this book is available from the Library of Congress

Typeset in Minion by Hewer Text UK Ltd, Edinburgh
Printed and bound by CPI Group (UK) Ltd, Croydon, CR0 4YY

In memory of L. D. Wainwright (1944–2024)

Steelmaker, historian, and father

Contents

Part I
THE EMERGENCE OF MARXIAN NATURAL HISTORY

Preface to Part I — 3
1. Marx Before Darwin — 15
2. Darwin and the Defeat of Teleology — 43
3. Marx after Darwin — 84

Part II
READING *CAPITAL* AS NATURAL HISTORY

Preface to Part II — 111
4. Labor, Nature, and Technology — 113
5. Population, Value, and Commodity Fetishism — 147

Part III
ELABORATIONS OF MARXIAN NATURAL HISTORY

Preface to Part III — 193
6. A Natural History of Capitalism — 195

| 7. Philosophical Implications of Marxian Natural History | 246 |
| 8. Prospect of an End | 285 |

A Marx-Darwin Timeline   318
Index   321

PART I
The Emergence of Marxian Natural History

Preface to Part I

We should have seen it coming, but some corners of history are hard to peer around. Like a seawall swept aside by a storm surge, planetary climate change is washing away some of our firmest illusions. The illusion that the natural world is an inert stage upon which the human drama unfolds, for instance: Gone for good. An ancient truth has reasserted itself: Human history is inseparable from nature. Like our past, the future is natural-historical.

This book is written for those who seek neither the comfort of a simple explanation for our planetary crisis nor ten easy steps to solve it. It offers, instead, a theoretical approach to the natural history of our present condition. I start from the proposition that no meaningful account of our crisis can ignore the fact that we, *Homo sapiens*, evolved—like every other species—within ecosystems and environments undergoing constant change. Seen in this light, the fact that humans are presently "adapting to climate change" (as is often said) is nothing new. Understood in Charles Darwin's terms, adaptation is ceaseless and inherent to all life. One of Darwin's great contributions was to help us understand that no animal, no plant, no species—no life—is a fixed thing, but only the capricious result of complex historical transformations. So, of course, humans are adapting. Such is life. What is truly novel today in our world is the rapid pace and scale of these changes. To appreciate these magnitudes requires a sense of proportion which is difficult to obtain, for we have no meaningful reference point for this moment in the natural history of our species.

In the preface to his most important work, *Capital*, volume I, Marx writes that he takes "the development of society's economic formation as part of natural history."[1] *The End* could be read as an extended commentary upon this strange statement of Marx's. It is strange because no one reads *Capital* as a work of natural history. There are neither tortoises nor finches. Nonetheless, I will take Marx at his word and credit him with establishing a means to locate capitalism within the earth's natural history, which is an achievement—I think there are grounds to call it a scientific discovery—that is essential for humanity today. To bear this out, I will elaborate what we may call the Marxian natural history standpoint in three steps. In part I, I explain what this standpoint is and how it emerged from Marx and Darwin. In part II, I show how this analytic operates in *Capital*. In Part III, I put this approach to work to answer fundamental questions facing humanity today: How did the capitalist form of society come into existence? What consequences might a Marxian natural-historical approach have for our understanding of the philosophy of history, human nature, and human consciousness? Finally, how might we address the planetary crisis politically?

I am hardly alone in posing such questions or attempting to answer them with Marx. We are presently experiencing a widespread renaissance in ecological Marxist thought. Important works have been written by, for instance, Paul Burkett, John Bellamy Foster, and Kōhei Saitō—the so-called *Monthly Review* school—and other ecological Marxists.[2] I have learned a great deal from this literature, and I regard this book as a contribution to it. Yet, by my reading, previous works of ecological Marxism have not given due emphasis to natural history. Moreover, the critique of

---

1 Karl Marx, Preface to the 1867 edition of *Capital*. In *Capital*, volume I, trans. Paul Reitter (Princeton: Princeton University Press, 2024), 8. Unless noted, all further references to *Capital*, volume I, are to this edition, Reitter's translation of the second German edition of 1872. One of the many virtues of this edition—the first English translation since the completion of the immense labor of the MEGA2 section on the *Capital* manuscripts—is to allow English readers to read, for the first time, the final German edition which Marx (as opposed to Engels) worked on. I thank Reitter for his steady support of this project.

2 "So-called" because the "school" is heterogenous and overlaps significantly with other approaches within ecological Marxism.

Marx's anthropocentrism needs to be confronted more effectively.³ To do so, we should follow Marx's lead and treat Darwin's approach to evolution as a means to study the economic forms of human society, beginning with capitalism. In this way, ecological Marxism can address the critique of anthropocentrism by taking as its object of analysis not humans per se, nor capital, but *evolving socio-economic formations*.

The Argument

My aim is to provide my readers with a natural-historical perspective on our planetary crisis. I use "planetary crisis" for the combined crisis we face today, which presents itself in multiple dimensions, including rapid global heating, the sixth great extinction, yawning economic inequality, and threat of world war. If we accept that the roots of this crisis lie in the capitalist form of society, where does Marx's thought lead us?

Confronting this question brings serious challenges. We are entering a phase of ecological and climate change without precedent in human history, yet the prospects for revolutionary change today are bleak. Facing this impasse, people everywhere are frantically clamoring for a way through the ruins of liberalism. Many, swayed by the ideologues of fossil capital, find the answer in reactionary authoritarianism.⁴ Others

---

3   Some have criticized Marx for his ostensible contribution to the separation of humans from nature. Hence, notwithstanding many erudite attempts to show that Marx was an ecological thinker, many people on the Left still find Marxism to be anthropocentric. The search for a more ecological and relational approach to grasping capitalism's ecological crisis often leads to "new materialism" or "monism": Philosophical approaches that deny any substantial distinction between humans and nature (see, e.g., the varied writings of Teresa Brennan, Donna Haraway, Bruno Latour, and Jason Moore). Although there is a good deal in his historical research that I admire, and my approach shares certain commonalities with the world-ecology framework, I reject the approach promoted by Jason Moore, as well as the (neo-physiocratic) reformulation of Marx's value theory around concepts like "negative value" and "cheap natures": See Moore, *Capitalism in the Web of Life: Ecology and the Accumulation of Capital* (New York: Verso Books, 2015). I settled my accounts with Latour long ago: Joel Wainwright, "Politics of nature: A review of three recent works by Bruno Latour," *Capitalism, Nature, Socialism* 16, no. 1 (2005): 115–122.

4   In *Climate Leviathan* (2018), Geoff Mann and I characterize this political formation as "Climate Behemoth"; Malm et al. (2021) use a more alliterative expression: "Fossil fascism." Different name, same monster. I thank Mann and Malm for their encouragement and inspiration.

rebel to stop business as usual: Stop Oil Now, Extinction Rebellion, Fridays for Future, and so on. Few on either side seem to believe that humanity can get out of this trap. We feel stuck; we are stuck. Many on the Left write of exhaustion and hopelessness.[5] Natural scientists place odds on the extinction of our species. The doomsday clock stands at ninety seconds to midnight.[6]

We cannot understand our present condition without a conception of natural history—and of how capitalism and climate change have brought humanity to a turning point within that history. To reset the doomsday clock it will be helpful to revisit Marx with Darwin and assess how we got here.

Consider global heating. Past climate change was a critical factor in the evolution of *Homo sapiens* and our becoming the dominant species on the planet. It follows that we must reconsider climate politics as an element of Earth's natural history. Grounding ourselves in natural history also provides a bracing technique for our psychological challenges: It can only strengthen our grip on our self-understanding as human beings living at a critical time within a distinctive socio-natural form. Such an approach requires drawing up a much broader balance sheet of forces than we are accustomed to. We need to draw up an entirely different ledger.

Marx remains an essential thinker for this task. In his early writings, he called for a natural history of humanity, but his argument remained highly abstract—a critique of the philosopher Hegel. Things changed after 1860 when Marx read Darwin's *Origin of Species*. Marx immediately recognized the significance of this book for his approach and, over the next few years, developed a distinctly powerful conception of the natural history of capitalism. Many Marxists since then have noted Marx's debt to Darwin, but it has been downplayed. An ecological, Darwinist Marxism did not fit with Marxism-Leninism. Marx could be a brutal critic and wrote with acid, even about those thinkers he learned a great deal from. He had his criticisms of Darwin too, particularly of his method. But Marx called Darwin's work "splendid," even "epoch-making."

Among other contributions, then, I hope to clarify how Marx's reading of Darwin influenced his understanding of natural history and

---

5   Hannah Proctor, *Burnout: The Emotional Experience of Political Defeat* (London: Verso Books, 2024).
6   Per *Bulletin of the Atomic Scientists*, thebulletin.org, February 21, 2024.

capitalism. His analysis can help us confront climate change as a political problem by bringing us back to fundamental questions: How are we to grasp what is distinctive about human society vis-à-vis nature? What, if anything, marks the distinctive historical events or processes that have reshaped these relations? Where, in short, do we stand today in the earth's natural history? How could Marx's natural-historical standpoint inform our understanding of the political economy of climate change and the potential for systemic change in response to the crisis? If ecological Marxism is to progress, it must answer such questions clearly.

Marx had, I think, the right approach: To resolve such questions through a natural-historical critique. Of course, no ultimate resolution is possible, because—to draw a term from Antonio Gramsci, one of those thinkers who rediscovered Marxian natural history circa 1932—a rigorous approach to Marxist analysis does not historicize events on the basis of pregiven categories but historizes the categories as well. The resulting method, which Gramsci calls "*absolute* historicism," has only one ultimate guarantor, which is that it is deployed in an ever-evolving world, roiled constantly by struggle.[7] That was one of Darwin's key insights; we should hold on to it.

On Natural History

Not long ago, natural history was the name of a discipline, a means of organizing knowledge. It encompassed a number of fields of science—geology, biology, geography, anthropology, history, and so on—that we now regard as distinct. Prior to their separation, natural history described a discipline concerning the history of nature and the role of

---

7 "It has been forgotten that in the case of a very common expression [i.e., historical materialism,] one should put the accent on the first term [—'historical'—] and not on the second [—'materialism'—] which is of metaphysical origin. The philosophy of praxis is the absolute 'historicism,' the absolute secularisation and earthliness of thought, an absolute humanism of history. It is along this line that one must trace the thread of the new conception of the world" (circa 1932: Antonio Gramsci, Q11§ 27, in *Selections from the Prison Notebooks* [New York: International], 465). One manifestation of our planetary crisis today is the growing recognition of the "earthliness" of all human thought, that is, the immanence of natural-historical processes in human reasoning. Gramsci's pairing of absolute historicism with absolute earthiness aligns with Adorno's insistence—also in 1932—upon a return to natural history: See chapter 7.

nature in history.⁸ Today, the elements of this discipline have been redistributed.⁹ Vestiges of the discipline remain in museums of natural history: Storehouses for dinosaur bones, mineral samples, and fading dioramas portraying early humans. Do such places have something to teach us about capitalism and the climate crisis?¹⁰

Before Darwin, one of the fundamental questions in natural history was the place of the human being: Most Enlightenment thinkers, following the Greeks, categorized the human being as distinct from other beings. The capacious style of thinking prompted by natural history opened the way to inquire into the nature of humans. Questions like "What is the natural history of human beings?" and "How do the interrelations of

---

8   If natural history is taken as a discipline, it is an odd one: Given its object, it is arguably an anti-disciplinary discipline, and an ancient one at that. While the English term "natural history" is a calque of the Latin *historia naturalis*, the tradition goes back much further; its history is too long and complex to recount here. A minimal genealogy would emphasize the pre-Socratic thinkers in Ionia, Aristotle's criticisms thereof, and the texts on *Theologia naturalis* (natural theology) by Augustine of Hippo, a.k.a. Saint Augustine. Natural history subsequently blossomed in the Islamic world as Greek and Latin sources were discussed alongside the Koran and scientific thought from across the Mediterranean (M. Sharif, ed., *A History of Muslim Philosophy*, volume II, book 5, chapter 25, "Natural History" [Lahore, Pakistan, 1963]). It is not correct, therefore, to claim that natural history is a European discipline. Natural history—as discipline and museum—came of age in Europe during the seventeenth and eighteenth centuries as early capitalist empires spread across the world, facilitating natural-historical exploration: See, for example, Susan Parrish, *American Curiosity: Cultures of Natural History in the Colonial British Atlantic World* (Durham: UNC Press, 2012); Anne Mariss, *Johann Reinhold Forster and the Making of Natural History on Cook's Second Voyage, 1772–1775* (New York: Lexington, 2019). By the time of Darwin's *Origin* (1859) and Marx's *Capital* (1867)—the high points of nineteenth-century natural history—the discipline was fracturing: See Daniel Worster, "History as natural history: An essay on theory and method," *Pacific Historical Review* 53, no. 1 (1984): 1–19.

9   Of course, work akin to natural history persists, albeit in distinct disciplines (geology, biology, etc.). Scholars in certain subdisciplines—evolutionary biology, environmental history, and my own, human-environment geography—sometimes claim their field as the rightful heir to the natural history tradition. I will not attempt to adjudicate these disciplinary issues.

10   What would natural history museums become, were they to genuinely address our planetary crisis? Beyond renouncing ties to fossil capital (often, such museums function as greenwashing machines as much as science education spaces), they would reorient along the lines of Marx's standpoint, teaching visitors the natural-historical implications of the capitalist form of society for humans and Earth, while cultivating imagination of post-capitalist, ecological society. For fertile thinking along these lines, see "The Natural History Museum," a project of artist Beka Economopoulos, at thenaturalhistorymuseum.org.

humans relate to human relations with Earth?" were prompted by the development of thought into the history of the earth and the different ways humans lived in different environments.

To be clear, I am not interested in defending a discipline, and I am not trying to redraw disciplinary boundaries. My aim is rather to retrieve and to recharge the tradition of natural-historical thinking to help us gain a stronger grasp on our planetary crisis. The basic value of the concept may be simply stated: In natural history, nature is subsumed into history, and history is subsumed into nature. From the standpoint of natural history, neither category has any absolute meaning apart from the other. Moreover, in principle, natural history knows no telos; it knows no cause, either. To affirm natural history today means to deny teleology, to constantly transpose nature into history and vice versa, and to affirm the possibility for a change in the form of any given structure.[11] Hence, I endorse the definition of Marxism once provided by Kōjin Karatani: "To describe how the relationships between nature and humans as well as among humans themselves transformed/developed throughout history."[12]

To read Marx as a natural historian at this moment of planetary crisis is to invite some pessimistic thoughts. But it is not as if, all hope of emancipation having been lost, all that remains is to paint a *memento mori* with a skull and a worn copy of *Capital*. Genuine pessimists do not write books—there would be no point. While I am deeply concerned about the planetary crisis, I remain hopeful about the prospects for humanity, assuming that we can escape the grip of our present social formation.

---

11 If "teleology" is understood in its etymological sense as the logos (reason) of telos (end), then *The End* is a study of teleology. But it is an antiteleological contribution that rejects arguments *from* ends. Claims that presuppose a specific ending in advance generally lead to truisms or trivialities. For instance, if one claims, "In the end, God will save us," the problems of the world are made trivial (and human agency is made irrelevant). If one claims, "In the end, God will either save us or not," we have a meaningless truism (and ditto). In the Western philosophical tradition before 1859, teleological argumentation was inseparable from one of the proofs of God's existence: Given that the world exhibits order, it must have a creator-designer; ergo, God exists. Many early readers of *Origin* (Marx included) recognized that Darwin had provided a nonteleological explanation for life's order, thereby undermining the teleological argument for God's existence.

12 Kōjin Karatani, *Transcritique: On Kant and Marx* (Boston: MIT Press, 2004), 7; cf., 18–19, 119–20, 127–30, 163–4, and 173. Karatani's reading of Marx and natural history—announced in a series of essays during the 1970s, later contributing to his modes of exchange theory—is a fundamental source of inspiration for this work.

Reactionary and conservative thinkers often criticize Marx and Darwin for rejecting teleology and challenging religious truth. The situation with liberal thinkers is more complex. Liberals celebrate Darwin (albeit abstractly) and dismiss Marx on political grounds, freighting both thinkers with misleading stereotypes:

Darwin = naturalizes the ethic of "survival of the fittest" and social inequality
Marx = economic determinist, loves the state, godfather of political tyranny

Whereas I claim:

Darwin historicizes variation of biological forms (species)
Marx historicizes variation of forms of human society (socio-economic formations)

The liberal stereotypes repress the essential commonality of Marx and Darwin and thereby devalue a critical analytical tool for explaining our world.

There is another reason that some on the Left do not talk about evolution or Darwin: Some of these ideas have come to be linked with a conservative, essentialist conception of race and sex. To avoid possible misunderstandings, let me say this: We must oppose racism, sexism, and biological essentialism. Are our strategies for opposing these enriched by disavowing Darwinism? No. Confronting a capitalist world divided by racism, sexism, and so on requires a perspective that is informed by the science of human evolution. Here is it important to recall that far too many people lack a foundational science education and that evolution in particular has never been accepted by many. There is a good reason that Marx and others on the Left immediately embraced Darwin's theory while the Right reacted against it: The idea had revolutionary implications for humanity's conception of our natural history.[13] Unfortunately, we have not yet cultivated those revolutionary implications.

---

13 David Stack, *The First Darwinian Left: Socialism and Darwinism, 1859–1914* (Cheltenham: New Clarion, 2003).

Preface to Part I

A Strange Idea

Shortly after the Nazi breakthrough in the July 1932 German federal election, with conservative reaction and cancer conspiring against his life, Sigmund Freud wrote some essays, published as *New Introductory Lectures on Psychoanalysis*. The collection ends with a lecture on the formation of one's *Weltanschauung*, or conception of the world. Freud examines the psychological factors that shape how we see the world and how our conception of the world makes us who we are.[14] After elaborating the contributions of his theory to this understanding, he considers two radical perspectives distinct from psychoanalysis—alternative viewpoints, competitors to his own. The first, anarchism, Freud dismisses. The second, he writes, "has to be taken far more seriously": This is "Karl Marx's investigations into the economic structure of society."[15] Freud's discussion of Marx is respectful. He acknowledges that Marx's ideas "have in our days acquired an undeniable authority." He even raises his criticism of Marx cautiously: Certain assertions by Marx, he writes, "have struck me as strange."[16] Marx made millions of assertions in his lifetime. Freud could have opened any one of the hundreds of texts Marx wrote and found an assertion that could be construed as strange. But the specific assertion that Freud uses as his first illustration is: "the development of forms of society is a process of natural history"—*this* is the claim of Marx's that struck Freud as strange.

A strange assertion, indeed! Made even stranger for appearing in the preface to *Capital*, a book that was not read *as* natural history (in Marx's day, nor Freud's). But appearances—as Marx and Freud both teach us—can be deceiving. If we can come to read *Capital* as a book that belongs on the same bookshelf as Darwin's *Origin*, then things about *Capital* become less strange, such as the theory of human overpopulation that constitutes *Capital*'s scientific conclusion (see chapter 5).

One of the major natural historical themes of *Capital* concerns the separation of humans from the land on which they live. While I agree

---

14 Sigmund Freud, *The New Lectures on Psychoanalysis*, trans. James Strachey (London: W. W. Norton & Co., 1965 [1932]). Although the book was published in 1933, Freud penned this essay in 1932.
15 Freud, *The New Lectures on Psychoanalysis*, 213.
16 Freud, *The New Lectures on Psychoanalysis*, 213.

with Kōhei Saitō that "Marx consistently bestowed a central role in his critique of modern society to the problem of the 'separation' of humans from the earth," this critique is only developed in *Capital*.[17] The reason for this is that "Marx was not thinking 'ecologically' when he first attempted to develop his critique of capitalism."[18] I concur. Yet, while Saitō demonstrates that Marx's thought became more ecological in the 1860s, he has not provided a compelling explanation for *why* this occurred.[19] He implies that it came about when Marx studied Justus von Liebig, leading Marx to use the concept "metabolism" more prominently. Yet Marx had already used "metabolism" for sixteen years before it took on a fundamentally new meaning in *Capital*.[20] Rather, I argue, what led Marx to think more ecologically in the 1860s was Darwin. Marx's early thought centers upon the relations between humans and nature. As I show in chapter 1, Marx's critique of Hegel centered precisely on these relations. However, Marx's critique of Hegel could not be

---

17 Kōhei Saitō, *Karl Marx's Ecosocialism: Nature, Capital, and the Unfinished Critique of Political Economy* (New York: Monthly Review Press, 2017), 258. I thank and acknowledge my debts to Saitō, not only for his scholarly contributions but for his comradely support of this project, including hosting me as a visiting researcher at the University of Tokyo (2023–24).

18 Saitō, *Karl Marx's Ecosocialism*, 253.

19 Saitō once told me that he did not emphasize Darwin's influence on Marx in his books because this topic had already been covered by Foster in *Marx's Ecology* (2000). While this position is understandable, Foster's book understated Darwin's influence on Marx, partly by emphasizing Marx's critique of Darwin—a pattern common in the Marxist literature.

20 Marx first used *Stoffwechsel* in March 1851: Saitō, *Karl Marx's Ecosocialism*, 70–1. Usually translated as "metabolism," I will also use "material exchange" or "material circulation," for, as Reitter explains:

> "Stoffwechsel" . . . could also be translated as "exchange of matter," or "material exchange" . . . The term "Stoffwechsel" gained currency in the 1830s and 1840s, when the German natural scientists Friedrich Tiedemann and Justus von Liebig used it to describe vital processes of substance conversion in physiology. The (Greek-derived) English term "metabolism," now the standard translation for "Stoffwechsel," wasn't employed widely until after 1900, and thus you could make the argument that it's an anachronistic rendering of Marx's use of "Stoffwechsel" in 1867/1872 . . . The French edition of *Capital* uses plain language for "Stoffwechsel": "la circulation matérielle." (*Capital*, 800.)

Treating capitalism as a socio-ecological form defined by definite patterns of material circulation is helpful for understanding *Capital* and appreciating Marx's claim to study capital in terms of natural history.

completed in the 1840s because of Marx's inability to conceive of human history in an evolutionary sense. Marx could not conceive of this for a reason that is easy for us to miss today: Darwin's explanation of evolution had not yet been published. It was only after Marx read Darwin that his thinking exhibited the ecological and world-historical character rightly celebrated by Saitō and other ecological Marxists.

But set Marx aside for a moment. During the early to mid-1860s—the period immediately following Darwin's great work—a flowering of natural-historical studies occurred across Europe, including Liebig but also many others.[21] This was a time of roiling debate not only of the validity of Darwin's hypothesis (a "hypothesis" since it was far from widely accepted) but also about the anatomical relationship between gorillas and humans, the morality of Darwin's implied critique of religious metaphysics, the implications of evolution for the philosophy of history, the role of sexuality and gender in primate and human evolution, and more. These debates are less intense today, but they continue.

My argument concerning Darwin's influence upon Marx is not widely accepted. Naomi Beck, for instance, writes that "Marx's objective could not have been further from Darwin's," that "the similarities between Darwin's ideas and [Marx's] ideological discourse were only superficial," and that "Darwin's theory fulfilled for [Marx] only the function of a pretext and was not in reality connected with [his] views."[22] David Stack finds that Marx represents "the death knell of social explanations rooted

---

21 For example, George Perkins Marsh, *Man and Nature* (Seattle: University of Washington Press, 2003 [1865]), a landmark study of the consequences of human activity on the earth; cf. C. N. Fraas's *Agricultural Crises and Their Remedies* (1866). In a letter to Engels on March 25, 1868, Marx praised Fraas's (1847) study, *Climate and the Plant World* (see Saitō, *Karl Marx's Ecosocialism*, 246–249). Saitō cites Marx's enthusiasm for Fraas as evidence of Marx's serious study of ecology and metabolism, and I concur. Yet, *pace* Fraas, Marx thought it "necessary that the harmony between civilization and nature should be realized by the conscious collective governance of the metabolism by the associated producers"; unfortunately "'as a bourgeois,'" Fraas "does not reach this point," wrote Marx (Saito, 249). Notably, the highest praise Marx offered Fraas's 1847 book was to call it "*Darwinist before Darwin*" (Saitō, 247)—implicit praise for Darwin, if misplaced; for it was Lamark that claimed that "newly emerged characteristics could be handed down to the next generation" (ibid.). Darwin's *Origin*, as I discuss in chapter 2, displaced this theory (these matters have grown more complex with the expansion of evolutionary theory after the modern synthesis).

22 Naomi Beck, "The *Origin* and Political Thought," in M. Ruse and R. Richards, eds., *The Cambridge Companion to the Origin of Species* (Cambridge, UK: Cambridge University Press, 2000), 295–391; 310, 311, and 313.

in natural history" and that his "use of Darwin, such as it was, was occasional [and] inconsequential."[23] Gareth Stedman Jones writes that Marx "accepted Darwin's importance" but was not "excited" by *Origin* and that Marx's "acknowledgments were somewhat backhanded."[24] Claims like these (which could be multiplied) provide an analytical mandate for this book.

Yet my desire to rethink Marx with Darwin derives less from academic debates than from our planetary crisis, which demands fresh approaches to old questions. By showing that Darwin influenced Marx in productive ways, I aim to strengthen our understanding of this crisis, which—like all problems rooted in our inability to imagine alternatives—should be met by creative questioning and collective experimentation.

---

23 Stack, *The First Darwinian Left*, 74–5. Notwithstanding my disagreement on this point, Stack's book (largely ignored in the recent literature on ecological Marxism) deserves close reading.

24 Gareth Stedman Jones, *Karl Marx: Greatness and Illusion* (Cambridge, MA: Harvard University Press, 2016), 567–8.

# 1

Marx Before Darwin

Marx in Paris

Our world proclaims certain principles—freedom, popular sovereignty, the dignity of all—that are not generally upheld. One might reasonably ask whether democratic governance or dignified livelihoods could be maintained in a world where the wealthiest 1 percent own half of everything, and when we peer into the future, it is difficult to envision the achievement of universal human rights on a planet wracked by flooding, fires, and war. Capitalist modernity is thus rent by a deep chasm between its ideals and its lived expression. We know these contradictions well, but we keep believing in freedom. We reaffirm the aspirations declared in the charter that established the United Nations in 1945, such as "universal respect for, and observance of, human rights and fundamental freedoms for all without distinction as to race, sex, language, or religion"; likewise those articulated in the UN Universal Declaration of Human Rights of 1948, which insists upon the "recognition of the inherent dignity and of the equal and inalienable rights of all members of the human family," including "the right to security in the event of unemployment."[1]

Historically speaking, these are recent and novel claims. They are fruit of Enlightenment philosophy, which was (to simplify) the result of

---

1 Accessed October 22, 2024, un.org.

social transformations in the seventeenth and eighteenth centuries produced by the emergence of capitalist social relations and battles surrounding the formal political emancipation of subaltern groups. The social changes that generated those ideas—culminating, among other things, in those UN declarations—were swift enough to generate confidence that the momentum would enable us to build a world of equality and collective emancipation. We are a long way from achieving those goals. There are grounds for skepticism that we may ever get there. Nevertheless, we can find ample testaments to the belief that we should build a world that rests upon the foundation of emancipation, equality, and dignity for all.

Karl Marx was a child of the Enlightenment. His thinking only makes sense if we remember that he shared this belief in equality and freedom. Marx made two great contributions to the Enlightenment tradition: First, to provide a historical critique of its presuppositions; second, to show that collective emancipation required the reorganization of social life. Capitalism, he demonstrated, had brought about huge advances in our productive capacities, scientific understanding, and the interconnection of peoples around the world—all to be celebrated. Yet while liberal thinkers proclaimed ideals of freedom and equality, bourgeois society fell far short in realizing them (Marx's book *Capital* explains why). Marx joined others is calling for the transformation of capitalist society; he sketched a new socio-economic form, communism, organized through networked associations of producers and consumers, to achieve these aims. When communists tried to construct such societies during the nineteenth century, they tended to make communes that were too small to withstand the entropic forces unleashed by capitalist modernity. When Marxist-Leninists tried to create communist societies in the twentieth century (first in the USSR), they created powerful states and armies to defend from attack by capitalist states. The result were state-centered societies that could not overcome the capitalist organization of economic life (state capitalism). This was never Marx's goal.

Today, we are in a position to return to Marx's Enlightenment ideals and critique of capitalism and to consider our prospects from a fresh perspective. Indeed, we must do so, for our planetary crisis shows that humanity has reached a dead end.

Hegel, Marx, and history

Marx's critique of capitalist society cannot be fully appreciated without considering his debts to Georg Wilhelm Friedrich Hegel (1770–1831). By emphasizing the social, historical, and intersubjective dimension of all human knowledge and experience, Hegel radicalized Kant's transcendental critique and generated a fertile conceptual language for philosophical reflection. Hegel's influence appears throughout Marx's writings, and the influence is often so fundamental as to justify claims that Hegel's philosophical project determined Marx's. Certainly, Marx's critical approach to philosophy and history reveals Hegel's abiding influence, and Marx's conceptions of abstraction, contradiction, and dialectic begin with Hegel. Nevertheless, Marx was not Hegel, and the student went much further than the teacher in explaining our world. While this book will not end the debates about Marx's debts to and departures from Hegel, by emphasizing the influence of Darwin to this story, I aim to contribute to them in one respect. I contend that a kernel of Marx's critique of Hegel's philosophy was that he found it to be teleological, which Marx could not accept.

Scholars do not agree as to whether Hegel was a teleological thinker. Hegel can be read to mean that history is the unfolding of Geist (spirit) toward itself without end (where the movement of consciousness adopts a sort of circular form), so, by one reading, there is no telos: By this reading, Hegel does not provide us with teleology, but science. I reject this, first, for what may be a naïve objection: I have never been able to understand Hegel's use of terms like "Geist" and "Absolute Spirit" without thinking that he really means God. More importantly for my book's argument, Marx's texts from the 1840s show that Marx found Hegel to be teleological and recognized that this was a problem. But here the plot thickens because many have accused *Marx* of teleological thinking.[2] I agree that Marx inherited elements of teleological

---

2 For example, philosopher Howard Williams argues that "Hegel's and Marx's views of history are very similar," for they share a set of assumptions, including "The development of human history falls into distinct stages ... represent[ing] a change in gravity in the location of historical development from East to West ... There is a distinct point of culmination where the higher level of society is achieved ... Both Hegel and Marx therefore take a teleological view of history. They believe that there is a purpose underlying the unfolding of events in world history": Howard Williams, "The end of

thinking from Hegel. Yet I contend that Marx overcame them in the 1860s as one element of a broader shift in his thinking about capitalism, history, and our world. To develop this claim, I will focus on Marx's departure from Hegel's philosophy of nature and history, for it is upon these grounds that Marx cleared a path toward natural history.³

The books by Hegel most closely associated with Marx are the *Phenomenology* (1807) and *Logic* (1812–16), abstract philosophical works in which (among other things) Hegel provides his conceptions of being, reason, dialectic, abstraction, and absolute idealism.⁴ These works influenced Marx (although how, and with what effect, remain a matter of much dispute); but I will emphasize Marx's engagement with Hegel's philosophy of history, introduced in Hegel's *Philosophy of Right* (1821) and enunciated through a series of lectures that Hegel delivered during the 1820s at the University of Berlin.⁵ After Hegel died in 1831, his audience split into competing schools of thought, with two poles: The left-and right-wing Hegelians. The debates surrounding the meaning of Hegel's thought were fundamental to Marx's intellectual

---

history in Hegel and Marx," *The Hegel-Marx Connection*, Tony Burns and Ian Fraser, eds. (London: Macmillan, 2000), 198–216, 198. *Pace* Williams, I agree with Mauricio Vieira Martins, who writes that while "in certain passages" of Marx and Darwin's texts "it is possible to locate . . . marks of a finalist conception," and therefore "an author who wants to present readers a finalist Marx or a teleological Darwin would partially be able to do so," such an "approach disregards what is fundamental: The enormous contribution each one made to break with the prevailing worldviews that projected finalist human categories into unintentional processes": Mauricio Vieira Martins, *Marx, Spinoza, and Darwin: Materialism, Subjectivity and Critique of Religion* (Berlin: Springer Nature, 2022), 206.

3   I discuss Marx's *political* critique of Hegel in Wainwright and Mann, *Climate Leviathan*, 190–3. See also Geoff Mann, *In the Long Run We Are All Dead: Keynesianism, Political Economy, and Revolution* (New York: Verso Books, 2017), chapters 5–7.

4   These are standard shortened names of Hegel's *Phenomenology of Spirit* (*Phänomenologie des Geistes*) and *Science of Logic* (*Wissenschaft der Logik*). I note in passing that—as Marx would have known—Kant and Hegel shared an abiding interest in natural history and astronomy: In 1755, Kant published (anonymously) his *Universal Natural History and Theory of the Heavens*; similarly, in 1801, Hegel wrote his doctoral dissertation on the motion of the planets.

5   G. W. F. Hegel, *Elements of the Philosophy of Right* (Cambridge, UK: Cambridge University Press, 1991 [1821]), §§ 341–360; G. W. F. Hegel, *The Philosophy of History* (New York: Dover, 1956 [1830-31]). These are not to be confused with Hegel's lectures on the history of philosophy: See G. W. F. Hegel, *Lectures on the History of Philosophy: 1825–26*, in three volumes (Oxford: Clarendon, 2009 [1825–26]). Hegel's lectures on the history of philosophy conclude by declaring his "standpoint": "The cognition of spirit, the knowledge of the idea of spirit" (vol. III, 212).

development; his early philosophical formation began as a Left Hegelian. It is impossible to understand Marx's standpoint with respect to natural history without considering Hegel's influence, therefore.

Hegel was one of the first modern thinkers to examine nature and history and to make them central to his philosophical project.[6] Among philosophers, Hegel is unusual for attributing meaning to history and his contention that the grasping of this meaning is an essential philosophical act. For Hegel, being is defined as a perpetually unfolding historical process; at each stage, being is in conflict with itself, negates itself, and thereby propels itself forward. But this historical process is not random, for Hegel discerned indications of progress, particularly in the development of reason.[7] Hegel posited that this development was tending toward a specific telos: A point at which reason would be able to know itself as Geist, a word that is usually (however imperfectly) translated as "spirit."[8]

Hegel discerned a tripartite distinction in historiography. He begins with an intuitive and familiar distinction: *Original* history, comprised of accounts written by direct observers; *reflective* history, in which observers reflect upon events to produce a narrative about the past and its relation to the present.[9] Then Hegel posits a third type, *philosophical* history,

---

6 G. W. F. Hegel, *Philosophy of Nature* (London: Allen and Unwin, 1970); Marina Bykova, ed., *Hegel's Philosophy of Nature* (Cambridge, UK: Cambridge University Press, 2024).

7 "The principle of *Development* involves also the existence of a latent germ of being—a capacity or potentiality striving to realise itself. This formal conception finds actual existence in Spirit; which has the History of the World for its theatre, its possession, and the sphere of its realisation" (Hegel, *Philosophy of History*, §61, 54). Hegel's views on history's development, though distinct, emerged in response to the debates in late nineteenth-century Germany concerning (among other things) the meaning of the French Revolution. (Hegel was an impressionable fourteen-year-old in Obergymnasium in 1784, the year that Kant published his "Ideas for a Universal History" and Herder published "Ideas for a Philosophy of History.")

8 "Geist" in Hegel typically means "spirit" in a manner akin to the colloquial expression, "the spirit of the times," but it can also mean "God." More imperfect translations are "mind" or "culture." (Whereas "mind" can feel appropriate where Hegel writes about particular manifestations of Geist, "culture" is fundamentally wrong: Indeed, its wrongness illustrates the problem of treating culture as a philosophical concept: See Qadri Ismail, *Culture and Eurocentrism* (London: Rowman & Littlefield, 2015).

9 Hegel, *The Philosophy of History*, §§1-13, 1–10. *Original* historians, in Hegel's terms, "simply transferred what was passing in the world around them, to the realm of representative intellect"; *reflective* history "is not really confined by the limits of the time to which it relates" and can encompass "criticism of historical narratives and an investigation of their truth and credibility."

which stems from the recognition that "the business of history"—the duty of documenting "what is and has been, actual occurrences"—is inherently set against philosophy, which is fundamentally about questioning the meaning of things, encompassing all that is and has been. If history means simply *recording what happened*, it is anti-philosophical and can contribute nothing to reflection. Conversely, philosophical reflection without knowledge of "what is and has been" will remain hopelessly abstract. One might think that the solution to this challenge is simply to mix history and philosophy, but Hegel's point is that, if the two are defined in opposition to each other, blending is already precluded. His philosophical history confronts this challenge through a definite set of propositions. First, philosophy brings to history the conception of "Reason [as] the Sovereign of the World; that the history of the world therefore, presents us with a rational process."[10] This proposition could be taken to mean that everything that happens is rational and therefore morally justified. Hegel's view is more complex. He sought to grasp history as reason unfolding, historically accounting for itself as the fulfillment of spirit:

> Reason is the *substance* of the Universe; viz. that by which and in which all reality has its being and subsistence . . . While it is exclusively its own basis of existence, and absolute final aim, it is also the energising power realising this aim; developing it not only in the phenomena of the Natural, but also of the Spiritual Universe—the History of the World. That this "Idea" or "Reason" is the True, *the Eternal*, the absolutely *powerful* essence; that it reveals itself in the World, and that in that World nothing else is revealed but this and its honour and glory— is the thesis which, as we have said, has been proved in Philosophy and is here regarded as demonstrated.[11]

This is Hegel playing absolute idealism in the key of philosophy of history. To put it formulaically, the history of reason's realization = realization of the history of reason. Reason's development culminates in a fulsome self-consciousness of spirit. One might describe this

---

10 Hegel, *The Philosophy of History*, §12, 9.
11 Hegel, *The Philosophy of History*, 9–10. The final sentence is a reference to his *Phenomenology*, published earlier.

conception of reason and history as dialectical, where the two comprise a unity of difference, defined through an immanent opposition, developing into itself as itself.

Hegel justifies philosophical history on the grounds that, among the three forms of historiography, it alone grasps historical reality at this level of understanding. Regardless of whether we agree that it could be executed in the fashion Hegel intended or not, his act of positing philosophical history was crucial for Marx; by doing so, Hegel called into question history's basic conditions of possibility. Even if we do not accept Hegel's elucidation of the relationship between the growing self-awareness of consciousness and the course of history towards its "absolute final aim," as "the True, the Eternal, reveals itself in the World," by thematizing the relationship of reason and history as a philosophical problem, Hegel made an important contribution.

Hegel's understanding of the task of philosophy had great influence on Marx's early self-conception. But Marx substituted social relations for Hegel's spirit and proposed to explain the immanent development of these relations through the contradictions of class struggle. In time, Marx's approach to history would cross over the three levels of Hegel's schema as he sought grounds for history's realization in communism.[12] This is not the only sense in which Marx could be said to develop Hegel's philosophy of history. Hegel's approach inspired Marx's conception of the history of humanity constituting itself through a developmental process of perpetual negation. As Marx wrote in 1844:

> The outstanding achievement of Hegel's *Phänomenologie* and of its final outcome, the dialectic of negativity as the moving and generating principle, is ... that Hegel conceives the self-creation of man as a process, conceives objectification as loss of the object, as alienation and as transcendence of this alienation;[13] that he thus grasps the essence of *labour*

---

12 My readings of *Capital* as natural history (part II) and the natural history of capitalism (chapter 6) treat Marx as a writer whose texts oscillate between critical-reflective and philosophical modes—using the one mode to interrupt the other.

13 The works of Marx, Darwin, and their contemporaries are riddled with anachronistic, gendered language. My strategy will be to quote selectively (to avoid the most offensive words) and to modify language for gender neutrality when it does not require significant changes to the meaning of the text. Nevertheless, some anachronistic, gendered language remains, since, to take this case, changing "man" to "individual," "humanity," or "person" would significantly distort Marx's meaning here. Marx's patriarchal language reflects a

and comprehends objective man . . . as the outcome of man's *own labour*. The *real, active* orientation of man to himself as a species-being, or his manifestation as a real species-being (i.e., as a human being), is only possible if he really brings out all his *species-powers*—something which in turn is only possible through the cooperative action of all of mankind, only as the result of history—and treats these powers as objects: And this, to begin with, is again only possible in the form of estrangement.[14]

Marx uses "species-being" here in contradistinction to Hegel's spirit: for Marx, it is species-being, not spirit, which is realized through history. Marx took this concept from Ludwig Feuerbach's 1841 critique of Hegel, *The Essence of Christianity*, in which humans are defined (in contrast to other animals) by our self-consciousness "as a member of a species."[15]

Still, the question remains to what extent Marx remained loyal to Hegel's thought. This question is often answered like this: While Marx initially adopted a left-wing interpretation of Hegel, he replaced the *cause* of historical change (from the contradictions of reason to class struggle) and the *object* of movement (from Geist to social relations). In doing so, Marx overcame Hegel's "idealism" to establish "dialectical materialism."[16] Later, Marx borrowed some of Hegel's language when writing *Grundrisse* and *Capital* (see part II). Without saying too much at this point, my contribution will be limited to clarifying an element of

---

genuine bias in his thought; it would be facile to address this solely by deleting anachronisms and changing pronouns. For the project of Marxian natural history, we must continue to examine this figure, the "man" of nineteenth-century philosophy, with Marxist-feminism. See, for example, Lise Vogel, *Marxism and the Oppression of Women: Toward a Unitary Theory* (New Brunswick, NJ: Rutgers University Press, 1983); Heather Brown, *Marx on Gender and the Family: A Critical Study* (Leiden: Brill, 2012); Silvia Federici, *Patriarchy of the Wage: Notes on Marx, Gender, and Feminism* (Binghamton, NY: PM Press, 2021). For reviews of recent debates, see Cinzia Arruzza, "Three debates in Marxist feminism," *The SAGE Handbook of Marxism* (London: SAGE, 2022), 1354–72; Frigga Haug, "Marxism-Feminism," in *Historical-Critical Dictionary of Marxism* (Leiden: Brill, 2023) 582–601.

14   Marx, Paris MS III (*Economic and Philosophic Manuscripts, 1844*), §XXIII; MECW III: 332–3 (cf. Gregor Benton's translation in Karl Marx, *Early Writings* [London: New Left Books], 385–6). See note 19.

15   The source is Feuerbach's (1841) *The Essence of Christianity*, but my quotation is taken (and lightly modified) from David McLellan, *Marx Before Marxism* (Middlesex: Penguin, 2022), 151.

16   Marx never used the expression "dialectical materialism," coined by Joseph Dietzgen. I reject the concept.

Marx's relationship to Hegel that has been downplayed: Marx's break with Hegel had more to do with nature and history than is usually recognized. By positing spirit as the subject of history, Hegel excludes what is usually called "nature," or natural processes, from his philosophy. In fairness, Hegel writes: "The phenomenon we investigate—Universal History—belongs to the realm of Spirit. The term 'World' includes both physical and psychical Nature. Physical Nature also plays its part in the World's History, and attention will have to be paid to the fundamental natural relations thus involved." Nevertheless, he continues: "*But Spirit, and the course of its development, is our substantial object.* Our task does not require us to contemplate Nature as a Rational System in itself . . . but simply in its relation to Spirit."[17] Marx could not accept this, for he saw humanity as part of nature. Reason is simultaneously natural and historical: It is an inherent quality of human species-being realized through humanity's practical activities. Hence to flip Hegel's dialectic around means adopting a natural-historical standpoint (see chapter 3).

Humans and Nature in Marx's Paris Manuscripts

Though he published nothing of significance, 1844 was a very good year for Karl Marx. A young journalist-scholar in Paris, he clarified his critique of Hegel and worked through literature on political economy, studying Adam Smith, Thomas Malthus, David Ricardo, and others.[18] Through this research and writing, Marx cleared the ground for what I will call "Marxian natural history."

---

17 Hegel, *Lectures on the Philosophy of History*, §19, my italics.

18 Marx kept up this research for several years, then set it aside to focus on his political work and journalism, later resuming his research into political economy in 1857 in London. In the decade 1838–1848, Marx wrote a considerable number of important works with relevance to this project: From his doctoral dissertation (*On the Difference Between the Democritean and Epicurean Philosophy of Nature*, written 1838–41), the "Debates Concerning the Law on the Theft of Wood" (1842), "On the Jewish Question" (1843), to the Paris manuscripts (see next note), the *German Ideology* (1845), the "Theses on Feuerbach" (1845), the "Manifesto of the Communist Party" (1848), abundant journalism, and so on. A vast secondary literature has been written on these texts. I will not try to review all this material; neither will I produce a biographical reconstruction of Marx's life and writings during this period nor a philology of any particular text. Since my aim is to stage my argument apropos natural history, I mainly discuss the Paris manuscripts and the *German Ideology*.

When Marx writes about nature and history in his Paris manuscripts, he often uses Hegelian expressions but develops his thinking upon different premises.[19] Marx begins from human interactions with nature: "The worker can create nothing without *nature*, without the *sensuous external world*."[20] This relationship with nature defines the species character of humanity: "Man is a species-being, not only because in practice and in theory he adopts the species (his own as well as other things) as his object, but—and

---

19   The manuscripts Marx wrote in Paris in 1844, which came to be called the *Economic and Philosophic Manuscripts*, were first published in German in 1932 (a date to which we shall return). The first English translation, by Martin Milligan, arrived in 1959 by the Foreign Languages Publishing House in Moscow; it would be republished, with editorial modifications by Dirk Struik, in 1975 in *Marx and Engels Collected Works*, volume III. A second translation, by Tom Bottomore, was published in 1963 in an affordable paperback edition that included other works from Marx's early years; a third translation, by Lloyd Easton and Kurt Guddat, followed in 1967; a fourth, by Gregor Benton, was published in 1974 in a collection with New Left Books/Penguin. These English editions vary not only in terms of translation style but source text and critical apparatus, raising complex philological issues (which I cannot discuss here). On the political history of the German source texts, see Marcello Musto, "The 'Young Marx' myth in interpretations of the economic–philosophic manuscripts of 1844," *Critique* 43, no. 2 (2015): 233–60, an excellent source which contains one small typo: The first English translation was published in 1959, not 1956. Given these issues, when citing Marx's Paris manuscripts, I will refer to the section (§) quoted in the original manuscripts, followed by the page in the two versions that are most widely read today: Benton, New Left Books/Penguin, 1974, and Milligan and Struik, MECW, volume III, 1975. Unless noted, quotations are drawn from the latter.

The multiplicity of translations from 1959 to 1975 speaks to the immense influence of the Paris manuscripts on left-wing thought during the 1960s and 1970s, an era of fierce debates over the status of this work in relation to *Capital*—particularly after 1965 when Louis Althusser proposed that Marx experienced an "epistemological break" (a concept Althusser drew from Bachelard) between the early, humanist, Hegelian Marx, and the late, scientific Marx in Althusser, *For Marx* (London: Verso, 2005 [1965]). Although I reject Althusser's hypothesis, I concur that Marx's thought *changed* and that he made a scientific discovery concerning "the history of 'social formations'" (Althusser, *For Marx*, 13); yet the course of Marx's discovery was more complex than Althusser suggests. While his thinking changed frequently, certain fundamental problems and themes reappear at different points in novel forms. Hence, even putting aside matters of intellectual historiography, "epistemological break" was always the wrong concept to apply to Marx. I agree with Lixin Han's contention that an important shift occurs after Paris manuscript I with Marx's renewed emphasis on social relations of production (and that this is tied to his 1844 *Notes on James Mill*): Han, *Studies of the Paris Manuscripts: The Turning Point of Marx* (Singapore: Springer, 2020). I would quibble with Han's general contention that Karl Marx "begins" in mid-1844, since even his doctoral dissertation expresses fundamental themes of his thought. While *The End* will bring no end to these debates, I hope it clarifies how Marx thought the interrelation of nature, history, and the capitalist social formation.

20   Marx, Paris MS I (1844), §XXIII; New Left Books/Penguin, 325; MECW III: 273.

this is another way of expressing [this relationship]—also because he treats himself as the actual, living species; because he treats himself as a *universal* and therefore a free being."[21] Like other species, humans survive and reproduce by transforming and consuming nature: "Physically man lives only on these products of nature, whether they appear in the form of food, heating, clothes, a dwelling, etc."[22] By laboring on and consuming the products of the earth, we are constantly reweaving the natural ties that constitute us. "Man *lives* from nature—i.e., nature is his *body*, with which he must remain in continuous interchange if he is not to die. To say that man's physical and mental life is linked to nature simply means that nature is linked to itself, for man is a part of nature."[23] We *are* nature, a form of nature. Humans became humans historically through practical activity. Our life activity—laboring upon Earth—and the material changes we make to our physical environment are aspects of one evolving, differentiated unity.

As a social species, our practical activities are inherently social. We come to exist—as individuals and as a species—as social beings. Moreover, the activity that (for Marx) defines us, our creative laboring activity, is always social. This socialness in no way negates our naturalness, our being part of nature, for we modify nature socially:

> The worker can create nothing without *nature*, without the *sensuous external world*. It is the material in which his labor realizes itself, in which it is active and from which and by means of which it roduces.
>
> But just as nature provides labor with the *means of life* in the sense that labor cannot *live* without objects on which to exercise itself, so also it provides the *means of life* in the narrower sense, namely the means of physical subsistence of the *worker*.
>
> The more the worker *appropriates* the external world, sensuous nature, through his labor, the more he deprives himself of *means of life* in two respects: Firstly, the sensuous external world becomes less and less an object belonging to his labor—a *means of life* of his labor; and, secondly, it becomes less and less a *means of life* in the immediate sense, a means for the physical subsistence of the worker.

---

21 Marx, Paris MS I (1844), §XXIV; New Left Books/Penguin, 327; MECW III: 275.
22 Marx, Paris MS I (1844), §XXIV; New Left Books/Penguin, 327; MECW III: 275.
23 Marx, Paris MS I (1844), §XXIV; New Left Books/Penguin (lightly modified), 328; cf. MECW III: 276.

I note in passing that Marx provides little justification for these claims—his theory of alienation—in the 1844 manuscripts. He posits them without substantive explanation. He did not provide such an explanation until *Capital* (see part II). Continuing:

> In these two respects, then, the worker becomes a slave of his object . . . The culmination of this enslavement is that it is only as a *worker* that he can maintain himself as a *physical subject* and it is only as a *physical subject* that he is a worker.[24]

Marx concludes this line of thinking with the claim that "every self-estrangement of man, from himself and from nature, appears in the relation in which he places himself and nature to men other than and differentiated from himself."[25] This is an odd claim, one that is worth parsing.

Marx is examining a pair of relationships which are, from his standpoint, distinct yet inseparable: Relations between humans and nature ("the relation in which he places himself and nature") and social relations ("men other than and differentiated from himself"). Because these distinct yet inseparable relationships will be important for my argument throughout the book, I propose to use some simplifying terminology here. I call the first relationship, human-nature relations, "HN," and the second, social relations, "HH" (human-human). Since humans are inherently part of nature, HH could be subsumed into HN. But in practice we do not equate HH and HN. We reason as humans; our consciousness is a human sort; we distinguish humans from other species and our social relationships (HH) from our relationships with the rest of nature (HN). Now, Marx sees these two relationships, HH and HN, as intertwined. This is not to say that, for Marx, HH = HN. Rather, HH and HN mutually condition each other. Whenever humans change how they relate with one another (HH)—for instance, in the ways they produce and reproduce their social lives—they change their relationship with the

---

24 Marx, Paris MS I (1844), § XXIII; New Left Books/Penguin, 325; cf. MECW III: 273. Marx adds: "We have until now considered this relationship only from the standpoint of the worker and later on we shall be considering it also from the standpoint of the non-worker"; Marx did not make good on this promise until *Capital*: See chapter 4.

25 Marx, Paris MS I (1844), §XXV; New Left Books/Penguin, 331; MECW III: 279.

local environment (HN). Conversely, every change in the relations between humans and nature (HN) has consequences for social relations (HH). I will symbolize this dialectical relation of constitutive mutual conditioning by HH×HN. For Marx, HH×HN is not a fixed thing, but an evolving ensemble of relations.

Even in 1844, Marx had reached another level of understanding. His argument is not merely that the relations HH and HN mirror one another perfectly, with each adjustment to one resulting in a spontaneous homeostatic adjustment in the other; rather, his conception of HH×HN came to encompass maladjustment, misunderstanding, and misrecognition. For it is not the *authenticity* of humanity's relationship to itself that appears in our relationship to nature, but our *self-estrangement*. Nothing is more human than our self-estrangement, our separation from ourselves as natural beings. In Marx's conception, human society formed via natural history, and the possibility exists for a sort of adequation of HH×HN in transcendental form—the "true resurrection of nature."

> [Human] activity and consumption, both in their content and in their *mode of* existence, are *social* activity and *social* consumption. The *human* essence of nature exists only for *social* man; for only here does nature exist for him as a *bond* with other *men*, as his existence for others and their existence for him, as the vital element of human reality; only here does it exist as the basis of his own human existence. Only here has his *natural* existence become his *human* existence and nature become man for him. *Society* is therefore the perfected unity in essence of man with nature, the true resurrection of nature, the realized naturalism of man and the realized humanism of nature.[26]

Clearly, humanity has not yet reached the "complete unity of man with nature." Our planetary crisis—rapid global heating, the sixth great extinction of species, despoilation of water, land, and atmosphere at every scale—suggest something like the opposite: Degradation and disunity. I intend these meditations on Marx to help us find a way forward.

~

---

26   Marx, Paris MS III (1844), §V; New Left Books/Penguin, 349–50; cf. MECW III: 298.

Marx is addressing an ancient question: If humans are natural, what distinguishes us from the rest of nature? His answer at this stage is deeply Hegelian: The essence of humanity is our self-estrangement from what we truly are. Instead of persisting in asking, "Are humans part of nature?" Marx poses a different question, one that Hegel did not ask: "Given that we are natural beings," Marx ponders, "how did this apparent separation between humans and nature emerge *historically*?" Pursuing *that* question would eventually lead Marx to a genuinely original standpoint.

Before explaining, we should note that Marx's conception of the differentiation of the human from other species—our "self-estrangement" from nature—also takes the shape of human *consciousness*. By Marx's reasoning, humans are a distinctive species because of our consciousness of our being what we are. While Marx is not especially precise about his definition of consciousness—in fairness, the concept is difficult to define—he makes an original move by positing that human consciousness mediates HH×HN:

> The animal is immediately one with its life activity. It does not distinguish itself from it. It is *its life activity*. Man makes his life activity itself the object of his will and of his consciousness. He has conscious life activity. It is not a determination with which he directly merges. Conscious life activity distinguishes man immediately from animal life activity. It is just because of this that he is a species-being. Or it is only because he is a species-being that he is a conscious being, i.e., that his own life is an object for him.[27]

These claims reappear in a substantially revised fashion in *Capital* (I will question Marx on this point in chapter 7). For now, I wish to emphasize that Marx conceived the human-animal distinction from a specifically human "conscious life activity" that determines "species-being." Two points deserve emphasis here: Marx's argument reflects a break from Hegel since for Marx human consciousness is equated with practical life activity (as opposed to the development of reason or

---

[27] Marx, Paris MS I (1844), §XXIV; New Left Books/Penguin, 328; MECW III: 276.

spirit). In Hegel, Marx writes, the "human character of nature and of the nature created by history" appear only "in the form that they are *products* of abstract mind and as such, therefore, phases of *mind—thought-entities*."[28] Consequently, humanity and nature "become mere predicates—symbols of this hidden, unreal man and of this unreal nature."[29]

Second, we must bear in mind that when Marx made these criticisms of Hegel, no coherent scientific theory of the evolution of human beings existed.[30] Marx was, so to speak, suspended between Hegel and Darwin. He anticipated the possibility of a unifying natural-historical scientific theory that would encompass the entire development of human existence vis-à-vis nature, consciousness, and our practical activities:

> The whole of history is a preparation, a development, for "*man*" to become the object of *sensuous* consciousness and for the needs of "man as man" to become [sensuous] needs. History itself is a *real* part of *natural history* and of nature's becoming man. Natural science will, in time, subsume the science of man, just as the science of man will subsume natural science: There will be *one* science.[31]

In a curious coincidence, Marx wrote these lines exactly when Darwin was writing his "1844 Essay" (see chapter 2).

---

28 Marx, Paris MS III (1844), §XVIII; New Left Books/Penguin, 385; MECW III: 332.
29 Marx, Paris MS III (1844), §XXX; New Left Books/Penguin, 396; MECW III: 342.
30 There were earlier ideas about species evolution (see chapter 2), but prior to Darwin's *Origin* (1859), these theories lacked coherence.
31 Marx, Paris MS III (1844), §IX, New Left Books/Penguin, 355; cf. MECW III: 303–4. One might respond that "one science" already exists: It is history, which encompasses everything. But as Marx observes: "Historiography only incidentally takes account of natural science, which it sees as contributing to enlightenment, utility, and a few great discoveries." Implied: For Marx, history has taken too little from the natural sciences, which have "intervened in and transformed human life all the more *practically* through industry and has prepared the conditions for human emancipation": Marx, Paris MS III (1884), §IX, New Left Books/Penguin, 355; cf. MECW III: 303.

Toward an earthly basis for history

Marx developed these ideas in new directions in *The German Ideology*, where he and Engels elaborate the implications of their critique of Hegel for understanding humans, nature, and history.[32] In a famous passage, Marx and Engels write that the "first premise of all human history, of course, is the existence of living human individuals. The first fact to be established, then, is the physical organization of these individuals and their consequent relationship to the rest of nature."[33] After specifying the sort of natural conditions that human groups must engage—they mention geology and climate—Marx and Engels add: "All historiography must proceed from these natural bases and their modification in the course of history through the actions of men."[34] Commenting upon this passage, György Márkus states that Marx's "later enthusiastic reception of the Darwinian theory can serve in all probability as an indication of his earlier theoretical predispositions in this matter."[35] I concur. Marx and Engels elaborate:

> [W]e must begin by stating the first premise of all human existence and, therefore, of all history, the premise, namely, that men must be in a position to live in order to be able to "make history." But life involves before everything else eating and drinking, habitation, clothing, and many other things. The first historical act is thus the production of the means to satisfy these needs, the production of material life itself. And indeed this is an historical act, a fundamental condition of all history, which today, as thousands of years ago, must daily and hourly be fulfilled merely in order to sustain human life ... Therefore in any interpretation of history one has first of all to observe this fundamental fact in all its significance and all its implications and to accord it its due importance. It is well known that the Germans [including

---

32 On the complex publication history of the *German Ideology* manuscripts, see Terrell Carver and D. Blank, *A Political History of the Editions of Marx and Engels's German Ideology Manuscripts* (London: Springer, 2014).
33 Marx and Engels, *German Ideology*, MECW V: 31.
34 Marx and Engels, *German Ideology*, MECW V: 31.
35 György Márkus, *Marxism and Anthropology: The Concept of "Human Essence" in the Philosophy of Marx*, trans. E. de Laczay and G. Márkus (Assen: Van Gorcum, 1978), 4.

Hegel—JDW] have never done this, and they have never, therefore, had an *earthly* basis for history.³⁶

To recapitulate: By 1843–45, Marx rejected Hegel's conception of nature and history *philosophically*, yet he had no *scientific* basis to critique its implicit teleology. At that time, then, Marx could only propose this abstract humanism of species-being and praxis (concepts that, retrospectively, we might characterize as "pre-Darwinian"). The farthest point that Marx could go in terms of his critique of Hegel was to posit a new unitary science: An as-yet-to-be-determined science of natural history that would take HH×HN as its object. Having clarified his standpoint, Marx set the matter aside.

Human Nature and Praxis: The "Theses on Feuerbach"

A common interpretation of Marx's difference from Hegel is that the former sought a material basis to grasp history, whereas Hegel was an idealist. The standard sources for this story—an oversimplification—are *The German Ideology* and Marx's critique of the Left-Hegelian philosopher Ludwig Feuerbach, whose critique of Hegel initially inspired Marx. The "Theses on Feuerbach" comprise a sort of culminating point that marks an epitome of self-clarification as well as a certain conceptual limitation. Let us consider them briefly.

The central concept of Marx's "Theses on Feuerbach" is praxis.³⁷ Consider these statements from Theses I, II, and VIII:

---

36 Marx and Engels, *German Ideology*, MECW V: 41–2; according to Carver and Blank, *Marx and Engels's 'German Ideology' Manuscripts* (London: Palgrave Macmillan, 2014), 62–5, this passage is written in Engels's hand. Marx adds a note to their manuscript here that says, enigmatically: "Hegel. Geological, hydrographical etc. relations. Human bodies. Needs, labor."

37 "Praxis" and its cognates ("praktisch") are used fourteen times, in seven of the eleven theses. Praxis—a transliteration of an ancient Greek word for action (πρᾶξις)—refers to almost any sort of human activity, any doing. Although the term appears occasionally in pre-Socratic Greek myth and philosophy, the foundation for modern use was laid by Aristotle in the *Nicomachean Ethics*, in which he categorizes human activity into three modes—praxis, poiēsis, and theōria—which could be interpreted as activity, production, and contemplation. For Aristotle, the first two transformed things directly, but poiēsis is defined by its end, production: The bringing of something into existence. Although praxis and poiēsis both entail human activity, Aristotle saw praxis as more distinct and

I. The chief defect of all previous materialism . . . is that things, reality, sensuousness are conceived only in the form of the object, or of contemplation, but not as human sensuous activity, practice [Praxis], not subjectively.

II. The question whether objective truth can be attributed to human thinking is not a question of theory but is a practical question. Man must prove the truth, i.e., the reality and power, the this-worldliness of his thinking in practice [Praxis].

VIII. All mysteries which mislead theory into mysticism find their rational solution in human practice and in the comprehension of this practice [Praxis].

Following Aristotle, Marx treats praxis as practical activity. Marx laments that prior materialist philosophers failed to grasp that "human sensuous activity," not objects, provides the basis of our knowledge. Marx, however, extends the classical usage here with one further step. He claims that those who seek to explain social relations (as he does) must therefore examine "human practice" as well as "the comprehension of this practice." The object of analysis becomes both practice and its understanding. Only by grasping "practice and . . . the comprehension of this practice," can philosophy *transform social circumstances* and thereby *realize itself*. In Thesis I, Marx criticizes Feuerbach for failing to recognize the "significance of 'revolutionary,' of 'practical-critical,' activity." Marx thus equates practical-critical activity (that is, praxis) with revolutionary politics. Thesis III concludes: "The coincidence of the changing of circumstances and of human activity or self-changing can be conceived and rationally understood only as revolutionary practice." I take this to mean that practices are revolutionary to the extent that the transformation of social relations coincides with the transformation of the circumstances that produce these social relations. For Marx in

---

virtuous than poiēsis, for praxis is always directed inward, and concerned with means (not solely ends-directed). And theōria, which emerges through contemplation of what exists (and these other modes of action upon it), is the highest mode of activity. Aristotle's praxis thus forms one element within an implied hierarchy of concepts that define human beings (as against non-human animals), with praxis a middle term between theōria and poiēsis (by implication, placing thinking above producing). This subsection draws from Joel Wainwright, "Praxis," *Rethinking Marxism* 34, no. 1 (2022): 41–62. The following quotation is taken from Marx (1845), "Theses on Feuerbach," MECW V: 6–8.

1844–46, social and economic relations shape consciousness. Through revolutionary activity, people may come to understand themselves differently and transform these economic conditions. Revolutionary praxis would characterize such "coincidence" of "changing circumstances" and social "self-changing." Thus for Marx, the critique of society must aim to explain actual social practices, to bring about the *comprehension* of practice—*through practice*. Where achieved, praxis would comprise the negation and overcoming (Aufheben) of philosophy.[38]

From whom or where did Marx obtain this concept of praxis? The standard answer is Hegel. By this view, Marx's conception of praxis is the result of his transposition of Hegel's logical categories into an ontology of social life.[39] While there is some truth here, Marx's use of praxis carries us back to Aristotle's conception of praxis (a term he adopts from Plato, whence the pre-Socratics). Aristotle's concern apropos praxis lies in elaborating a theory of how to live well: Eudaimonia (flourishing). For Aristotle theōria is superior to praxis since contemplation of the act always follows from activity; theōria could be seen as praxis's full elaboration. Yet Aristotle's reasoning here is more complex. For one thing, he saw theōria as a specific kind of activity, one that he distinguished from but also included within praxis. What distinguishes theōria ultimately is its end, or *telos*, the correct grasping of existence—something Aristotle implies is possible only for the gods.[40]

---

38 There is an ambiguity concerning the relationship between the two meanings of praxis that I have drawn from the theses. If these two meanings are indeed found in the theses (as I believe they are) the ambiguity results from the distance between "sensuous practice" as the substance of Marx's analysis, on one hand, and his conception of praxis qua revolutionary activity, on the other. The former is the basis of Marxist *materialism*; the latter defines a Marxist *political aim*. Notably, both are valorized. Although the two conceptions of praxis may be distinct in the theses, to Marx they are both good.

39 On the Hegelian conception of praxis, see Richard Bernstein, *Praxis and Action: Contemporary Philosophies of Human Activity* (Philadelphia: University of Pennsylvania Press, 1971). Andrew Feenberg writes that Marx's philosophy of praxis hinges upon "the idea that history, properly understood, has ontological significance. As a philosopher of praxis, Marx did not choose between an ontological and a historical interpretation of the social categories; he chose both": Feenberg, *The Philosophy of Praxis: Marx, Lukács, and the Frankfurt School* (New York: Verso Books, 2014), 7.

40 Although the term "praxis" is commonly used by many on the Left today, its meaning is often ambiguous. The prominence and ambiguity of the term stem from Marx's usage. "Praxis" was not central to Marx's writings. It is fundamental in three

Thesis VI

One—only one—of Marx's eleven theses on Feuerbach says anything about history or human nature. It is the sixth:

> VI. Feuerbach resolves the religious essence into the human essence. But the human essence is no abstraction inherent in each single individual. In its reality it is the ensemble of the social relations. Feuerbach, who does not enter upon a criticism of this real essence, is consequently compelled:
>
> 1. To abstract from the historical process and to fix the religious sentiment as something by itself and to presuppose an abstract—isolated—human individual.
>
> 2. Essence, therefore, can be comprehended only as "genus," as an internal, dumb generality which naturally unites the many individuals.[41]

Restated: Because Feuerbach fails to see that the essence of humanity is expressed in "the ensemble of the social relations," his thought leads him to "abstract from the historical process" that constitutes this ensemble of relations. Marx says that if one wants to understand humanity, one must grasp the total ensemble of human social relations as historical, in other words, without reducing human nature to a list of traits.[42] For Marx, "the ensemble of social relations" must be grasped *as* "historical process"; the ostensibly general "natural unity" of individuals is an illusion. We find here a recapitulation of the principle that HH×HN is not a fixed thing, but an evolving ensemble of relations.

---

texts, all written in 1844–46: His critique of Hegel, the "Theses on Feuerbach," and *The German Ideology*. Then Marx basically dropped it. This disappearance of the concept praxis from Marx's writing after 1846 is somewhat mysterious and, so far as I am aware, has not been adequately explained. I offer a hypothesis in chapter 4.

41 Marx, "Theses on Feuerbach" (1845), in *Karl Marx Early Writings* (New Left Books/Penguin: London, 1974), 423.

42 As Vanessa Wills observes, "according to a Marxist approach to ethics, we ought to do that which promotes human flourishing. It follows that . . . we must know something of what it is to flourish *as* a human being . . . However, if 'human nature' is the sort of thing that can be resolved into a discrete list of fixed traits . . . then Marxist theory offers no such thing": Wills, *Marx's Ethical Vision* (Oxford: Oxford University Press, 2024), 46.

During the 1960s, a prominent school of thought, led by Louis Althusser, argued that Marx rejected the concept of human nature. In Althusser's reading, Marx was an anti-humanist. To be precise, Althusser claimed that Marx *became* an anti-humanist, after breaking from his early Hegelian thinking in around 1845. For sake of argument, let us say that Marx's early thought is "humanist" because it takes a self-conscious species-being, *humanity*—not God, nor Geist, nor this or that king—as the subject of history. Was Althusser right to claim that Marx became an anti-humanist? No.[43] Althusser's argument for Marx's anti-humanism hinges upon a narrow reading of Thesis VI: He reads Marx's statement "the human essence is no abstraction inherent in each single individual" to mean that Marx believes that there is no human essence, ergo, no human nature. The validity of Althusser's claim therefore hinges on how we interpret Thesis VI. Geras analytically parses the three plausible interpretations:

Human nature ("the human essence") is . . .
i. *conditioned by* the ensemble of social relations.
ii. *manifest in* the ensemble of social relations.
iii. *determined by* the ensemble of social relations; therefore, the concept of human nature has no meaning.[44]

Geras argues persuasively that Marx meant interpretation (i) and probably (ii), but definitely not (iii). Marx was not a determinist and did not deny human nature. He sought an approach to human nature that neither abstracted from individuals nor defined humans religiously, but explained human nature historically through study of the "ensemble of the social relations" in different times and places. (I return to this in chapter 7.)

Today any discussion of these matters must address the human

---

43 Norman Geras, *Marx and Human Nature: Refutation of a Legend* (New York: Verso Books, 2016 [1983]); cf. McLellan, *Marx Before Marxism*.

44 Geras, 46. These are my renderings of Geras's statements. While I find much to praise in Geras's study, I should acknowledge that it gives—for my purposes—too little emphasis to natural and historical processes. I would also quibble with Geras's discussion of Marx's conception of "nature as man's creation," which Geras reads from a post-Darwinian vantage, though Marx's statements about "nature as man's creation" predate his reading of the *Origin of Species* by around fifteen years.

relationship with nature. So, to Marx's Thesis VI we should add the following: The "ensemble of social relations" (HH) is always already mediated by social relations with nature (HN). A restatement of Thesis VI therefore might be to say that human nature is conditioned by and manifest in the ensemble of socio-ecological relations. Formulaically: Human nature = HH×HN.

Marx Contra Hegel in *The German Ideology*

Marxian natural history is implicitly a critique of Hegel's teleology. Hegel conceives the development of reason as culminating in spirit's self-recognition. Hegel's philosophy provides Marx with a conception of nonlinear historical progress which he would recast to theorize the emergence of capitalism and its eventual overcoming as communism. Yet Marx's inheritance from Hegel on this point was complicated by the inherent complexity of Hegel's philosophy. Let us return briefly to Hegel's lectures on *The Philosophy of History*, where Hegel summarizes his perspective:

> Divine Wisdom, i.e., Reason, is one and the same in the great as in the little; and we must not imagine God to be too weak to exercise his wisdom on the grand scale. Our intellectual striving aims at realizing the conviction that what was intended by eternal wisdom, is actually accomplished in the domain of existent, active Spirit, as well as in that of mere Nature. Our mode of treating the subject is, in this aspect, a *Theodicaea*—a justification of the ways of God . . . so that the ill that is found in the World may be comprehended, and the thinking Spirit reconciled with the fact of the existence of evil. Indeed, nowhere is such a harmonizing view more pressingly demanded than in Universal History.[45]

Hegel conjures the prospect of philosophical history and universal history, therefore, as a "justification of the ways of God." *Theodicaea* (theodicy), a term from Christian theology, characterizes attempts to explain God's wisdom in the face of all of our world's cruelties. Theodicies answer questions like: If a loving God created this world,

---

45 Hegel, *The Philosophy of History*, §17; 15.

why is there so much evil? Given that our world is so unjust, can we be sure it is actually God's creation? When, at the outset of his lectures, Hegel says that his "mode of treating" the philosophy of history is *Theodicaea*, he has promised us a teleological mode of reasoning about history. Let us turn to the conclusion of his lectures to find their ultimate end:

> Philosophy concerns itself only with the glory of the Idea mirroring itself in the History of the World. Philosophy [is concerned with] the recognition of the process of development which the Idea has passed through in realizing itself—i.e., the Idea of Freedom, whose reality is the consciousness of Freedom and nothing short of it. That the History of the World, with all the changing scenes which its annals present, is this process of development and the realization of Spirit—this is the true *Theodicaea*, the justification of God in History. Only this insight can reconcile Spirit with the History of the World—viz., that what has happened, and is happening every day, is not only not "without God," but is essentially His Work.[46]

Hence the story told by Hegel's philosophical history is that the course of world history = development toward Spirit's realization = God's work.

Marx knew this story well. A radical Enlightenment thinker, he was unpersuaded by Hegel's theodicy. Marx respected Hegel's genius but sought to bring him back down to Earth:

> Criticism has plucked the imaginary flowers on the chain not in order that man shall continue to bear that chain without fantasy or consolation but so that he shall throw off the chain and pluck the living flower. The criticism of religion disillusions man, so that he will think, act, and fashion his reality like a man who has discarded his illusions and regained his senses, so that he will move around himself as his own true sun . . . It is therefore the *task of history* . . . to establish the *truth of this world*. It is the immediate *task of philosophy*, which is in the service of history, to unmask self-estrangement in its *unholy forms* once the *holy form* of human self-estrangement has been

---

46 Hegel, *The Philosophy of History*, 457

unmasked. Thus the criticism of heaven turns into the criticism of Earth.[47]

While the character of Marx's critique of Hegel never changed after the 1840s, its *expression* and Marx's *standpoint* toward history continued to evolve as Marx learned more about nature and society from the natural sciences.

The antithesis of nature and history is created

Marx's deepest critique of Hegel's theodicy-cum-philosophy of history is that it abstracts away from the practical and earthy qualities of human life. For Marx, history begins with humans producing a means to live on Earth (HN), the social relations of humans (HH), and their complex interrelations (HN×HH). Marx contends that by abstracting away from this, Hegel—like many thinkers before and after him—generated an artificial, ideological separation of nature and history:

> In the whole conception of history up to the present this real basis of history has either been totally neglected or else considered as a minor matter quite irrelevant to the course of history. History must, therefore, always be written according to an extraneous standard; the real production of life seems to be primeval history, while the truly historical appears to be separated from ordinary life, something extra-superterrestrial. *With this the relation of man to nature is excluded from history and hence the antithesis of nature and history is created.* The exponents of this conception of history have consequently only been able to see in history the political actions of princes and States, religious and all sorts of theoretical struggles, and in particular in each historical epoch have had to *share the illusion of that epoch.*[48]

---

47   Marx, *A Contribution to the Critique of Hegel's Philosophy of Right*, trans. R. Livingstone and G. Benton (London: New Left Books/Penguin, 1975 [1844]), 244–5; cf. MECW III: 176.

48   Marx and Engels, *German Ideology*, MECW V: 55, italics modified. Marx draws out an ethical implication through an eviscerating critique of an article by Rudolph Matthäi. Marx warns against treating ahistorical nature as such as a basis of, or criteria for, social ethics: "We should be only too pleased to believe that 'all the social virtues' of our true socialist are based 'upon the feeling of natural human affinity and unity,'" as proposed by Matthäi, "even though feudal bondage, slavery, and all the social

Marx claims that the theoretical legitimation of an artificial separation of humans from nature—the failure to recognize HH×HN as the ground of historical analysis—is the root cause of a view of the world in which history means nothing but "the political actions of princes and States, religious and all sorts of theoretical struggles," which in turn generates illusions particular to each epoch.[49] Marx's argument is that Hegel—like other thinkers of his time—inhabited an illusion in which human society and its history could be grasped apart from our practical, worldly relations. This error was motivated, on one hand, by the emerging bourgeois social relations which encouraged this illusion, and on the other hand, by a Christian theological drive to subtract the human body, its earthiness and desires, from nature and history:

> Christianity wanted to free us from the domination of the flesh and "desires as a driving force" ... because it regarded our flesh, our desires as something foreign to us; it wanted to free us from determination by nature only because it regarded our own nature as not belonging to us.[50]

Of course, Christianity is not alone here. It seems to me that all the universal religious traditions—by which I mean those which have succeeded in organizing political spaces by subordinating social differences under an encompassing spiritual framework (Hinduism, Buddhism, Judaism, Islam, and so on)—have found a powerful tool in teaching the separation of the "merely existing" human-body-in-the-world and the "true" spirit behind it. The consequence of this teaching is to divide human life, to separate us from our natural existence and bodily desires. The latter are effectively denaturalized and repressed:

---

inequalities of every age have also been based upon this 'natural affinity.' Incidentally, 'natural human affinity' is an historical product which is daily changed at the hands of men" (*German Ideology*, MECW V: 479, my italics). This passage is cited by Cornell West, *The Ethical Dimensions of Marxist Thought* (New York: Monthly Review Press, 1991), 74–101, in an illuminating discussion of Marx's adoption of radical historicism.

49 Marx illustrates his Left-Hegelian style with his insistence that each period is governed by an "illusion" specific to its conception of what it is, historically, basically an adaptation of Hegel's claim about spirit. Marx turns Hegel's "Geist"—the spirit of an age—into an "illusion" specific to it. Around 1932, Gramsci will revise Marx's "illusion" as "hegemony."

50 Marx and Engels, *German Ideology*, MECW V: 254.

"For if I myself am not nature, if my natural desires, my whole natural character, do not belong to myself," but are merely an epiphenomenal trace of an ideal, a material effect of the presence of something like a soul, then it follows that "all determination by nature—whether due to my own natural character of to what is known as external nature—seems to me a determination by something foreign, a fetter, compulsion used against me."[51] One effect of the religious denial of an evolving human nature—of our being at once natural and historical beings—is to inculcate the sense that every limit we encounter in the world is an external force set against us (a "determination by something foreign").

Separation from nature legitimates religious authority. An illustration may be useful here. For me to believe that a drought can be fixed by my making a sacrifice to a rain god, I must believe that the drought is something that is connected to my spirit by that god. This relation—between me and the rain god, which is to say, between me and myself as a natural-historical being—is required for there to be a mediator, a priest, between me and the god, that is, between me and nature.[52] The insertion of that mediator is a socio-historical event bound up with human relations with nature: Religion is cause and effect of HH×HN. Fortunately for humanity, Marx argues, religion finds its historical limit when human nature rebels against these constraints in the form of desire: "Christianity has indeed never succeeded in freeing us from the domination of desires."[53]

---

51 Marx and Engels, *German Ideology*, MECW V: 254.

52 By implication, the fundamental, enabling condition of all fetishism is therefore the separation of humanity from nature. On fetishism, see chapter 5.

53 If communism = human emancipation, then communism presupposes the social liberation of desire. Marx's natural history may therefore provide a means for conceiving demands for the destruction of sex- and gender-based repression. This would require a sympathetic reading of passages in Marx such at this: "The immediate, natural relationship of human being to human being is the relationship of man to woman. In this natural species-relationship the relation of man to nature is immediately his relation to man, just as his relation to man is immediately his relationship to nature, his own natural condition. Therefore this relationship reveals in a sensuous form ... the extent to which the human essence has become nature for man or nature has become the human essence for man. It is possible to judge from this relationship the entire level of development of mankind": Marx, Paris MS III (1844), New Left Books/Penguin, 347; MECW III: 295–6. That last expression, *die ganze Bildungsstufe des Menschen*, could be translated as "degree of human education," that is, transformation.

What underpins these theological ideas in everyday life? What gives religion its power? The answer is that it provides us with an encompassing explanation of what we are and where we stand in terms of nature and history.[54] Marx writes that the concept of Creation is "very difficult to dislodge from popular consciousness. The fact that nature and man exist on their own account is incomprehensible to it, because it contradicts everything tangible in practical life"; nevertheless, Marx speculates, the development of natural sciences will dissolve these popular prejudices over time. For the doctrine of the "creation of the earth has received a mighty blow" from "the science which presents the formation of the earth, the development of the earth, as a process, as a self-generation. *Generatio aequivoca* [spontaneous generation] is the only practical refutation of the theory of creation."[55]

Marx here correctly predicted the development of a scientific theory to explain the self-generation of nature into forms of life—including the evolution of humanity itself—which would transform popular conceptions about human nature, religion, and human history. Marx made this prediction in the mid-1840s, when the natural sciences generated evidence for geological time scales and a deeper understanding of the formation of the earth, but no adequate theory to explain the evolution of species. There was at the time a great deal of speculation but no scientific basis for rebutting Creation stories with anything better than the vague concept of spontaneous generation of species. In 1844–45, the furthest point that Marx could reach was to write:

> You can reply: "I do not want to postulate the nothingness of nature, etc. I ask you about its genesis, just as I ask the anatomist about the formation of bones, etc." But since for the socialist man the entire so-called history of the world is nothing but the creation of man through human labor, nothing but the emergence of nature for man, so he has the visible, irrefutable proof of his birth through himself, of his genesis. Since the real existence of man and nature has become evident in practice, through sense experience, because man has thus

---

54 Freud reached similar conclusions. I discuss his critique of religion, and of Marx, in chapter 8.
55 Marx, Paris MS III (1844), §X; New Left Books/Penguin, 356; MECW III: 304–5.

become evident for man as the being of nature, and nature for man as the being of man, the question about an alien being, about a being above nature and man—a question which implies the admission of the unreality of nature and of man—has become impossible in practice.[56]

Marx saw that socialism would require a theory of "the history of the world ... [and] the creation of man": A natural history of humanity.

Although some Marxists have found in these lines evidence of a fully-fledged, dialectical scientific theory, I regard them as question begging. For, in my view, it was not sufficient for Marx to claim "visible, irrefutable proof of his birth through himself, of [humanity's] genesis," and hence to reject the existence of God ("a being above nature and man") without anything more concrete than these philosophical objections to Hegel. Reading Marx's Paris manuscripts from the vantage of the present, it is clear that his thought at this stage was missing something specific: A scientific framework for explaining the natural-historical becoming of the human species. But we should not judge Marx harshly for this lack, since no one had such a thing in 1844. Except Darwin.

---

56   Marx, Paris MS III (1844), §XI; New Left Books/Penguin, 357; MECW III: 305–6.

2

Darwin and the Defeat of Teleology

Darwin's Discovery

Few scientists are as famous as Charles Darwin. There are things about him that are commonly known: He was appointed as the naturalist for a world-spanning voyage on the HMS *Beagle*, during which he developed his theory of evolution; he subsequently proved that theory; although Darwin's ideas raised the ire of religious thinkers, he remained a faithful Christian. Yet none of those statements are true. Darwin developed his theory—of descent through natural selection, not *evolution*—in *London*; he never *proved* that theory; late in life, Darwin became an agnostic.[1]

I begin with these points (elaborated below) not to quibble, but to emphasize something important about discussions of Darwin. Fate and fame have burdened him with a false sense of familiarity. If we want to understand what Darwin accomplished, how this influenced Marx, and its meaning for us today, we must be willing to slow down and revisit some well-trod ground.

My aim is to clarify Darwin's achievements, why his ideas were revolutionary, and what Marx gained from them. For the sake of my argument, I distinguish between (a) what Darwin wrote in the early editions

---

1 Darwin uses the verb "to evolve" only once in *Origin*, in the book's final line: "from so simple a beginning endless forms most beautiful and most wonderful have been, and are being, evolved" (quoting the 1859 and 1860 editions read by Marx).

of *Origin* read by Marx, (b) texts that Darwin wrote later (his 1871 study of human evolution, *The Descent of Man*, for example), and (c) all that we know today about evolution.[2] This chapter focuses on (a). Apropos (c), we have learned so much about the science of evolution since 1859 that it is difficult to appreciate from the present just how great a leap Darwin made. Darwin had no knowledge of genetics. The rediscovery of Mendel's pea plant experiments in variation and inheritance was still decades off when the debates surrounding the implications of Darwin's theory spread around the world. Darwin's theory was not widely accepted until the 1940s, with the reconciliation of his conceptions of natural selection and fitness with formal models of genetic heredity and change—the "modern synthesis."[3] Recent research into epigenetics (among other topics) has generated proposals for a new synthesis (but no standardized name).[4] The theory of evolution is still evolving.

If we have learned so much since 1859, one might wonder, Does Darwin still matter? The answer to that question depends on how we answer another: What did Darwin do that was genuinely important and novel?

There is no scholarly consensus on the answer to these questions.[5] By my reading, Darwin made three fundamental achievements, each with

---

2   Darwin tinkered with *Origin* between 1859 and 1872, introducing significant changes and publishing six distinct editions. Since Marx initially read *Origin* in late 1860, he most likely read the second edition; a few years later, he acquired and read the German translation, by H. G. Bronn, of Darwin's second edition. Therefore, in this chapter, unless otherwise noted, I quote from Darwin's second edition, drawing pagination from the excellent variorum edition prepared by Barbara Bordalejo for Darwin Online, darwin-online.org.uk, and commenting upon significant changes as appropriate. On evolutionary thought since Darwin, cf. Ernst Mayr, *One Long Argument: Charles Darwin and the Genesis of Modern Evolutionary Thought* (Cambridge, MA: Harvard University Press, 1991); S. J. Gould, *The Structure of Evolutionary Theory* (Cambridge, MA: Harvard University Press, 2002), $I; and the essays in part III of M. J. Hodge, ed., *The Cambridge Companion to Darwin* (Cambridge, UK: Cambridge University Press, 2009).

3   The ideas underlying the synthesis emerged earlier: Cf. Ronald Fisher, *The General Theory of Natural Selection* (Oxford: Oxford University Press, 1930); J. B. S. Haldane, *The Causes of Evolution*, (London: Longmans Green, 1932).

4   For one proposal, see Michael Skinner and Eric Nilsson, "Role of environmentally induced epigenetic transgenerational inheritance in evolutionary biology: Unified Evolution Theory," *Environmental Epigenetics* 7, no. 1 (2021), dvab012.

5   The literature on Darwin and his theory is far too vast to summarize. My assessment of his achievements draws from sources cited below. I thank Mary Evelyn Tucker for supporting a fellowship on the theory of evolution which provided an early workshop

some influence upon Marx's critique of political economy. First, as is well known, Darwin assembled the first coherent theory of species change and natural selection. This achievement required two earlier, but less well-recognized, breakthroughs. On one hand, Darwin proposed that the differences between individuals (upon which natural selection works) arises through *chance*. He therefore identified uncertainty as a fundamental dynamic within natural history. On the other hand, by providing an explanation for how changes in forms of life occur that required no plan nor end point, Darwin destroyed teleology in natural history.[6] Darwin laid the groundwork for these achievements in 1837–1844 then waited until 1859 to publish *Origin*. The conditions of Darwin's breakthrough and his delay in publishing his world-shaking ideas are instructive for reading *Origin*, so I will comment upon them before turning to his famous book.

Darwin, natural historian

Darwin was born February 12, 1809, at the Mount, his family home in Shrewsbury, England. His family was affluent: His father was a doctor, and both parents came from relatively wealthy families. Darwin was expected to follow his father into medicine, so, in 1825, he went to the University of Edinburgh Medical School with his brother Erasmus to begin training. Yet Charles was an indifferent medical student, skipping lectures to explore the countryside and cultivating a beetle collection. His father ultimately redirected his son toward Cambridge to study divinity (Darwin was open to becoming a "naturalist parson"). Nevertheless, his years of exploring at Edinburgh (1825–1828) had left a mark. In 1826, Charles took a course in natural history. Though he found it "dull," he met people who would have important influences upon his life, such as the philosopher James Mackintosh and the natural historian William MacGillivray ("I had much interesting natural-history talk with him").[7] Darwin also joined

---

for some of these ideas. I thank Martina Ramirez for teaching me the basics of evolutionary theory.

6 Not all Darwin scholars agree. For a contrary reading of Darwin (and Marx), see Stanley Edgar Hyman, *The Tangled Bank: Darwin, Marx, Frazer, and Freud as Imaginative Writers* (New York: Atheneum, 1962).

7 Darwin, *The Autobiography of Charles Darwin: 1809–1882* (New York: W. W. Norton & Co., 1993 [1887]), 52, 53.

the Plinian Society—a natural history study group that exposed Charles to materialism, a philosophical school that sought scientific, non-religious explanations of processes of natural history.[8] He later later recorded in his autobiography that these conversations "had a good effect on me in stimulating my zeal."[9]

In his autobiography, Charles includes one sentence in passing describing another influence:

> By the way, a negro lived in Edinburgh, who had travelled with Waterton and gained his livelihood by stuffing birds, which he did excellently; he gave me lessons for payment, and I used often to sit with him, for he was a very pleasant and intelligent man.[10]

This "intelligent" man was named John: Because he had been enslaved on Edmonstone's Guyana plantation, his name is recorded as John Edmonstone. Little is known about John or the lessons he gave to Darwin, but the emerging consensus is that they were important for Darwin's development—not only for what John taught about taxidermy, but for what he likely shared with Darwin about the geography and animals of South America.[11] Since Darwin came from a line of anti-slavery abolitionists, they likely discussed John's experience with slavery, too.

Three years of study at Cambridge followed. Darwin records that his "time was wasted" there, but, as at Edinburgh, he gained a number of important relationships and studied major works in the tradition of natural history:[12] "I read with care and profound interest [Alexander von] Humboldt's [1814–1825] *Personal Narrative*. This work and Sir J. Herschel's

---

8 On Darwin's materialism, cf. Edward Manier, *The Young Darwin and His Cultural Circle* (Boston: Reidel, 1978), chapter 4; Michael Ruse, *The Darwinian Revolution* (Chicago: University of Chicago Press, 1979); Mauricio Vieira Martins, *Marx, Spinoza, and Darwin: Materialism, Subjectivity, and Critique of Religion* (London: Springer Nature, 2022).

9 Darwin, *Autobiography*, 50; cf. Adrian Desmond and James Moore, *Charles Darwin* (Oxford: Oxford University Press, 1991), 31–4.

10 Darwin, *Autobiography*, 51.

11 Adrian Desmond and James Moore, *Darwin's Sacred Cause: How a Hatred of Slavery Shaped Darwin's Views on Human Evolution* (Chicago: University of Chicago Press, 2014). I recommend the Natural History Museum's website on John Edmonstone: Nhm.ac.uk.

12 For example, John Stevens Henslow, who facilitated Darwin's appointment as naturalist to the *Beagle*, and Leonard Jenyns, with whom Darwin "had many a good walk and talk . . . about Natural History": Darwin, *Autobiography*, 67.

[1830] *Introduction to the Study of Natural Philosophy* stirred up in me a burning zeal to add even the most humble contribution to the noble structure of Natural Science."[13] When the offer came to join the voyage of the *Beagle*—a trip assigned to confirm geographical and hydrographical data around coastal South America—Darwin found an outlet for all his zeal. The *Beagle*'s captain, Robert FitzRoy, recounts the circumstances:

> Anxious that no opportunity of collecting useful information during the voyage, should be lost; I proposed to the Hydrographer that some well-educated and scientific person should be sought for who would willingly share such accommodation as I had to offer, in order to profit by the opportunity of visiting distant countries yet little known ... [After some consultations, this led to Darwin], a young man of promising ability, extremely fond of geology, and indeed all branches of natural history.[14]

This retrospective explanation elevates Darwin's scientific role aboard the *Beagle*. Darwin's chief assignment was to take meals with FitzRoy, who—abiding by the rules governing British sea voyages of that era—never socialized with the men under his ship's command.[15]

Their voyage lasted five years. Our image of Darwin on that famous voyage is dominated by finches, tortoises, fossils, and bones. Certainly, Darwin observed animals and collected specimens. But he mainly spent those years reading and thinking. The *Beagle* carried an impressive library with at least four hundred volumes "housed in bookcases in the poop cabin at the stern of the ship, which was also Darwin's cabin. Thus, Darwin lived and worked in the *Beagle* library for nearly five years."[16]

---

13 Darwin, *Autobiography*, 51.

14 Admiralty's instructions for the HMS *Beagle*, quoted in Darwin, *A Naturalist's Voyage: Journal of Researches into the Natural History and Geology of the Countries Visited During the Voyage of HMS Beagle: With Maps and Illustrations* (London. Penguin, 1989 [John Murray (1882; 1839)]), 380.

15 On the *Beagle*, Darwin "did devote his attention to natural history. But he was brought on board for another purpose": To take meals with Captain FitzRoy, who needed someone to talk for sake of his mental health. The ship's surgeon, Robert McCormick, "originally held the official position of naturalist": Stephen Jay Gould, "Darwin's sea change," in *Ever Since Darwin: Reflections in Natural History* (New York: W. W. Norton & Co., 1977), 28.

16 Quoting John van Wyhe and the Darwin Online Correspondence Project, where the library is painstakingly pieced together: Darwin-online.org.uk. Roughly

This was the site of Darwin's greatest education. To say the least, he made the most of the opportunity. He first gained fame with the publication in 1839 of his *Journal of Researches into the Geology and Natural History of the Various Countries Visited by HMS* Beagle.[17]

This brief biography allows me to underscore a point that will take on greater significance in chapter 3. Although Darwin is today called a scientist, biologist, or evolutionary theorist, he never used those terms to describe himself.[18] Rather, he called himself a "naturalist," shorthand for "natural historian." Darwin inherited the natural history tradition that encompassed Buffon, Cuvier, Humboldt, Lyell, Agassiz, and others. He absorbed this tradition not only through school and his readings on the *Beagle*, but also through his family; his grandfather Erasmus Darwin, an influential natural historian, was an early advocate of the idea that species arise from other species (he claimed that "the strongest and most active animal should propagate the species, which should thence become improved").[19]

Given the breadth of research in natural history during this nineteenth century and the dissemination of such ideas, it is extremely likely that someone other than Darwin would have put together a theory of descent by natural selection if he had not done so. Indeed, Alfred Russel Wallace arrived independently at some of Darwin's core ideas, and other naturalists were examining species change.[20] Although Darwin made

---

one-third of the books in the *Beagle*'s library concern natural history, one-third are travel books, one-third other topics. Notably, the only work of political economy on the ship was Harriet Martineau's (1832) *Illustrations of Political Economy*, a popular representation of Adam Smith influenced by French positivism.

17 Darwin, *Voyage of the* Beagle.

18 "The most influential biologist to have lived," for example: John Bowlby, *Charles Darwin: A New Life* (New York: W. W. Norton & Co., 1992), 1. Darwin did not call himself a "scientist," a gender-neutral term only entering into use, but "man of science"—and that rarely.

19 Erasmus Darwin, *Zoonomia, or, The Laws of Organic Life: In Three Parts*, volume 1 (Boston: Thomas & Andrews, 1809 [1794–96]).

20 On June 18, 1858, Wallace (1823–1913) shared his paper, "On the Tendency of Varieties to Depart Indefinitely from the Original Type," with Darwin. Darwin and Wallace publicly presented their papers in parallel (see note 37); then Darwin hurried to complete and publish *Origin*. Wallace's theory, while sharing numerous points with Darwin's, was not as well-developed (on the question of the cause of variation, for example). Wallace's legacy as a theorist of evolution was also tarnished by his subsequent metaphysical turn. On Wallace, see Andrew Berry, ed., *Infinite Tropics: An Alfred Russel Wallace Anthology* (New York: Verso Books, 2003).

undeniable contributions, it could be claimed that he was fortunate to receive practically sole credit for the discovery of descent by natural selection.[21] To appreciate how Darwin arrived first at his theory, the place to begin is not the Galapagos Islands, but his unpublished London texts.

Darwin's unpublished writings, 1837–44

In the spring of 1837, shortly after his return to London, Darwin filled a series of private research notebooks, identified today by letters (Notebook B, C, M, and so on). Darwin writes incomplete sentences with abundant abbreviations, "frenetic" punctuation, and "spelling peculiarities"; some entries are crossed out (but still legible), suggesting that Darwin changed his mind.[22] The texts suggest that as Darwin was working out his theory, he was overwhelmed by the strangeness of the ideas and their implications.

Consider an illustration: The (relatively legible) page 18 from Notebook B, which Darwin started writing in July 1837 and which includes some of his earliest statements about natural selection. There Darwin writes: "Each species changes / does it progress. / Man gains ideas. / the simplest cannot help becoming more complicated; & if we look to first origin then must be progress."[23] He is struggling to determine appropriate language to describe the changes that species undergo and whether this constitutes "progress." Darwin asks, as a species changes, does it progress? *Progress* was a loaded term: A central tenet of Victorian liberalism (and British imperialism) was that science, reason, and capital would all contribute to the

---

21 I would not say that Darwin was "lucky," as some have; he paid a considerable price—not only in terms of the labor time spent on his research, but also in terms of his personal health, which failed during the voyage of the *Beagle* and remained poor in the decades he worked out his theory. The literature on his (not well-diagnosed) medical problems has reached no conclusion, but on one widely held view, the cause was mental and psychological more than physical; the stress of elaborating his theory seriously disrupted his health. Another (not mutually exclusive) view concerns the death of his mother in 1817, when Charles was eight: The critical event of his youth. John Bowlby contends that the anxiety and depression caused by this event were subsequently piqued by worry over his theory: Bowlby, *Darwin: A New Life*.

22 Editorial note in Darwin, *On Evolution*, T. Glick and D. Kohn, eds. (Indianapolis: Hackett, 1996) 49.

23 From "Notebook B," in Darwin, *On Evolution* 18. (David Kohn has "there must be progress.")

betterment—the *progress*—of society. Observing that "the simplest [forms of life] cannot help becoming more complicated," if we consider the "first origin" of forms of life on Earth, then from that vantage, it could be said that we see a sort of progress in natural history.[24]

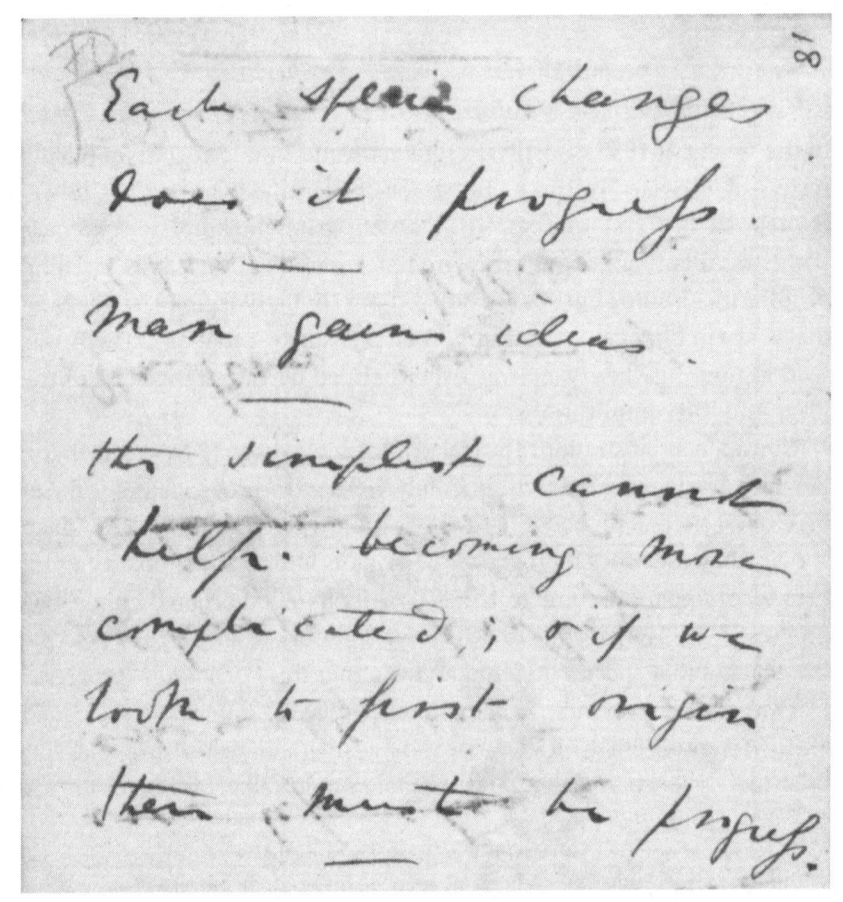

Figure 2.1: Darwin writes, "Each species changes / does it progress. / Man gains ideas. / the simplest cannot help becoming more complicated; & if we look to first origin then must be progress." Selection of page 18 from Darwin's Notebook B. Cambridge University Library, DAR 121. Printed with permission of Cambridge University Library.

---

24 Notably, Darwin interjects humans into this line of thought ("we gain ideas: Is that not progress?") but does not answer this question. On progress in Darwin's theory, see Curtis Johnson, *Darwin's Dice: The Idea of Chance in the Thought of Charles Darwin* (Oxford: Oxford University Press, 2015), 62, 162.

Still, the character of this progress—a greater complication of forms of life through time—is highly qualified in Darwin's research notebooks. What seems to become more complex are species, but species are complex phenomenon. Since Darwin's most important book is entitled *On the Origin of Species*, readers would expect that Darwin provides a clear definition of species, but the opposite is true. Darwin puzzled over the meaning of species his whole life, never offering a final definition (and the discussion has continued to the present).[25] The conclusion that Darwin reached at the time of his notebooks was that species was a valid concept for distinguishing between forms of life, and yet no ultimate definition of a species was possible: All species change through time, and every species comes from another, so the distinctions between and boundaries around species are fluid. Darwin invites us to grasp a species at once as a thing, a real phenomenon, but also as a spatio-temporal process. In chapter 2, for instance, after surveying the differences found within species and among subspecies, Darwin writes that "varieties cannot be distinguished from species—except, first, by the discovery of intermediate linking forms, and secondly, by a certain indefinite amount of differences between them."[26] In lieu of fixed species, Darwin substituted "varieties" marked by "intermediate linking forms" and "indefinite" differences. Further, he announces that *all* varieties are "incipient species"; it follows that any given species is nothing but the historically temporary form of some particular variety of a previously existing species. To put this in philosophical terms, we could say that Darwin introduced a dialectical conception of species that both sustains the concept (species qua form of *being*) while constantly emphasizing its *historical becoming*.

This thinking also suggested to Darwin a novel spatial image portraying the natural history of species and their interrelations. Species should not be seen as independent beings portrayed alongside one another, as in so many previous figures; nor as beings unfolding along a line, progressing from one stage to another; but arrayed as branches of a tree of life (see Figure 2.2). Darwin's famous "tree" sketch in Notebook B—

---

25 For recent views on the species concept and speciation, see James Sobel et al., "The biology of speciation," *Evolution* 64, no. 2 (2010): 295–315; cf. the papers in the special issue of *Journal of Evolutionary Biology* 35, no. 9.

26 Darwin, *Origin*, fifth edition (1869), 69.

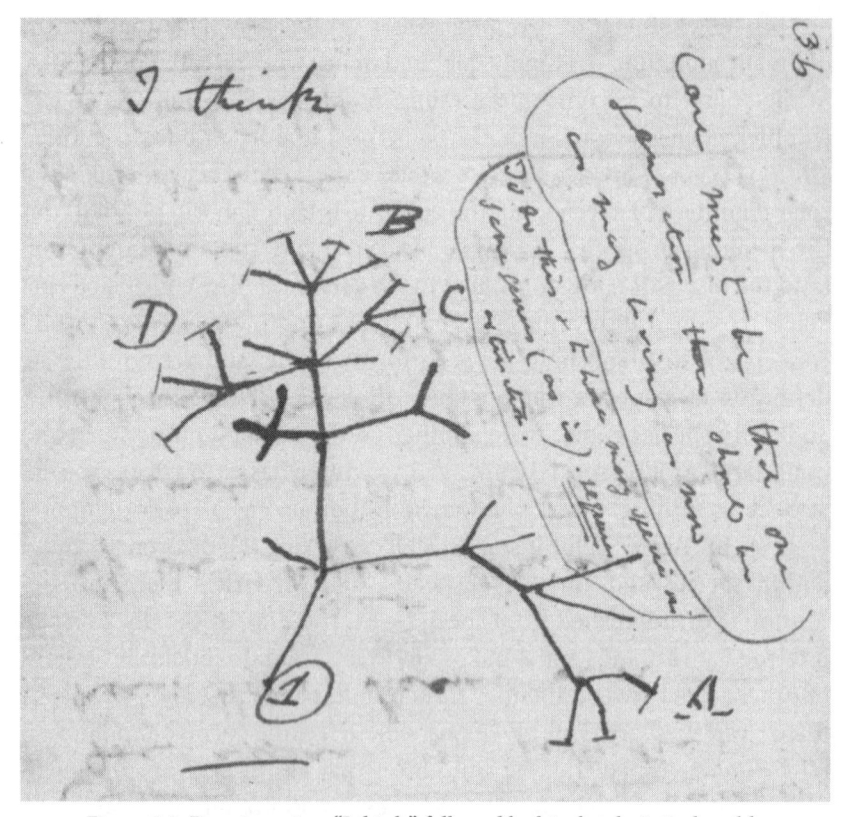

Figure 2.2: Darwin writes, "I think," followed by his sketch. Later he adds two notes in the upper right corner. The first says, "Case must be that one generation then should be as many living as now." The second: "To do this & to have many species in the same genus (as is) *requires* extinction." Selection of page 36 from Darwin's Notebook B. Cambridge University Library, DAR 121. Printed with permission of Cambridge University Library.

re-presented in *Origin* with a more formal, linear diagram—is a rare instance of a figural representation that captures a creative leap of imagination and a world-historical refiguring of humanity's place in nature. It expresses a new conception of life forms that do not progress linearly from one stage to another, but in which chance and natural selection produce new branches, which in turn vary with themselves—sometimes generating new species, eventually expiring.

The diagram helps us grasp why Darwin avoids terms in *Origin* that imply progress. Darwin describes species change more neutrally as "descent with modification." Even "to evolve" was suspect, since it implies unfolding toward some end. Darwin's thought was driving toward the end of such ends.²⁷ He only uses "evolve" once in *Origin*, in the final line (discussed below). By contrast, Darwin frequently employs the verb "to adapt" and its cognates; his conception of adaptation was undoubtedly one of his greatest achievements. Darwin correctly recognized a significant difference in meaning between the two words—a difference that is commonly elided today, when the words "evolve" and "adapt" are often treated—wrongly—as synonyms. Both words have Latin roots in verbs (to evolve, to adapt) that became nouns (evolution, adaptation).²⁸ In Latin, to adapt means to *fit*; to evolve means to *unfold*. The meaning of adapt is more "other" oriented and relational: The being that adapts is adapting to, fitting with, some *other* being or beings. To evolve, by contrast, is a process by which a being unfolds as *itself*; such a process is more "self"-oriented and typically implies forward progression.²⁹ Comprehending this distinction is crucial to appreciating Darwin's wariness about "evolution" and clarifies the fundamental gap between his conception of species change and that of Jean-Baptiste de Lamarck (1744–1829), an early proponent of species transmutation. Lamarck claimed that characteristics acquired during life were passed along to offspring. By contrast, in *Origin* individual organisms do not evolve

---

27 Stephen Jay Gould, "Darwin's dilemma," in *Ever Since Darwin*, 34–45.

28 This paragraph draws upon the entries for "evolution" and "adaptation" from the *Oxford English Dictionary*, second edition (Oxford: Oxford University Press, 1989).

29 The earliest biological use of "evolution" in English dates to around 1670 ("a gradual and Natural Evolution and growth of [insect] parts"); the first references to a *theory* of biological evolution occur almost one century before Darwin published *Origin*. The concept "evolution" and its application to society predates Darwin. Debates about social evolution were a key feature of the episteme in Paris during the 1840s, when Marx was there, filling his earliest notebooks on capitalism (see chapter 1). The OED cites Harriet Martineau's 1853 abridged translation of Comte's *Positive Philosophy* (1830–42, volume II, vi): "The elements of our social evolution are connected, and always acting on each other." (Martineau was the author of the only volume on political economy in the *Beagle* library.) "Evolution" would later come to be a synonym with "development" in English: Both words share a root (*volvere*) referring to the action of unfolding something; both have come to mean a directed unfolding, as when we say that a story evolves from one plot point to another. Of course, "development" would go on to play a fundamental role in the world-encompassing hegemony of capital: Joel Wainwright, *Decolonizing Development: Colonial Power and the Maya* (Oxford: Wiley, 2011).

to fit their environment.[30] They simply *are* a certain way; consequently, they happen (or do not happen) to be likely to survive and have offspring. Whereas one could speak of the evolution of an idea, or the evolution of a song, in Darwin's theory, an individual member of a species does not evolve. And adaptation occurs at the level of the population or species.[31]

The manuscripts of 1842–44

Darwin labored privately to refine his ideas. In 1842, when he and his wife Emma bought Down House and left central London for the sake of his health, Darwin started writing in more carefully articulated prose while preparing a preliminary "sketch" of natural selection. By July 1844, he had completed a manuscript of approximately 52,000 words, which he gave to Emma with instructions that it should be edited and published if he were to die before his book was completed. This 1844 essay contains several formal qualities that recur in Darwin's *Origin* edition of 1859. He begins with variation under domestication, then considers variation in nature; then, before introducing the concept of natural selection, Darwin introduces the image of a transcendent Being to invite the reader to imagine a force that could discern minute differences between individual organisms:

---

30 In 1844, Darwin mocked Lamarck's ideas: "I am almost convinced (quite contrary to opinion I started with) that species are not (it is like confessing a murder) immutable. Heaven forfend me from Lamarck nonsense of a 'tendency to progression' 'adaptations from the slow willing of animals' &c,—but the conclusions I am led to are not widely different from his—though the means of change are wholly so—I think I have found out (here's presumption!) the simple way by which species become exquisitely adapted to various ends": Darwin, Letter to J. D. Hooker, January 11, 1844, accessed via the online Darwin Correspondence Project.

31 Darwin's conception of adaptation was incomplete, however, and has generated some confusion. In the first instance, *Origin*'s treatment of adaptation often seems to presuppose a conceptual separation of species from their environment. Biological beings—populations or individuals of a species—are discussed as if they are subjected to competition but with no effect upon their environment. Yet such competition is never limited only to changes in the frequencies of one or another species, and the existence of any number of beings has some effect upon environment. The contemporary expression, "evolving ecosystem," is a post-Darwinian concept that incorporates abiotic factors into evolutionary theory.

Let us now suppose a Being with penetration sufficient to perceive the differences in the outer and innermost organization [of individual organisms] quite imperceptible to man, and with forethought extending over future centuries to watch with unerring care and select for any object the offspring of an organism produced under the foregoing circumstances; I can see no conceivable reason why he could not form a new race (or several were he to separate the stock of the original organism . . . on several islands) adapted to new ends.[32] As we assume his discrimination, and his forethought, and his steadiness of object, to be incomparably greater than those qualities in man, so we may suppose the beauty and complications of the adaptations of the new races and their differences from the original stock to be greater than in the domestic races produced by man's agency . . . With time enough, such a being might rationally . . . aim at almost any result.[33]

Darwin uses a similar passage to the same effect in *Origin*. In both texts, Darwin invites us to conceive nature metaphorically as an agent—a (masculine) Being, akin to an animal breeder, using the tools of artificial selection—to appreciate a power capable of generating new species from existing ones.

Although he never published this essay, 1844 was a very good year for Darwin. He worked out several ideas that would eventually appear in his masterpiece, *Origin*. Still, Darwin remained dissatisfied with his theory. In the passage just quoted, we can see one of the complexities. While, for literary effect, Darwin figures natural selection here as a "Being" that "perceives" and "selects," he knew it was no such being, but a complex, unguided, causal process, and one that only acts because of differences which preexist its operation, for reasons independent of natural selection.[34]

---

32 By "race," Darwin refers to a variety of any given species. As used by Darwin, "race" is not a human-specific concept.

33 Darwin, *On Evolution*, 103–4.

34 Johnson, *Darwin's Dice*, 58. The merits of the version of the theory written in 1844, and the broader question of when exactly Darwin developed his theory, have generated lively academic debate with no signs of imminent resolution. Adrian Desmond and James Moore (*Charles Darwin*, 316) present a representative critique of Darwin's 1844 theory, in which natural selection appears to operate like an "automatic feedback loop" that corrects temporary imbalances between beings and their

Darwin delayed publication of his theory for several reasons. He wanted to provide himself time to develop it further while gaining expertise with some specific genus. For the next decade, then, Darwin set his theoretical work aside to conduct empirical research on barnacles.[35] In late 1854, he returned to his theory and, in 1856, decided to write up his theory as a book, to have been entitled *Natural Selection*.[36] But before completing that book, on June 18, 1858, Darwin received the fateful letter from Wallace, writing Darwin from Indonesia to share his thoughts on the evolution of species. Darwin was stunned by the proximity between his theory and Wallace's and proposed that the two present their ideas together.[37] Then Darwin hunkered down at Down House to write *Origin*. While I presume my reader is familiar with the book's general argument, a concise statement on the text, and how it works, is necessary for my broader argument.

Descent with Modification by Means of Natural Selection

On the Origin of Species by Means of Natural Selection, or the Preservation of Favoured Races in the Struggle for Life was published on November 24, 1859. The first print run sold out immediately, the second shortly

---

environment. Whereas a few scholars accept Darwin's word that he "clearly conceived" "the theory" by "about 1839" (citing Darwin's *Autobiography*, 124.), or by 1842–44, when he wrote his transmutation notebooks, most Darwin scholars support the view that his theory consolidated in 1855–57: See, for example, Janet Browne, *The Secular Ark: Studies in the History of Biogeography* (New Haven: Yale University Press, 1983), 193–4), who argues that "over a period of about twelve years," that is, between 1844 and 1856, Darwin "surrendered the greater part of his 1842/1844 theory of transmutation"; hence the 1856 version "seems like a different theory." Further support for the latter is provided by Derek Partridge, "Darwin's two theories, 1844 and 1859," *Journal of the History of Biology* 51, no. 3 (2018): 563–92. This is not the place to parse the debate. Suffice to say that a major point of contention is the question of what exactly constitutes "Darwin's theory" and whether it can be conceived as a unitary phenomenon (for an argument to the contrary, see Ernst Mayr, *One Long Argument*).

35 This period is examined minutely in Rebecca Stott, *Darwin and the Barnacle* New York: W. W. Norton & Co., 2003).

36 The same period during which Marx wrote his 1857–58 notes on capitalism (*Grundrisse*): See chapter 4.

37 Their papers were read out *in abstentia* to the Linnean Society of London on July 1, 1858, then published in that body's journal.

thereafter. Darwin was instantly famous. He was also under attack and working on edits for a second edition (published in 1860).

Origin proved successful not only because Darwin got the big idea right, but because of the text's composition. He wrote it with style. Its structure, form, and prose are clear, the arguments explained with elegant illustrations.

Consider its opening moves. In light of the book's remarkable ambition, one might think *Origin* would begin with an expansive philosophical statement about natural history or a warning about the radical implications of the argument to follow, and—given scientific norms—a review of the existing literature. The first edition did no such thing.[38] It begins with pigeons. Chapter 1, "Variation Under Domestication," in which we read about Darwin's participation in the London Pigeon Clubs and his discovery that, for instance, "the trumpeter and laugher, as their names express, utter a very different coo from the other breeds" and that "the period at which the perfect plumage is acquired varies, as does the state of the down with which the nestling birds are clothed when hatched."[39] No reader of these pages has ever doubted that Darwin had observed pigeons closely. Their effect is to defamiliarize that object of everyday urban life, the humble pigeon, to show the extraordinary variation across its populations. Darwin's point is made in simple but powerful fashion: Artificial selection can generate a plethora of changes and variations—so much so that we could even speak of distinct species:

> Altogether at least a score of pigeons might be chosen, which if shown to an ornithologist, and he were told that they were wild birds, would certainly, I think, be ranked by him as well-defined species. Moreover, I do not believe that any ornithologist would place the English carrier, the short-faced tumbler, the runt, the barb, pouter, and fantail in the same genus.[40]

---

38 Darwin added a brief (and somewhat misleading) "Historical sketch" to the third edition (1861).
39 Darwin, *Origin* (1860), 21–3.
40 Darwin, *Origin* (1860), 22–3. This is the sort of style of argumentation that irritated Marx, who criticized Darwin's "crude English method": But see Chapter 3, note 5.

Nevertheless, Darwin clarifies, for all these differences, they share a common ancestor:

> Great as the differences are between the breeds of pigeons, I am fully convinced that the common opinion of naturalists is correct, namely, that all have descended from the rock-pigeon (*Columba livia*) ... how, for instance, could a pouter be produced by crossing two breeds unless one of the parent-stocks possesses the characteristic enormous crop?[41]

Beginning with the artificial generation of pigeon heterogeneity allows Darwin to stage an elaborate metaphor in chapter 2, "Variation Under Nature." Given that humans have been able to produce such extraordinary forms of variation in species with crude tools of artificial selection, just imagine, Darwin asks, what sorts of variation could be generated naturally by the sort of "Being" Darwin postulated in his 1844 essay. Once the reader has recognized the importance and diversity of forms of difference among (and within) plants and animal species that are modified by humans, Darwin draws our attention to the same diversity among natural species.[42] Given the diversity of forms of life and the gradations between these forms, what mechanism acts to cohere life into *these* forms and not some other ones?

Darwin answers this question in chapter 3, "Struggle for Existence," and chapter 4, "Natural Selection."[43] As noted, many thinkers before Darwin had hypothesized that species were not fixed (including his grandfather, Erasmus Darwin). But Darwin did not merely suggest that species change through time. Along with Wallace, Darwin proposed

---

41 Darwin, *Origin* (1860), 22–3. From the vantage of our contemporary evolutionary theory, we could quibble that Darwin's question here is based on a mistaken premise—because of the phenomenon called transgressive segregation: Loren Rieseberg et al., "Transgressive segregation, adaptation and speciation," *Heredity* 83, no. 4 (1999): 363–72—but we can let this pass.

42 Darwin's complex theorization of divergence and diversity in *Origin* remains a point of debate: David Kohn, "Darwin's keystone: The principle of divergence," in M. Ruse and R. Richards, eds., *The Cambridge Companion to the Origin of Species* (Cambridge, UK: University of Cambridge, 2009), 87–108.

43 Regrettably, in the fifth edition (1869), Darwin changed the title of chapter 4, adding a subtitle, "Or the Survival of the Fittest" (discussed below).

natural selection as a *mechanism* for changes in species. In the second paragraph of chapter 3, Darwin asks:

> How is it that varieties, which I have called incipient species, become ultimately converted into ... distinct species ...? How do those groups of species, which constitute what are called distinct genera, and which differ from each other more than do the species of the same genus, arise? All these results ... follow from the struggle for life. Owing to this struggle for life, any variation, however slight, and from whatever cause proceeding, if it be in any degree profitable to an individual of any species, in its infinitely complex relations to other organic beings and to external nature, will tend to the preservation of that individual, and will generally be inherited by its offspring. The offspring, also, will thus have a better chance of surviving, for, of the many individuals of any species which are periodically born, but a small number can survive. I have called this principle, by which each slight variation, if useful, is preserved, by the term of Natural Selection, in order to mark its relation to man's power of selection.[44]

Thus did Darwin present his central argument in the first editions of *Origin*. In the fifth edition, published February 10, 1869, he added the following sentence at this point in the text: "The expression often used by Mr. Herbert Spencer of the Survival of the Fittest is more accurate, and is sometimes equally convenient."[45] To put it generously, Darwin's decision to champion the expression "survival of the fittest" from Herbert Spencer (1820–1903) was an unfortunate concession to its popularity (thanks to Spencer's promotion of it as a distillation of Darwin's *Origin*). Darwin partly used the phrase to counteract a sense of teleology that some readers detected in his theory: If survivors are merely more "fit," that in itself implies neither beginning nor end.[46] But "survival of the fittest" took on a life of its own and would henceforth be linked to Darwin's name. (We will return to Spencer and his infamous expression shortly.)

---

44 Darwin, *Origin* (1860), 61–2.
45 Darwin, *Origin* (1869), 72.
46 Johnson, *Darwin's Dice*, 73, 121.

A history of the world imperfectly kept

Darwin's presentation of the mechanism of natural selection as the cause of speciation was undoubtedly a major achievement. There is, however, more to *Origin*. Before elaborating on these, a brief aside is due concerning geology.

Darwin's conception of the descent of species emphasizes that natural selection works slowly, with minute variations accumulating through many generations. Producing different species from natural processes therefore takes a long time. This is where the advances by natural historians from two generations before Darwin were critical: The discovery of geological time meant that millions of distinct species could emerge slowly—Earth was really old.[47] Darwin was influenced by Charles Lyell (1797–1875), whose "uniformitarian" geology proposed that landscape features could be explained as the effect of natural-historical processes operating uniformly across time and space.

Uniformitarian geology introduced a complication for Darwin's theory. If natural-historical processes (both geological and biological) were relatively slow and uniform, what explains the apparent jumps or breaks we find in the fossil record? Darwin used the imperfections of the geological record to respond to a criticism he anticipated—that we should expect to find a complete line of evidence for the evolution of species in the fossil record:

> I look at the natural geological record, as a history of the world imperfectly kept, and written in a changing dialect; of this history we possess the last volume alone, relating only to two or three countries. Of this volume, only here and there a short chapter has been preserved; and of each page, only here and there a few lines. Each word of the slowly-changing language, in which the history is supposed to be written, being more or less different in the interrupted succession of chapters,

---

47 Today we estimate that life has existed on Earth for 3.7 billion years. This was unknown in Darwin's time. But by the mid-nineteenth century it was estimated that life forms had existed for at least hundreds of millions years—long enough for slow processes of speciation to occur. On the discovery of geological time, see Stephen Jay Gould, *Time's Arrow, Time's Cycle: Myth and Metaphor in the Discovery of Geological Time* (Cambridge, MA: Harvard University Press, 1988).

may represent the apparently abruptly changed forms of life, entombed in our consecutive, but widely separated, formations.[48]

Hold on to that last line: "abruptly changed forms of life, entombed in our consecutive, but widely separated, formations." It may provide a useful metaphor for human life under capital.

Chance, selection, and adaptation

I presume that my readers know that *Origin* provides an explanation of the mechanism of natural selection, so I will not recount the details of chapters 4 and 5. I do wish to stress, however, three important points that are often lost.

First, in Darwin's exposition, no species was a fixed thing but only the capricious and temporary result of ongoing evolutionary processes. Darwin showed that adaptation was inherent and ceaseless for all life forms.[49] Adaptations come about as the result of natural selection over time by the environment and interactions with other organisms (including humans). Not all environmental conditions that populations of a species are exposed to and not all organisms with which they interact select for an adaptation.[50] Then, what triggers selection that can result in

---

48 Darwin, *Origin* (1860), 311. Prior to *Origin*'s publication, Darwin had already received awards for his geological research into volcanos, coral reef formation, and his *Fossil Mammalia*: Sandra Herbert and David Norman, "Darwin's geology and perspective on the fossil record," in M. Ruse and R. Richards, eds., *The Cambridge Companion to the Origin of Species*, 129–52.

49 By "adaptations," biologists refer to those characteristics of a species (or a population within a species) that allow individuals to survive in the environment in which they live and with the organisms with which they interact.

50 Since his aim in *Origin* was to explain how species evolve, not to discuss environmental change, Darwin sometimes writes of the adaptation of species to environment as a unidirectional phenomenon, as if natural selection acts on a static stage, leaving out of discussion the constant changes to the environment caused by the actions of living species. In reality, we cannot separate the evolution of populations from changes in the natural environment and vice versa. Some evolutionary biologists have developed the concept of "niche construction" as a counterpart to Darwinian "adaptation of species": See Richard Lewontin, *The Triple Helix: Gene, Organism, and Environment* (Cambridge, MA: Harvard University Press, 2000); Lev Jardón, "Construcción de nicho y causalidad: Algunas implicaciones en el estudio de la agrobiodiversidad desde la praxis," in *Biofilsofías para el Antropoceno: La Teoría de Construcción de Niche desde la Filosofía de la Biología y la Bioética* (Mexico City:

adaptation? We have a stronger answer to this question today than was available to Darwin: The trait or characteristic must be *variable* in the population, the variation must affect the organism's *fitness*, and it must have a *genetic* basis.[51] Consider an example. Imagine a farming community experiencing climate change. Suppose their maize plant populations are subjected to warmer temperatures and reduced precipitation. Let us assume the height of a given maize plant varies across a population, and smaller plants tend to survive warmer and drier conditions, thereby producing more seeds (higher fitness); if variation in height, is, in part, controlled by genes, then we would expect selection to result in a relatively larger proportion of short plants in coming years.[52]

Second, Darwin recognized that *a potential fitness advantage does not determine life outcomes*. Darwin recognized that natural selection and random chance coincide. Consider two baby birds in a nest, siblings of the same mother. One may well be faster and stronger and therefore seem more "fit" than the other, yet nevertheless fail to survive to maturity due to bad luck: A hawk that swoops down to the nest will not take the time to examine its prey. The "less fit" sibling may be more lucky and produce more offspring (and therefore have higher fitness).

Third, Darwin claims that natural selection acts upon variation within populations: Because there is variation, selection occurs. But this presupposes some sort of prior answer to the question "Why do variations occur?" The careful reader of *Origin* will find that Darwin provides no simple, clear answer to this question. In the early editions (such as those that Marx would have read), Darwin strongly implies that the

---

UNAM, 2021), 159–183. I thank Jardón for his insights here and encouragement of this project.

51   Fitness describes the ability of an organism to survive and to reproduce (produce progeny). The more progeny an individual contributes to the next generation, the higher their fitness. Fitness can only be measured *ex post facto*, after reproduction or non-reproduction. The fitness of an individual organism that does not reproduce is measured as zero. Darwin believed that *every* feature of *every* species either conferred today, or at one point in time conferred, some fitness advantage. We now know that this is not true. Random mutations can generate phenotypic expressions which are non-advantageous and yet persist in a population.

52   This is a simplification, for in reality, no single characteristic of any species (like height in my illustration) can predict fitness. Myriad variable factors influence the fitness of maize under climate change: Brian Pace, Hugo Perales, Noelymar Gonzalez-Maldonado, and Kristin Mercer, "Physiological traits contribute to growth and adaptation of Mexican maize landraces," *Plos one* 19, no. 2 (2024): e0290815.

answer is that variations occur randomly, as a result of *chance*.⁵³ Given the limits of his knowledge, this was a brilliant answer and a positive change from his private notebooks of the 1840s.⁵⁴ Yet it was neither as conceptually clear nor as scientifically solid as Darwin would have liked, and it opened the door to various criticisms.

Since variation and selection are frequently conflated, let me restate the distinction in different terms. Darwin correctly identified that species change by natural selection operating upon variations that affect fitness. But where do these variations come from? Darwin's answer in *Origin* is that variations emerge naturally, through chance. This constitutes a distinct and separate argument from his identification of natural selection as the mechanism that acts upon variations. It was also the more original claim; no one had previously made this leap. It constituted a leap because it required the imagination and courage to unmoor variation from any metaphysical ground. In retrospect, it may seem obvious that novel variations are an effect of chance, particularly now that we have learned about genetic mutation, gene flow, and random genetic drift.⁵⁵ Yet Darwin knew nothing of these processes.⁵⁶ As Curtis Johnson notes, Darwin confronted two distinct questions:

---

53   Darwin was wary of stating this thesis directly, since he knew what it would mean for his religious readers. Thus, at the outset of chapter 5, he qualifies himself: "I have hitherto sometimes spoken as if the variations . . . had been due to chance. This, of course, is a wholly incorrect expression, but it serves to acknowledge plainly our ignorance of the cause of each particular variation," *Origin* (1860), 131.

54   Darwin had earlier sketched causal explanations for the origins of variation, but gave up on them: "Early in his career he put the burden on 'divine creation,' but in time he came to regard that as a non-answer": Johnson, *Darwin's Dice*, 12.

55   These three distinct processes are all random in some respect. Briefly: Genetic mutation generates genetic variations; gene flow moves genetic variation between populations; and genetic drift can change frequencies of genetic variation within populations. On the latter, see Russell Lande, "Natural selection and random genetic drift in phenotypic evolution," *Evolution* (1976): 314–34. On gene flow and climate change, see Kristin Mercer and Hugo Perales, "Evolutionary response of landraces to climate change in centers of crop diversity," *Evolutionary Applications* 3, nos. 5–6 (2015): 480–93. I thank Mercer for her support of this project and assistance with this chapter in particular.

56   That Darwin knew nothing of genetics makes his breakthrough more remarkable. Perhaps self-reflection upon the role of chance events in his life opened Darwin to the prospects of random variation as a factor in evolution. For, as he writes in his *Autobiography* (76–7): "The voyage of the *Beagle* has been by far the most important

[Q1] what causes modification of species (or populations), and [Q2] what causes variations among individuals[?] Usually Darwin was focused on "natural selection" as the *primary* (though not the only) cause of modification of species. But natural selection cannot act without something to act upon—actual organisms in all of their variety. It is thus a short mental step to steer the question to the causes of variation.[57]

The "mental step" is so short that—like a *coupé* in dance—it is easy to miss. Yet this step, which had never been taken before, proved decisive for Darwin's other breakthrough—his destruction of teleology.

The End of Teleology

Stephen Jay Gould reminds us that, in Europe during the first half of the nineteenth century, "evolution was a very common heresy."[58] What was genuinely heretical was to claim, as Darwin did, that natural selection acts upon variations that emerge by chance and that this process unfolds in a nonteleological fashion.

Darwin's achievement here had historical precedents. Pre-Socratic philosophers like Empedocles of Akragas speculated that forms of life come from other forms, without end, but this line of thought was largely lost after Aristotle criticized it. Aristotle rejected the view that chance could explain the variation we see in nature.[59] As Kōjin Karatani explains, while there were theories of evolution in Ionia, the "epoch-making significance" of Darwin's *Origin* was its radical capacity "to reject teleology of any kind … whether aware or not, Darwin

---

event in my life and has determined my whole career; yet it depended on so small a circumstance as my uncle offering to drive me thirty miles to Shrewsbury, which few uncles would have done … I have always felt that I owe to the voyage [my] first real training … [for] I was led to attend closely to several branches of natural history": No ride with friendly uncle, no *Beagle* voyage, no *Origin*.

57 Johnson, *Darwin's Dice*, 141. Although implied in early editions of *Origin*, Darwin failed to specify that natural selection was not the *cause* of variation until the fifth edition.
58 Stephen Jay Gould, *Ever Since Darwin*, 23.
59 Aristotle, *Physics*, Book II.

recuperated Ionian thought, which had long been suppressed by Aristotelian thinking."[60]

Darwin did not study ancient philosophy, but he knew what he had done; he knew his theory presented a radical challenge to conventional thinking. This explains why Darwin delayed the publication of his theory for fifteen years. Though personal dissatisfaction with his theory contributed to his delay, Darwin's early notebooks demonstrate that he was seriously concerned with the philosophical and religious implications of his scientific theory. He saw that

> the primary feature distinguishing his theory from all other evolutionary doctrines was its uncompromising philosophical materialism. Other evolutionists spoke of vital forces, directed history, organic striving, and the essential irreducibility of mind—a panoply of concepts that traditional Christianity could accept in compromise, for they permitted a Christian God to work by evolution instead of creation. Darwin spoke only of random variation and natural selection.[61]

Darwin's affirmation of inexplicable chance as the cause of variation caused a collision with the concepts of God, design, and end, or telos.

It may be useful to provide a brief example of a teleological conception of natural history. Consider the first two theses of Immanuel Kant's "Idea for a Universal History from a Cosmopolitan Point of View":

> All natural capacities of a creature are destined to evolve completely to their natural end.

> In man (as the only rational creature on earth) those natural capacities which are directed to the use of his reason are to be fully developed only in the race, not in the individual.[62]

---

60 Kōjin Karatani, *Isonomia and the Origins of Philosophy* (Durham: Duke University Press, 2017), 66.
61 Stephen Jay Gould, *Ever Since Darwin*, 25.
62 Kant (1784), in *On History*, trans. L. W. Beck (Indianapolis: Bobbs-Merrill, 1963), theses 1 and 2.

Kant's problematical reasoning returned in different guise in Hegel and, with further modifications, in some texts by Marx. The general idea shared by these thinkers is that certain things have an essence that is destined to be developed to its natural end. Kant's object of analysis was the use of reason by humans; Hegel's object was reason's development as spirit; before reading Darwin, Marx's object was the possibilities inherent to a social order organized by a particular mode of production. Consider this famous line from Marx's preface to his *Contribution to the Critique of Political Economy*, written only a year before he read Darwin:

> No social order is ever destroyed before all the productive forces for which it is sufficient have been developed, and new superior relations of production never replace older ones before the material conditions for their existence have matured within the framework of the old society.[63]

Darwin's theory challenges such teleological reasoning.

Darwin's theory bore implications for religious faith as well as our belief in human superiority vis-à-vis other species. This was obvious to Darwin. As he wrote in one of his earliest notebooks:

> Thought (or desires more properly) being hereditary.—it is difficult to imagine it anything by structure of brain hereditary . . . love of the deity [God, is an] effect of organization, oh you Materialist! . . . Why is [human] thought being a secretion of brain, more wonderful than gravity a property of matter? It is our arrogance, our admiration of ourselves.[64]

Darwin's theory would effectively replace an essentialist conception of nature with historicist naturalism: Nature remains the ontological ground of history, but it is itself subsumed to history. Such an approach undermines traditional, conservative appeals for order in the name of nature. No longer could one claim that stability was natural. Showing

---

63 Marx, *Contribution to the Critique of Political Economy* (New York: International Publishers, 1970 [1859]), 21.

64 Darwin, "Notebook C," in *On Evolution*, 71.

that human life resulted from natural processes implied that there was nothing permanent about our existing social condition. Human reason—that celebrated keyword of the Enlightenment—could even be understood as certain brain "secretions," an effect of material exchanges between the human body and the natural environment.

Darwin was more circumspect when discussing these philosophical thoughts with colleagues. Consider this famous letter that Darwin wrote to his interlocutor, the botanist Asa Gray. Gray admired Darwin's work enormously and agreed with him about natural selection but could not accept that mere chance generated the variations that it worked upon. He wrote Darwin to encourage his friend to clarify for his audience that behind all the variations lay laws that governed life, laid down by God. Darwin replied:

> With respect to the theological view of the question; this is always painful to me.— I am bewildered . . . But I own that I cannot see, as plainly as others do, & as I [should] wish to do, evidence of design & beneficence on all sides of us. There seems to me too much misery in the world. I cannot persuade myself that a beneficent & omnipotent God would have designedly created the Ichneumonidæ [a family of parasitoid wasps] with the express intention of their feeding within the living bodies of caterpillars, or that a cat should play with mice. Not believing this, I see no necessity in the belief that the eye was expressly designed. On the other hand I cannot anyhow be contented to view this wonderful universe & especially the nature of man, & to conclude that everything is the result of brute force. I am inclined to look at everything as resulting from designed laws, with the details, whether good or bad, left to the working out of what we may call *chance*.[65]

Here lies Darwin's anti theodicy. God need not be vindicated for creating parasitic wasps, for there is no need to invoke God to explain their sting. Chance and natural selection are sufficient.

---

[65] Darwin, Letter to Asa Gray, May 22, 1860, accessed from the online Darwin Correspondence Project.

What did Darwin do?

Despite what we are often told about the way to do science, Darwin's breakthrough was not the result of ingenious methods nor the production of especially insightful data. His methods were conventional and while Darwin presents abundant empirical evidence in *Origin*, it is plainly secondary to the analytical argument—data provide illustration, not proof.[66] Darwin's theory also differs from a conventional conception (Popperian, for example) of what a scientific theory is supposed to look like (hypothetico-deductive and predictive). And while most contemporary presentations of the theory rely on probabilistic descriptions (statistical likelihood), Darwin made no attempt to learn statistics (such as it existed at the time), nor formalize his theory mathematically. So, what exactly did Darwin do? Simply put, he creatively changed an existing line of questioning to present a new answer to an old puzzle.

Earlier, I noted that Darwin is fortunate to receive sole credit for establishing the theory of evolution and natural selection, since others (like Wallace) were on the path to doing so. Then again, we do not always give sufficient credit to thinkers who can shift a conversation by posing a good question, which is what Darwin did. To appreciate this, consider Aristotle. After some pre-Socratic philosophers developed materialist theories of the origin of species, Aristotle criticized them in his *Physics*. Forms in nature, he reasoned, could not result from pure chance. Why, he asked, do we see the species that we do? Why are there bats, but no flying cats? Aristotle reasoned that nature is a cause that operates for some purpose. (His reasoning dovetailed neatly with the monotheistic traditions to come.)

Darwin did not so much rebut Aristotle as change the question. Alan Garfinkel puts it elegantly:

> Darwinian biology simply does not answer Aristotle's question ["why do we see species x?"—JDW]. The scientific advance that Darwin made can partly be seen as a rejection of *that* question and the

---

66 Although Darwin's insights into natural selection were inspired by empirical observations, the development of his theory was not empiricist in the philosophical sense. This is a quality that *Origin* shares with Marx's *Capital*.

substitution of a different question, namely: *Given* that a species comes to exist (however it does), why does it continue to exist or cease to exist? That is precisely not the question of the origin of species but rather why species survive.[67]

The question Darwin answered with his theory of natural selection is why species change and become other species. In this sense, Darwin's greatest book was misnamed. It could have been called *On the Becoming of Species from Other Species*, though I admit this title is not so euphonious. The point is that Darwin's breakthrough was accomplished not by generating original data, or by mathematical reasoning, or by making brilliant predictions, but by shifting from one standpoint to another. (As I explain in part II, the same can be said of Marx's *Capital*.)

Darwin needed more than a good question; he needed evidence, too. Yet Darwin knew that he could not prove his theory to his readers by showing direct evidence of the evolution of species. He needed to present the theory in a logical form and then use multiple lines of evidence to bolster his claims as well as he could. Darwin employed three sorts of empirical evidence to illustrate that evolutionary processes change species. The first, discussed earlier, is his use of artificial selection as a metaphor for natural selection. Since we can see the direct results of artificial selection, Darwin only needed to convince his reader that natural selection works in a fashion akin to artificial selection. Second, Darwin uses evidence from paleontology to great effect. *Origin* features illustrations of previously existing forms of species (particularly invertebrates) to show morphological changes in species over geological time.[68] Third, biogeography. Inspired by his travels on the *Beagle*, Darwin provides examples of species that vary across space, such as the finches and tortoises of the different Galapagos Islands, in geographical patterns that imply adaptative variation across environmental gradients.

---

67 Alan Garfinkel, *Forms of Explanation: Rethinking the Questions in Social Theory* (New Haven: Yale University Press, 1982), 9–10.

68 Warren Allmon, "Darwin and palaeontology: A re-evaluation of his interpretation of the fossil record," *Historical Biology* 28, no. 5 (2016): 680–706. As noted, this line of evidence introduced a problem for Darwin: He could present no complete evolutionary sequence for a well-known species. Darwin explained this away by acknowledging gaps in the fossil record.

Notably, Darwin's theory of natural selection explains observable biological conditions—variation in species—without making any pretense to the sort of specific predictions often made in physics and chemistry. *Origin* therefore demonstrates that the scientific capacity for *explanation* and *prediction* are independent qualities—and that explanation without predictive capacity is virtuous and no less scientific.[69] Darwin's theory shows us "that one cannot regard explanations as unsatisfactory when they do not contain laws, or when they [cannot] enable the event in question to have been predicted"; it follows that the "impossibility of a Newtonian revolution in the social sciences . . . is not fatal to their status as sciences."[70] A theory that provides satisfactory explanations of *social* processes need not be formal or predictive to be scientifically valid and useful.

On the reception of Origin

The publication of *Origin* was a major event of social and political significance.[71] In retrospect, given the strength of religious thinking at the time (and Darwin's hesitation to defend his arguments in the public sphere), it is remarkable that so many orators and writers leaped to defend *Origin* in England during the 1860s.[72] Eric Hobsbawm

---

69 Michael Scriven, "Explanation and prediction in evolutionary theory: Satisfactory explanation of the past is possible even when prediction of the future is impossible," *Science* 130, no. 3374 (1959): 477–82. This is not to say that Darwin's theory *prevents* prediction, only that the character of Darwinian predictions is inherently open-ended. Consider Darwin's definition of natural selection: "This preservation of favourable variations and the rejection of injurious variations, I call Natural Selection" (*Origin*, 2nd edition, 81). This is a general prediction: "favourable variations" will be preserved while "injurious" ones will not. But this prediction must remain open (both in the sense of temporality and probability), because Darwin recognized that we cannot predefine "favourable" versus "injurious" variations. Notably, after reading *Origin*, Marx never attempted to define ahistorical criteria for the survival of social classes or social practices.

70 Scriven "Explanation and prediction in evolutionary theory," 477. Marx would have recognized this quality of Darwin's *Origin*—a paradigm of a natural science which was explanatory but not strictly predictive.

71 On the global reception of *Origin*, see Thomas Glick, ed., *The Comparative Reception of Darwinism* (Chicago: University of Chicago Press, 1988).

72 Even today, the teaching of Darwin's ideas in the USA is fiercely contested, with conservatives insisting upon "balancing" the teaching of evolutionary biology with the (non-scientific) concept of God's creation. I invite the reader to search the term

hypothesizes that the willingness of the so-called evolutionists (as defenders of *Origin* were called) to counterattack versus religious critics of Darwin—and the rapidity of their victory (in Europe at least)—marked a "happy conjuncture of two facts[: 1] the rapid advance of a liberal and 'progressive' bourgeoisie and [2] the absence of revolution. The challenge to the forces of tradition grew stronger, but it no longer seemed to imply social upheaval."[73] Assuming that Hobsbawm is correct—and I believe he is—then we can appreciate why Darwin's theory proved divisive for European Marxists in the decades after *Origin*'s publication: It was associated with bourgeois and liberal thinkers (beginning with Darwin himself) and seen as one element of the architecture of liberal hegemony in capitalist society. By the time that the Bolsheviks stormed the Winter Palace in 1917, an unfortunate ideological distinction between evolutionary and revolutionary theory had been deeply etched.

The reception of *Origin* varied notably by geography, social class, and political ideology. In the UK, it generated a sensational debate over its implications for religion. If forms of life evolved in response to natural selection, and the underlying differences were generated by chance, there was no need for a transcendent divinity.[74] While such concerns were also present among conservative thinkers in the USA, the early fate of *Origin* played out differently, for it was readily embraced by anti-slavery abolitionists—including many Christian evangelicals—who recognized that *Origin* implied the natural unity of all human beings. Yet while *Origin* therefore came to be (rightly) associated with abolition and anti-racism in the USA, Darwinism there was also converted into an "ideology of militant capitalism"—social Darwinism.[75]

"Darwinism" at amazon.com (US edition): Most of the best-selling books are still religious criticisms of Darwinism—another distinction that Darwin shares with Marx. For a defense of natural history and account of anti-evolutionary politics in US education, see Prosanta Chakrabarty, *Explaining Life Through Evolution* (Boston: MIT, 2023).

73   Eric Hobsbawm, *The Age of Capital* (London: Penguin, 1996 [1975]), 260.

74   Nor divine revelation, as Darwin explained in an efficient letter: "I am sorry to have to inform you that I do not believe in the Bible as a divine revelation, & therefore not in Jesus Christ as the son of God. / Yours faithfully": Darwin, Letter to Frederick McDermott, November 24, 1880.

75   Hobsbawm, *The Age of Capital*, 261. This bifurcated initial reception of *Origin* still complicates discussions of Darwin on the Left in the USA.

Malthus, Darwin, Spencer, and "Social Darwinism"

Darwin knew that his theory applied to humans as much as other species and that his readers would ask themselves how our species had evolved. Yet he made a deliberate decision not to discuss human evolution in *Origin*, except for one remark: "Psychology will be based on a new foundation, that of the necessary acquirement of each mental power and capacity by gradation. Light will be thrown on the origin of man and his history."[76] These predictions have been borne out.[77] Darwin's book on human evolution, *The Descent of Man, and Selection in Relation to Sex*, was not published until 1871, with a companion study of the evolution of expressions of emotions the following year.[78] The temporal gap between *Origin* and *Descent* created a vacuum in Victorian England: Every member of the reading public held the tools to speculate upon human evolution, but, for twelve years, Darwin did not present his own research. Nature abhors a vacuum; book markets, too. For scientific, religious, ideological, and commercial reasons, many writers rushed to fill the gap.[79]

At the center of this discourse stood the Neanderthal (a skull was found in 1856 in Germany) and the Western gorilla of equatorial Africa (*Gorilla gorilla*). These were displayed—initially via artifacts, particularly skulls, then live gorillas—in various museums and spectacular

---

76 Darwin, *Origin* (1860), 489. Regrettably, in the sixth edition, Darwin changed this line to make a fateful prediction: "Psychology will be securely based *on the foundation already well laid by Mr. Herbert Spencer*." Thankfully, Darwin got this wrong.

77 For a recent assessment, see Peter Richerson, Sergey Gavrilets, and Frans de Waal, "Modern theories of human evolution foreshadowed by Darwin's *Descent of Man*," *Science* 372, no. 6544 (2021): Eaba3776.

78 Darwin, *The Descent of Man, and Selection in Relation to Sex* (London: John Murray, 1871); Darwin, *The Expression of Emotions in Man and Animals* (New York: Penguin, 2009 [1872]). Both texts are available in multiple editions at the excellent Darwin Online website. In *The Expression of Emotions*, Darwin notes that "as long as man and all other animals are viewed as independent creations, an effectual stop is put to our natural desire to investigate as far as possible the causes of Expression as to every other branch of natural history," 23.

79 "Darwin's caution and long hesitation . . . all failed. When he left man out of the *Origin*, his enemies immediately assumed that Darwin really included man, and attacked him for doing so. Meanwhile, whether he wanted it or not, his allies included the most radical thinkers of his day": Howard Gruber, *Darwin on Man: A Psychological Study of Scientific Creativity* (Chicago: University of Chicago, 1981), 72.

productions, inviting the public to judge firsthand the claims of those who argued that humans did or did not "come from the apes."[80] Today we know that humans did not evolve from *Gorilla gorilla*, of course, but that was how the matter was framed at the time. Culturally, the decade of the 1860s was partly defined by this great debate—the so-called gorilla controversy—particularly in England, where Marx was then living.[81]

Even without writing a single paragraph on the human species, Darwin's *Origin* invited the reader to subsume human history within natural history and thus to abolish, in Hobsbawm's words, "the sharp line between natural and human or social sciences."[82] This was a welcome change, but it opened the door to right-wing ideas, such as social Darwinism. Social Darwinism could be defined by two claims: First, that "success in an unconstrained free market economy proves an individual's biological fitness to survive in the struggle for existence"; second, that "constraints on free markets and charitable practices will result in the growing inability of a nation, race, or the whole species to preserve the fit," and therefore should be avoided or eliminated.[83] The second claim predates Darwin—Malthus and Ricardo provided earlier versions—yet proved especially persuasive and endurable when combined with the

---

80   The first live gorilla was brought into Europe in 1876 and died one year later. "Procured in Gabon . . . for the German African Society, the gorilla was passing through Liverpool en route to Berlin"; the gorilla, nicknamed Pongo, "caused quite a stir during its brief stay on Merseyside. The *Illustrated Police News* was excited by 'actual demonstration of the "missing link" between the human and animal creation . . . the nearest known approximation of the human form.'" After dying in November 1877 in London, obituary notices "were carried in British and German newspapers, with references to Pongo cropping up in pantomimes and other stage shows for a decade or so afterwards": Alexander Scott, "The 'missing link' between science and show business: Exhibiting gorillas and chimpanzees in Victorian Liverpool," *Journal of Victorian Culture* 25, no.1 (2020): 1–20, 1–2.

81   On the controversy, see Thomas Huxley, *Evidence as to Man's Place in Nature*, in *The Major Prose of Thomas Henry Huxley*, ed. Alan Barr (Athens, GA: University of Georgia Press, 1997 [1863]), 20–153; Joel Mandelstam, "Du Chaillu's stuffed gorillas and the savants from the British Museum," *Notes and Records of the Royal Society of London* 48, no. 2 (1994): 227–45; Amanda Hodgson, "Defining the species: Apes, savages and humans in scientific and literary writing of the 1860s," *Journal of Victorian Culture* 4, no. 2 (1999): 228–51; also see previous note, above, Scott, "'Missing Link.'"

82   Hobsbawm, *The Age of Capital*, 258.

83   David Depew and Bruce Weber, "The fate of Darwinism: Evolution after the modern synthesis," *Biological Theory* 6 (2011): 89–102, note 1.

first claim. In the face of our planetary crisis, many powerful people are becoming more social Darwinist and neo-Malthusian. It would help to reexamine the interrelations between Darwin, Herbert Spencer, and the Reverend Thomas Malthus (1766–1834).

Few intellectual debts have been vetted so thoroughly as Darwin's to Malthus. The starting point is recorded with precision by Darwin in his *Autobiography*: On September 28, 1838, he recounts, he read Malthus's *Essay on the Principle of Population* (sixth edition). Encountering Malthus's assertion that the human "population, when unchecked, goes on doubling itself every twenty-five years, or increases in a geometrical ratio" helped to clarify, for Darwin, that all species produce more offspring than can survive until the next successful reproduction.

Malthus writes that the main factor that distinguishes humans from other animals "is the power which he possesses of very greatly increasing" our means of subsistence. This is the reason, he explained, human numbers were increasing during his time. Nonetheless, Malthus reasons: Humanity's power in this respect "is obviously limited" by scarcity, specifically of land ("the great natural barrenness of a very large part of the surface of the earth") and of

> the decreasing proportion of produce which must necessarily be obtained from the continual additions of capital applied to land already in cultivation. It is, however, specifically with this diminishing and limited power of increasing the produce of the soil, that we must compare the natural power of mankind to increase, in order to ascertain whether . . . the natural power of mankind to increase must not, of absolute necessity, be constantly retarded by the difficulty of procuring the means of subsistence; and if so, what are likely to be the effects of such a state of things.[84]

Malthus's concern was that, if social welfare provisions allowed for humans to avoid natural checks—e.g., if poor people are fed and housed thanks to labor organizing and welfare policies—then the poor would multiply unnaturally beyond any reasonable limit. Herbert Spencer, not Malthus, connected these ideas to Darwin,

---

84 Thomas Malthus, *An Essay on the Principle of Population* (London: Penguin, 1970 [1798]), 225.

generating social Darwinism.[85] From there it was a small step to, for example, Andrew Carnegie's claim that science proved that the state in a capitalist society should not help working people: "While the law of competition may be sometimes hard for the individual, it is best for the race, because it ensures the survival of the fittest in every department."[86]

Darwin recognized that Malthus was not only wrong about his data; he also recognized that Malthus's generalizations and ratios were unsupported and invalid. Darwin moreover explicitly rejected the *goal* of Malthus's book. Recall that Darwin never claimed that evolution meant improvement. His conception of fitness does not imply development or progress: It simply describes the adaptation of species to a given environment (and to the presence of other species: For example, Darwin often describes technical changes in species' bodies which allow them to catch prey or escape from predators).

Nevertheless, as noted, Darwin cited Spencer at a crucial point in *Origin*. Inspired by Malthus, Spencer summarized Darwin's *Origin* with the expression "survival of the fittest," which Darwin in turn used in later editions of his book.[87] This was unfortunate. Spencer's implicit claim, "the fittest are those which survive," has three serious faults. First, it is tautological, at least if "the fittest" is defined as "those which survive." Second, since fitness is shaped by myriad factors, identifying "the fittest" in any given population is complex and difficult. Third, as noted earlier, it is untrue to claim (as one often hears) that "the fittest individuals

---

85 Specifically in Spencer's *First Principles*, a flawed work rushed into print in 1862, in which Spencer "redefined his own version of evolution, first outlined in articles published in the 1850s. To this he added views sharpened by his reading of Darwin ... [Spencer] seems to have believed that his views were vindicated in Darwin's writings, although he differed from him on several grounds ... Much of what was ultimately attributed to Darwin was the result of philosophical shifts expressed in one form or another by Spencer. At the very least, most of what Spencer proposed about directional development and progress, although basically Lamarckian or environmentalist in thrust, was conflated in people's minds with the Darwinian impetus": Janet Browne, *The Power of Place* (New York: Knopf, 2002), 184–5.

86 Andrew Carnegie (1889), cited in Hiram Caton, "Getting our history right: Six errors about Darwin and his influence," *Evolutionary Psychology* 5, no. 1 (2007): 147470490700500106; cf. John White, "Andrew Carnegie and Herbert Spencer: A special relationship," *Journal of American Studies* 13, no. 1 (1979): 57–71.

87 Gregory Claeys, "The 'survival of the fittest' and the origins of social Darwinism," *Journal of the History of Ideas* 61, no. 2 (2000): 223–40.

will survive and reproduce," because *chance* plays a major role in life and reproduction. (Recall the two bird siblings in the nest.) These points render Spencer's popular expression meaningless. Thus, Darwin weakened *Origin* by inserting it during revisions for the fifth edition.

The ideological consequences were also severe, and—as is evident in light of our planetary crisis—have not yet been overcome. Citing Spencer at a crucial point in *Origin* undoubtedly elevated the status of social Darwinism. Hobsbawm notes that one of Darwin's achievements was that his theory "ratified the triumph of history over all the sciences"; unfortunately, however, thanks to Spencer's influence, "'history' in this connection was generally confused by contemporaries with 'progress.'"[88] By citing Spencer, Darwin contributed to the misrecognition of one of his greatest achievements—the destruction of teleology in natural history. This misunderstanding of Darwin's conception of descent with modification caused by natural selection as *progress* was one of the most important events in the history of late nineteenth-century ideas. Its consequences still haunt our world.

Was Darwin a social Darwinist?

Reading Malthus's *Essay on the Principle of Population* helped to inspire Darwin's theory. Yet Malthus was hardly the first to recognize something like a "struggle for existence" in nature and Darwin's theory is by no means limited to the recognition of it. Malthus helped Darwin to see that populations tend to produce larger numbers of offspring than can survive to reproduction. In this narrow sense, it can be said fairly that "Darwin was Malthusian"—but this is saying little, for Darwin had many sources of inspiration (not only scholarly ones). Certainly, none of Darwin's major insights owe anything to Malthus.[89] Moreover, Darwin departed from Malthus's theory by undermining its teleological presuppositions. How Darwin understood this related to the evolution of *Homo sapiens* and to the form of human societies is another matter entirely. From the vantage of contemporary biology, the issue is moot:

---

88 Hobsbawm, *The Age of Capital*, 258.
89 As Marx recognized: See chapter 5.

Among the advantages of twentieth-century Darwinism is that fitness and unfitness are not defined by particular qualities, such as cunning aggression in the case of the supposedly fit or congenital degeneracy in the case of the unfit . . . Nor in twentieth-century Darwinism does natural selection occur only at or near Malthusian limits. Unfortunately, however, these technical reformulations of the science of Darwinism are, after almost a century, still largely unknown in popular culture.[90]

Social Darwinism is incompatible with contemporary evolutionary theory.

Many scholars claim that Darwin rejected what came to be called "social Darwinism" and there is considerable evidence to support this view. For instance, after *Origin* was published, Darwin wrote a letter complaining: "I have received in a Manchester newspaper rather a good squib, showing that I have proved might is right and therefore that Napoleon is right, and every cheating tradesman is also right."[91] Nevertheless, however much he may have regretted this sort of misrepresentation of his ideas, Darwin did not explicitly and publicly condemn social Darwinism, and many writers in his day treated his theory and Spencer's as two versions of the same idea. When Darwin died in 1882, the *New York Times* obituary summed up his thought in these lines:

> The central principle . . . of Mr. Darwin's system is "natural selection," called by Herbert Spencer "the survival of the fittest" . . . which results inevitably from "the struggle for existence." It is a law and fact in nature that there shall be the weak and the strong. The strong shall triumph and the weak shall go to the wall. The law, though involving destruction, is really preservative.[92]

Darwin was buried as the patron saint of natural law and order.

---

90 Depew and Weber, "The fate of Darwinism," 89–102, note 1.
91 Darwin, Letter to Charles Lyell, 1859, in *The Life and Letters of Charles Darwin*, volume II, 62, cited in Peter Singer, *A Darwinian Left: Politics, Evolution, and Cooperation* (New Haven: Yale University Press, 2000).
92 *New York Times*, April 21, 1882, 1, cited in Caton, "Getting our history right."

In fairness, we cannot blame Darwin for what others wrote about him when he died. And Darwin was motivated to show the universal commonality of all human beings. It is sometimes said that as a white European man his thinking was marked by racist attitudes typical to his society. While it is true that everyone inherits the prevailing ideas of their time and one can find statements in some of Darwin's texts that appear racist from the vantage of the present, he was raised by a family of committed abolitionists and from his youth advocated for the recognition of the unity of all humanity. By one reading, Darwin's drive to explain evolution stemmed from a desire to undermine the racist notion that humans were created as separate races by God.[93] Since all humans have evolved from common ancestors, then we comprise one family and should treat each other equally and respectfully. This was Darwin's anti-racist argument—perhaps the strongest one ever made.

But matters are more complex because, privately, Darwin gave some support to those who used his arguments to make claims for social class hierarchy and against workers' organization. For instance, on July 26, 1872, he wrote a letter to a colleague in which he justifies repression of workers's attempts to organize to improve their conditions:

> I much wish that you would sometimes take occasion to discuss an allied point . . . namely the rule insisted on by all our Trades-Unions, that all work-men,—the good and bad, the strong and weak,— sh[oul]d all work for the same number of hours and receive the same wages. The unions are also opposed to piece-work,—in short to all competition. I fear that Cooperative Societies, which many look at as the main hope for the future, likewise exclude competition. This seems to me a great evil for the future progress of mankind. — Nevertheless under any system, temperate and frugal workmen will have an advantage and leave more offspring than the drunken and reckless.[94]

---

93 Desmond and Moore, *Darwin's Sacred Cause*.

94 Darwin, Letter to Heinrick Fick, 1872, quoted in Richard Weikart, "A recently discovered Darwin letter on social Darwinism," *Isis* 86, no. 4 (1995): 609–11. Weikart concludes that Darwin "clearly linked economic success with selective fitness and thought his theory supported individualist economic competition."

This is an unambiguously bourgeois application of social Darwinist ideas to condemn labor organizations and cooperativism as "a great evil" impeding the "progress of mankind."[95]

So, while Darwin was an antiracist, he was a social Darwinist. Not only did he (knowingly or not) employ metaphors from political economy to describe nature in *Origin*;[96] what is worse, he applied his theory to political economy to justify class inequality. Taken together, Darwin contributed to naturalizing capitalism qua class competition.[97] If we are going to use Darwinian ideas to make sense of our place in natural history, then we will need to pair him with a more critical thinker on political economy.

---

95   Apropos Darwin's class: "Supported by a family fortune . . . Darwin was content to become a thoroughly respectable Victorian gentleman. He put away his *Beagle* shotguns, cast a discerning eye over his investments, and began to participate in the growing sense of national prosperity. He had no need to seek employment. Like many others in his circle, he was free to pursue his interests, in his case a magnificent obsession with natural history": Browne, *The Power of Place*, 4.

96   As Marx and Engels complained in their letters on Darwin: See, for example, Engels's statement:

> Of the Darwinian theory I accept the *theory of evolution* but only take Darwin's method of proof (struggle for life, natural selection) as the first, provisional, and incomplete expression of a newly-discovered fact . . . All that the Darwinian theory of the struggle for existence boils down to is an extrapolation from society to animate nature of Hobbes' theory of the *bellum omnium contra omnes* and of the bourgeois-economic theory of competition together with the Malthusian theory of population. Having accomplished this feat (the absolute admissibility of which . . . I contest, especially where the Malthusian theory is concerned), these people proceed to re-extrapolate the same theories from organic nature to history, and then claim to have proved their validity as eternal laws of human society. The puerility of this procedure is self-evident . . . (Engels's letter to Lavrov, November 12, 1875. MECW 45: 107–8).

While Engels's critique of direct extrapolation from biology to human history is entirely valid, his claim that "the Darwinian theory . . . boils down to . . . extrapolation from society to animate nature of Hobbes' theory . . . and of the bourgeois-economic theory" is at best a polemical oversimplification. For a counterargument to Engels's claim, see Scott Gordon, "Darwin and political economy: The connection reconsidered," *Journal of the History of Biology* 22, no. 3 (1989): 437–59.

97   On this point, among the scientists with whom Darwin is frequently compared, Einstein provides the clearest counterpoint: Albert Einstein, "Why Socialism?," *Monthly Review* 57, no. 1 (2005); cf. Joel Wainwright, "Climate change, capitalism, and the challenge of transdisciplinarity," in *Geography of Climate Change* (New York: Routledge, 2013), 270–8.

There Is Grandeur in This View of Life

Darwin concludes *Origins* by presenting an image of life as an "entangled" riverbank, alive with species:

> It is interesting to contemplate an entangled bank, clothed with many plants of many kinds, with birds singing on the bushes, with various insects flitting about, and with worms crawling through the damp earth, and to reflect that these elaborately constructed forms, so different from each other, and dependent on each other in so complex a manner, have all been produced by laws acting around us.[98]

Darwin's proposal of a standpoint of an "entangled" world anticipated the concept of ecology. Ernst Haeckel coined this word, "ecology," in 1866. It is no coincidence that this word was invented seven years after *Origin* by one of Darwin's leading advocates. While natural historians had already established methods for studying the histories of landscapes and the interconnections between different species (both basic elements of ecological theory), until Darwin promulgated his theory of the descent of species with modification through natural selection, there was no coherent explanatory framework through which to interpret ecological relations between species and no way to conceptualize ecosystems comprised of evolving interrelations of multiple species and environmental factors. All ecology is evolutionary, and all evolution is ecological. Our world is indeed entangled.

It can also be explained by reference to natural laws. Darwin's search for laws of the transmutation of species did not lead him to formal, predictive laws (as with Newtonian physics), but it certainly helps us to understand natural-historical processes—which are "laws" to the extent that they allow us to explain interrelations retrospectively, if not to predict specific outcomes. In the final lines of *Origin*, Darwin summarizes the laws he thought govern speciation:

---

98 Darwin, *Origin* (1860). I cite the 1860 edition here because that is the version that Marx first read. Darwin made modest changes to this paragraph in later editions, but no substantive changes to the section quoted here: See darwin-online.org.uk.

Darwin and the Defeat of Teleology

> These laws, taken in the largest sense, being Growth with Reproduction; Inheritance which is almost implied by reproduction; Variability from the indirect and direct action of the external conditions of life, and from use and disuse; a Ratio of Increase so high as to lead to a Struggle for Life, and as a consequence to Natural Selection, entailing Divergence of Character and the Extinction of less-improved forms. Thus, from the war of nature, from famine and death, the most exalted object which we are capable of conceiving, namely, the production of the higher animals, directly follows. There is grandeur in this view of life, with its several powers, having been originally breathed by the Creator into a few forms or into one; and that, whilst this planet has gone cycling on according to the fixed law of gravity, from so simple a beginning endless forms most beautiful and most wonderful have been, and are being, evolved.[99]

Many have commented on the peculiar tone of these lines from the conclusion of *Origin* and their distinct blend of Victorian-era themes: Scientific law, war, character, improvement, and so on. What does it all add up to? For some readers, Darwin sought to defend his critique of teleology with soothing references to the "grandeur" from the forms "breathed by the Creator";[100] others contend that we have proof here of a full-blown appeal to teleology.[101] A broader consideration of Darwin and ideology is helpful here. With his secularization of Christian ideals, criticisms of slavery, presumption of law and order, and critique of proletarian organization, we find in *Origin* the crystallization of the consciousness of a scientifically-minded, Victorian-era liberal. What especially distinguishes Darwin's vision at the conclusion of *Origin* is his celebration of the interrelatedness of all life forms. Herein lies the origin of ecological thinking—appearing first in a liberal guise.

But there is another insight to derive from Darwin's entangled bank, so to disembark from *Origin*, I offer another hypothesis, which is that Darwin was specifically inspired here by the conclusion of natural

---

99 Darwin, *Origin* (1860), 490.
100 Darwin, *Origin* (1860), 490. Darwin removed the reference to the "Creator" in the third edition of 1861.
101 For the former view, see Johnson, *Darwin's Dice*; for the latter, Hyman, *The Tangled Bank*. Thierry Hoquet provides a historical parsing of this literature in part II of *Revisiting the Origin of Species: The Other Darwins* (New York: Routledge, 2018).

historian James Hutton's (1788) *Theory of the Earth*, a founding text of geology.[102] Hutton established the "uniformitarian" school of which Darwin was a resolute student. Hutton proposed that the physical processes creating landforms were uniform in time and space; hence, to understand Earth's natural history, we must examine processes presently at work and extrapolate back in time. Hutton's *Theory* concludes with lines intended to reassure the reader that, although his theory requires that our sense of time be radically expanded—with disorienting consequences for human subjectivity—all will be well:

> We have the satisfaction to find, that in nature there is wisdom, system, and consistency. For having, in the natural history of this earth, seen a succession of worlds, we may from this conclude that, there is a system in nature . . . [I]t is vain to look for anything higher in the origin of the earth. The result, therefore, of our present enquiry is, that we find no vestige of a beginning, no prospect of an end.[103]

Hutton's conclusion offers no telos except the satisfaction of finding "wisdom, system, and consistency" in nature. To find our place in natural history, we need look no higher than the earth. Here is the standpoint Darwin arrived at in London circa 1844.

Today, the expression "prospect of an end" resonates differently than in Hutton's day. Our planetary crisis signals that the prospect of an end for humanity is all too real. Here was another achievement of Darwin's: To recognize that speciation implies extinction. Look carefully at Figure 2.2; you will see that, after writing "I think," and sketching his tree, Darwin added additional language in the upper-right corner: "Case must be that one genus then should have as many living as now / to do this & to have species in same genus (as is) requires *extinction*." The becoming of species means the end of species. As Darwin would write in *Origin*: "We need not marvel at extinction; if we must marvel, let it be

---

102   James Hutton, *Theory of the Earth; or an Investigation of the Laws Observable in the Composition, Dissolution, and Restoration of Land upon the Globe*, Earth and Environmental Science Transactions of the Royal Society of Edinburgh 1, no. 2 (1788): 209–304. Hutton (1726–1797) was an interlocutor with David Hume and Adam Smith. In his *Investigations of the Principles of Knowledge*, Hutton anticipated aspects of evolution, but his deism prevented him from grasping that species emerge from one another.

103   James Hutton, *Theory of the Earth*, 304.

at our own presumption in imagining for a moment that we understand the many complex contingencies on which the existence of each species depends."[104]

It is a small step for a natural historian to ask: How long will humans last? When will be our end?

To grasp these existential questions—to conceive of human extinction in a nonteleological, natural-historical fashion—is a puzzle that I claim Marx "solved" after reading Darwin, marking his ultimate departure from Hegel. What is at stake is not a matter of settling philosophical scores. Today, we need such a standpoint to discern the prospect of an end—but not the end of our species. Rather: The prospect of an end to *rule by capital*.

---

104   Darwin, *Origin* (1860), 322.

3
Marx After Darwin

Marx in the Natural History Museum

Writing in Paris in 1844, Karl Marx made a bold prediction: "There will be *one* science."[1] He did not name this science, but we can sense what he was thinking:

> Man is not merely a natural being: He is a *human* natural being ... And as everything natural must *come into being*, so man also has his process of origin in *history* ... *History is the true natural history of man*. (We shall return to this later).[2]

---

1 Marx, Paris MS III (1844), §IX; New Left Books/Penguin, 355; cf. MECW III: 304.

2 Marx, Paris MS III (1844), §XXVII; New Left Books/Penguin,Penguin, 391, italics lightly modified; cf. MECW III: 337. Apropos these lines, Yusuke Akimoto writes, "If Marx wishes to preserve the possibility of admitting any difference between human beings and nature, then it would be quite inappropriate to look for a teleological claim within [his] thesis of 'fully developed naturalism, equals humanism.' It only appears to be teleological, because Marx operates ... from a viewpoint in which the relationship between human beings and nature is conceived of as a totality of natural history": Akimoto, "Marx's philosophy on natural history," in *Natural Born Monads: On the Metaphysics of Organisms and Human Individuals*, eds. Andrea Altobrando and Pierfrancesco Biasetti (Berlin: De Gruyter, 2020), 137–52, 145. I concur. Yet, to recapitulate, Marx could not realize his conception of this relationship (or of humanity as natural-historical) until reading Darwin.

Marx's parenthetical comment, "(We shall return to this later)," stood guard for two decades, bracketing these matters as Marx lived through many adventures: He met Engels; had children and raised a family with Jenny; wrote heaps of notes and journalism; got kicked out of France; wrote the *Manifesto of the Communist Party*; cheered as the revolutionary tide of 1848 rose, and wept as it fell; moved to London, where he filled a stack of notebooks about political economy; got sick again and again; then, one day in late 1860, convalescing from another bout of illness, opened *The Origin of Species*.[3]

He was far from alone in doing so. People were discussing Darwin's book everywhere, debating its implications for humanity. This was, without a doubt, the major intellectual event during the period that Marx wrote *Capital*. I concur with Eric Hobsbawm's assessment: "If any single scientific theory is to represent the advances of natural science and was recognized as crucial" during the years that Marx worked out his critical political economy of capitalism, "it was the theory of evolution, and if any one figure dominated the public image of science it was . . . Charles Darwin."[4]

Darwin provided Marx with a feeling of intellectual vindication. Immediately upon reading *Origin*, Marx wrote to Engels, declaring that the book "contains the natural-historical foundation of our outlook."[5]

---

3 According to the editors of MEGA2, the text of Darwin's that Marx owned and quoted in *Capital* was: Charles Darwin (1863), *Über die Entstehung der Arten im Their-und Pflanzen-Reich durch natürliche Züchtung, oder Erhaltung der vervollkommneten Rassen im Kampfe urn's Daseyn. Nach der 3. engl. Aufl* (From the third English edition), translated and with commentary by H. G. Bronn. However, as previously noted, Marx read Darwin's *Origin* in late 1860, a year after initial publication. From this and other clues, I surmise that Marx read Darwin's second edition in English (in 1860) and reread *Origin* (third edition) when he acquired the German translation in 1863.

4 Eric Hobsbawm, *The Age of Capital* (London: Penguin, 1996 [1975]), 253.

5 Marx, Letter to Engels, December 19, 1860, MECW 41: 232, in which Marx's words (ist dies das Buch, das die naturhistorische Grundlage für unsere Ansicht enthält) are translated, "This is the book which, in the field of natural history, provides the basis for our views." Note that MECW 41 has naturhistorische as a noun, whereas Marx uses "natural-historical" as an adjective that modifies "foundation" (cf. note 16). At the start of that sentence, Marx qualifies his praise for Darwin's book: "Though developed in an unsophisticated English manner . . ." ("Obgleich grob englisch entwickelt . . .": MECW [41: 232] has "Although developed in the crude English fashion"). This expression has often been cited by those emphasizing Marx's methodological and philosophical differences from Darwin. Granted that Darwin's *style* differs

A few weeks later, he wrote to Lassalle that Darwin's *Origin* "is most important and . . . provides a basis in natural science for the historical class struggle . . . For the first time, 'teleology' in natural science is not only dealt a mortal blow but its rational meaning is empirically explained."[6]

Marx wrote these letters midway between the start of his notebooks on political economy and the completion of *Capital*, volume I.[7] He read Darwin at a critical point, therefore, in the development of his theory of capitalism.[8]

---

from Marx's: That is how I understand Marx's criticism of *Origin* here (an instance of Marx's German chauvinism). We should also bear in mind that Marx is repeating to Engels here what Engels wrote earlier to Marx: Darwin's *Origin*, Engels wrote, "is absolutely splendid. There was one aspect of teleology that had yet to be demolished, and that has now been done . . . One does, of course, have to put up with the *crude English method*" (Engels to Marx, December 11 or 12, 1859; MECW 40: 551, my italics). So let us not overemphasize Marx's critique of Darwin's "grob englisch entwickelt"; the essential words are those that follow: Ist dies das Buch, das die naturhistorische Grundlage für unsere Ansicht enthält.

6   Marx, Letter to Lassalle, January 16 1861. MECW 41: 246–7.

7   His notebooks, first published in 1939–41, came to be called *Grundrisse*: Marx, *Grundrisse*, trans. Martin Nicolaus (London: New Left Books/Penguin Books, 1973).

8   This much—and no more—is agreed upon by the numerous scholars who have written on Marx and Darwin. We are presently in the third distinct phase of this literature. After a raft of early assessments in the late nineteenth and early twentieth century (on which see David Stack, *The First Darwinian Left: Socialism and Darwinism, 1859–1914* [London: New Clarion Press, 2002]), and a postwar lull, scholars of 1960s and 1970s generated numerous reassessments: Gerald Runkle "Marxism and Charles Darwin," *The Journal of Politics* 23, no.1 (1961): 108–26; Stanley Edgar Hyman, *The Tangled Bank: Darwin, Marx, Frazer and Freud as Imaginative Writers* (New York: Atheneum, 1962); Alfred Schmidt, *The Concept of Nature in Marx* (New York: Verso Books, 2014 [1962]); Erhard Lucas, "Marx und Engels Auseinandersetzung mit Darwin," *International Review of Social History* 9, no. 3 (1964): 433–69; Dieter Groh, "Marx, Engels, und Darwin: Naturegesetzliche Entwicklung oder Revolution? Zum Problem der Einheit von Theorie und "Praxis," *Politische, Vierteljahresschrift* 8, no. 4 (1967): 544–59; Guntram Knapp, *Der Antimetaphysysische Mensch: Darwin, Marx, Freud* (Stuttgart: Klett, 1973); Valentino Gerratana, "Marx and Darwin," *New Left Review* 82 (1973): 60–82; Enrique Ureña, "Marx and Darwin," *History of Political Economy* 9, no. 4 (1977): 548–59; Sven-Eric Liedman, *The Game of Contradictions: The Philosophy of Friedrich Engels and Nineteenth-Century Science* (Chicago: Haymarket, 2022 [1977]); Terence Ball, "Marx and Darwin: A reconsideration," *Political Theory* 7, no. 4 (1979): 469–83; and Paul Heyer, *Nature, Human Nature, and Society: Marx, Darwin, Biology, and the Human Sciences* (Westport: Greenwood Press, 1982 [a revised 1975 PhD dissertation]).

Darwin's *Origin* awakened a profound awareness of natural-historical becoming: Species have only evolved into their present form from some other form, indeed from other species; all beings are expressions of historical processes (always involving a degree of random chance). The implication is that these processes are universal, operating upon the ancestors of puffball mushrooms and finches and every other being on Earth. Species have therefore no essence outside of natural history. Since humans are a species, it follows that the forms of human society—how people organize themselves, basically—are essentially natural and

---

After another lull during the 1980s—a decade of retreat for Marxist scholarship—our third phase is marked by its liberation from the distortions of Stalinism and the dedication myth: Angus Taylor, "The significance of Darwinian theory for Marx and Engels," *Philosophy of the Social Sciences* 19, no. 4 (1989): 409–23; Helena Sheehan, *Marxism and the Philosophy of Science: A Critical History* (London: Verso Books, 2017 [1993]); Paul Nolan, *Natural Selection and Historical Materialism* (Watford: Glenfield Press, 1993); Richard Weikart, *Socialist Darwinism*, PhD dissertation, University of Iowa, 1994; Ball, "Marx and Darwin: A reconsideration," 469–83; Stephen Jay Gould, "A Darwinian Gentleman at Marx's Funeral," *Natural History* 108, no. 7 (1999): 32–40; J. B. Foster, *Marx's Ecology: Materialism and Nature* (New York: MR Press, 2000); Alan Carling and Paul Nolan, "Historical materialism, natural selection and world history," *Historical Materialism* 6 (2000): 215–64; Doyne Dawson, "The marriage of Marx and Darwin?," *History and Theory* 41, no. 1 (2002): 43–59; David Harvey, "On the deep relevance of a certain footnote in Marx's *Capital*," *Human Geography* 1, no. 2 (2008): 26–32; Paul Thomas, *Marxism and Scientific Socialism: From Engels to Althusser* (London: Routledge, 2008); Naomi Beck, "The *Origin* and Political Thought," in M. Ruse and R. Richards, eds., *The Cambridge Companion to the Origin of Species* (Cambridge, UK: Cambridge University Press, 2009), 295–391; Gareth Stedman Jones, *Karl Marx: Greatness and Illusion* (Cambridge, MA: Harvard University Books Press, 2016); J. B. Foster and Paul Burkett, *Marx and the Earth: An Anti-critique* (Leiden: Brill, 2016); Mike Davis, *Old Gods, New Enigmas: Marx's Lost Theory* (New York: Verso Books, 2018); J. B. Foster, *The Return of Nature: Socialism and Ecology* (New York: Monthly Review Press, 2020); Mauricio Vieira Martins, *Marx, Spinoza and Darwin: Materialism, Subjectivity and Critique of Religion* (New York: Springer, 2022).

Space does not permit a review of this literature. Suffice to say that—notwithstanding all this—there is more work to do. If the central contribution of the first phase was to stage a confrontation between Darwin and Marx, and the second to dispel various myths from the first, it remains to incorporate the wisdom of this literature into ecological or natural-historical Marxism. For instance, two of the outstanding contributions of our third period—Paul Burkett, *Marx and Nature: A Red and Green Perspective* (Chicago: Haymarket, 2014 [1999]); and Kōhei Saitō, *Karl Marx's Ecosocialism: Capital, Nature, and the Unfinished Critique of Political Economy* (New York: Monthly Review Press, 2017)—do not cite Darwin or discuss his influence upon Marx's shift to an ecological perspective.

historical.⁹ *Origin* also shows that every species and individual influences every other in a web of historically evolving relationships. Taken together, *Origin* invited readers to interpret the forms adapted by species as effects of natural-historical processes and ecological interrelations.¹⁰ Marx accepted this invitation. Darwin led Marx to reconceptualize his analysis of capital from a standpoint that emphasizes natural history and social form. Marxian natural history was born.¹¹

Consider when Marx wrote *Capital*. As noted in chapter 2, Darwin left his analysis of humans out of *Origin*. He published *Origin* in 1859 and *Descent* in 1871. In the twelve years between these epochal books, the world erupted in debate over the implications of Darwin's theory for humanity. A publishing boom met the reading public's appetite to speculate on the burning question of the natural history of humans as a species. Some of these works were good, some were terrible; most were forgotten. *Capital* volume I was written and published in the heat of this debate, in the weight of this pause before *The Descent of Man*. It deserves

---

9 Marx and other contemporary readers encountered *Origin* with no knowledge whatsoever of genetics, DNA, and epigenetics. What distinguishes humans from other species, Darwin would explain, was the character of human consciousness: Unlike horses and finches, we are aware of our place in natural history. Marx, we have seen, already agreed with this, but his presentation of the point in *Capital* shifted in ways that suggest Darwinian influence: See part II.

10 Some Marxists contend that Darwin's descriptions of organismal adaptation to natural environment was so bifurcated—the species active, the environment passive—as to inhibit an ecological perspective: For example, Richard Lewontin, *Biology as Ideology: The Doctrine of DNA* (Toronto: House of Anansi, 1991). While there is a basis for Lewontin's critique of Darwin from the vantage of contemporary evolutionary theory (though I note in passing that he does not actually cite Darwin in the aforementioned texts where he levels this critique), it should not mislead us from what is essential: Darwin's breakthrough facilitated an evolutionary-ecological conception of the life-world, thereby displacing an older conception—rooted in religion—that life forms were created within a harmonious, balanced nature: See Anneliese Griese, "Karl Marx und Friedrich Engels über das Verhältnis von Gesellschaft und Natur," *Beiträge Zur Marx-Engels-Forschung* 26 (1989): 70–82.

11 In recent years, the perspective I am defending has been called "ecological Marxism" or "eco-socialism." I have no objection to these terms. Yet, if we are talking about *Capital*, "ecological" is arguably less appropriate than "natural-historical." Either way, we must address Darwin. Haeckel coined the term "ecology" in 1866 (as Marx was finishing *Capital*) under inspiration from Darwin's theory. Marx derived similar inspiration as Haeckel, but developed another direction for inquiry: Whereas Haeckel's evolutionary ecology concerns biological life generally, Marx's natural history concerns the economic forms of human society.

recognition as the most important of the books written after *On the Origin of Species* that responded to it by helping to make sense of the natural history of human societies.

Consider where Marx wrote *Capital*. Denied peace in Germany, France, and Belgium, Marx was a political refugee in London. He fled to Britain not only because it was a relatively liberal society with a healthy community of German exiles and because his best friend Engels was in Manchester, running the family business.[12] London was the center of the capitalist world—the ideal place to observe capital in motion, with bourgeois society fully formed and on display. Moreover, that bourgeois society had produced excellent institutions to facilitate research. Marx worked on *Capital* in one of them, the reading room of the British Library within the British Museum, just down the street from where, a few years earlier, Darwin had written his notebooks.[13] The British Museum was London's natural history museum. Bloomsbury was a center of the world debate on Darwin's theory (if not where it was born).[14]

---

12   Rosemary Ashton, *Little Germany: Exile and Asylum in Victorian England* (London: Faber & Faber, 2013). Marx directly participated in debate surrounding Darwin's *Origin*: "Wilhelm Liebknecht, the 1848 veteran and founder of the SPD, fondly recalled attending six of [Thomas Huxley's 'Lectures to the Working Men'] with Karl Marx, then staying up all night excitedly discussing Darwin. The whole Marx household ... was caught up in the great debates": Mike Davis, *Old Gods, New Enigmas*, 115.

13   In 1839–1842, while writing his transmutation notebooks, Darwin lived at 12 Upper Gower Street (site of the present Darwin building—his residence was destroyed in World War II), which lies just north of the British Museum reading room where Marx wrote *Capital*, and a short walk to Marx's home at 28 Dean Street. Cf. Joe Cain, "Darwin in London," *The Linnean* 30, no. 2 (2014): 13–21; Asa Briggs and John Callow, *Marx in London* (London: Lawrence & Wishart, 2007). Darwin and Marx were both buried in London: Marx in Highgate Cemetery, near Hampstead Heath, where he would take restorative walks with his family; Darwin in Westminster Abbey, next to Isaac Newton Darwin's position in death, as in life, reflects higher social status; still, Marx's final resting place is the more congruent. For Darwin's great idea challenges the Church's theology, and, while Darwin may have admired Newton, the nature and implications of their scientific achievements were profoundly different.

14   In the 1860s these institutions were unified at the present location of the British Museum. The museum made a major contribution to Darwin's breakthrough: The zoological keeper, Edward Grey, "agreed to lend Darwin the Museum's entire cirripede collection, much of it uncatalogued ... The whole national collection at his fingertips": James Costa, *Darwin's Backyard: How Small Experiments Led to a Big Theory* (London: W. W. Norton & Co., 2017), 63. This marked a turning point in Darwin's research into barnacle speciation and sexual reproduction, arguably the critical empirical element of his theoretical development during the period of the delay of publication (1844–1856).

Take the when and where together and we have the setting of the production of *Capital*: Marx wrote it in a museum of natural history amidst the vociferous debate on the meaning of *Origin* for humanity's natural history. The conclusion is inescapable: Among the many ways that we can read *Capital*, it must be read in part as a contribution to the 1860s publishing boom on the natural history of human societies. But do not take my word for it. Marx emphasized this point himself.

Marx's Standpoint In *Capital*

Marx wrote an enigmatic paragraph at the end of the preface to the first German edition of *Capital*. Since this paragraph is central to my argument (and Marx's language is rather peculiar), I provide two translations and then comment upon the passage. Here is Ben Fowkes's translation:

> To prevent possible misunderstandings, let me say this. I do not by any means depict the capitalist and landowner in rosy colours. But individuals are dealt with here only in so far as they are the personifications of economic categories, the bearers [*Träger*] of particular class-relations and interests. My standpoint, from which the development of the economic formation of society is viewed as a process of natural history, can less than any other make the individual responsible for relations whose creature he remains, socially speaking, however much he may subjectively raise himself above them.[15]

Here is Paul Reitter's:

> To prevent possible misunderstandings, let me say this: I don't paint the figures of the capitalist and landlord in rosy colors—far from it.

The Natural History Museum was opened in 1881 in South Kensington. Legally it remained part of the British Museum until 1963 and was still called the British Museum (Natural History) until 1992. The British Library moved to St. Pancras in 1973. In recent years, the greatest annual conference of Marxist research, organized by the journal *Historical Materialism*, has been organized at SOAS, in the heart of Bloomsbury, walking distance from Darwin's erstwhile London home and the sites of Marx's work on *Capital*. (See also previous note.)

15   Marx, *Capital* (New York: Penguin, 1976), 92.

But individual persons play a role here only insofar as they are the personifications of economic categories, or the bearers of particular class relations and interests. My approach treats the development of society's economic formation as part of natural history, as that type of process, and no other approach does less to make the individual responsible for conditions that he remains a creature of socially, however much he manages to transcend them subjectively.[16]

Let me restate this in the form of a few propositions, starting with Marx's statement that his "approach treats the development of society's economic formation as part of natural history."

This is the only place in Marx's writings on capitalism where he declares his "standpoint."[17] He says his standpoint is that of a natural historian studying economic forms of human society. To recapitulate his language more literally, Marx declares that, from his standpoint, he examines natural-historical processes that cause the development of socio-economic formations.[18] By implication, his object of analysis in *Capital* is dual: He says he will examine [a] socio-economic formations and [b] the natural-historical processes that create [a]. The standpoint from which Marx will scrutinize this dual object could be called the natural history of socio-economic formations. I call it Marxian natural history for short.

I dwell on this passage from the preface, and these basic propositions, for two reasons.

First: This is definitely not the standard view of Marx's approach in *Capital*. Although Marx makes this claim right at the outset of the book,

---

16   Marx, *Capital* (Princeton: Princeton University Press, 2024), 7–8 (unless otherwise noted, all subsequent references to *Capital* I are to this edition, a translation of the second German edition of *Capital*: Incidentally, the version Marx sent to Darwin). Though Marx uses "natural-historical" here as an *adjective* that modifies the noun "process" (einen naturgeschichtlichen Prozeß), Reitter translates this passage using "natural history" as a *noun* ("part of natural history, as that type of process"), since, as he explained to me, we no longer use "natural-historical" as an adjective in colloquial English, and he sought to match the register of the source text. I have no quarrel with Reitter, but since I contend that we must become reaccustomed to the adjective "natural-historical"—common enough in Darwin's and Marx's time—throughout this book I emphasize that, from Marx's standpoint, *socio-economic forms* are the result of *natural-historical processes*.

17   Marx's term is "Standpunkt."

18   Marx's words are "naturgeschichtlichen Prozeß" and "ökonomischen Gesellschaftsformation."

in a place (the preface) where it cannot be missed, I know of practically no readings of *Capital* that emphasize these points.[19] Rather, in the popular understanding—which I believe to be a serious error, with unfortunate political consequences—Marx's approach in *Capital* is supposedly that of a "dialectical materialist" studying a "mode of production." But Marx never described his thought with the first term (indeed he never even wrote the words "dialectical materialism") and although Marx does refer to capitalism as a "mode of production" in *Capital*, he frequently complements this concept with a more general one, "social form" (or "social formation," or "economic social formation," as in this sentence in the preface).[20] What exactly is this? We must read *Capital* to obtain a fulsome answer, but a concise definition is "a way of organizing human lives." Basically, Marx thought that every social formation was comprised of multiple kinds of production, with one predominant and others subordinate.[21] *Capital* aims to explain capitalism as a social formation, that is, as the present form in which human lives are organized.[22]

---

19 The second—largely forgotten—English translation of *Capital*, Eden and Cedar Paul's contribution to Everyman's Library, makes no reference to natural history (as noun or adjective): "Inasmuch as I conceive of the development of the economic structure of society to be a natural process, I should be the last to hold the individual responsible for conditions whose creature he himself is, socially considered": Marx, *Capital*, volume II, trans. Eden and Cedar Paul (New York: Dutton, 1930), 864.

20 "The wealth of societies dominated by the capitalist mode of production appears in the form of an 'enormous accumulation of commodities'": Marx, *Capital*, 13. Note that, at the outset of his study, the object of analysis is not mode of production per se but the form in which its social wealth appears. Mode of production is not a straightforward category in *Capital*.

21 On the relationship between mode of production and social formation, Marx wrote in his notebooks on political economy (*Grundrisse*): "In all forms of society there is one specific kind of production which predominates over the rest, whose relations thus assign rank and influence to the others. It is a general illumination which bathes all the other colours and modifies their particularity. It is a particular ether which determines the specific gravity of every being which has materialized within it": *Grundrisse*, Nicholas translation, 107. Social formation translates Gesellschaftsformen. See Tony Burns, "Marx and the concept of a social formation," *Historical Materialism* 32, no. 3, 158-87.

22 Marx was not only concerned with one social formation. It is just that in *Capital* he could only attempt to explain capitalism. Marx failed to complete even this analysis.

Second: This is the moment in *Capital* at which Marx signals his debt to Darwin most clearly. Darwin was the world's best-known natural historian in 1867. If Marx's standpoint is to study the development of economic formations in light of natural history, what exactly constitutes this "development"?[23] Marx is definitely not talking about "economic development," as this term became widely used in the twentieth century (and *Capital* is not a book about development policy). What develops is neither a pregiven national economy nor a capitalist economy: Marx is not referring to an economy at all as if it were a thing. Rather, Marx says he will examine the historical development—becoming and logic—of economic social *forms*.[24] That is his object; his intent is to explain the laws of the unfolding of these forms. Recall that Darwin said that his aim was to explain the laws of the unfolding of new forms of species. When we take these statements in parallel, the influence of Darwin's *Origin of Species* on Marx's project in *Capital* is unmistakable. Following Darwin, Marx recognized that—like all other species—humans emerged through descent with modification. Such is our natural history. Marx wanted to explain how this natural-historical becoming extended down to the present day—requiring attention not only to natural selection but to social change (driven, for Marx, by class struggle).[25] In short, *Capital* aims to explain *how human societies organize themselves today*.

In *Capital*, Marx observes that while humans have always *labored*—the general activity of laboring is therefore universal and transhistorical—the specific social *form* through with labor is organized, and the sort of social and production relations this entails, vary considerably

---

23 Marx's term is Entwicklung, which would commonly be translated as "evolution" (both then and today). Indeed, in MECW 35 this passage is translated with "evolution" ("My standpoint, from which the evolution of the economic formation of society is viewed as a process of natural history"), underscoring Marx's Darwinian tone.

24 That is, "der ökonomischen Gesellschaftsformation." Elsewhere in the preface, where he comments on the lack of an equivalent in political economy to the study of chemical reagents, Marx states that his aim is the analysis of economic forms for which abstraction takes the place of chemistry experiments ("der Analyse der ökonomischen Formen kann außerdem weder das Mikroskop dienen noch chemische Reagentien. Die Abstraktionskraft muß beide ersetzen").

25 Recall the opening line of the *Manifesto of the Communist Party*: "The history of all hitherto existing societies is the history of class struggles": Marx and Engels, *Manifesto of the Communist Party* (London: Penguin, 2002 [1848]), 219.

through time and space. What most clearly distinguishes the capitalist epoch from previous ones is the existence of a substantial class of people, whom Marx named the "proletariat," who produce a living by selling the commodity labor power. This class produces the social surplus in a form that Marx will call "surplus value." While the exploitation of this class through the expropriation of surplus value is distinct to capital, the very possibility of class exploitation is ancient. Following the writings of classical political economists like Adam Smith and David Ricardo, Marx will call the thing which gives capitalist society its form, "value." Capitalist society is organized by the value form.[26] This is not so much an economic fact as an interpretation of the natural history of the organization of human societies. As Kōjin Karatani emphasizes apropos the passage from the preface under discussion:

> The "economic categories" mentioned here signify the forms of value. Who are capitalists and proletariats is determined by where the individuals are placed: In either relative form of value or equivalent form. This is totally irrespective of what they think . . . In *Capital*, there is no subjectivity.[27]

Yet, if there is no subjectivity in *Capital*—I agree with Karatani—there are indeed social relations and historical processes that function by designating or assigning specific class roles, and social burdens, to particular people. Unlike Marx's treatment of the subject ca 1844–46, in which he emphasizes the concepts species-being and praxis, in *Capital* Marx emphasizes this form-determination of historical processes of social subjection. That is what is meant by his statement that "the individual" cannot be held "responsible for relations whose creature he socially remains." Instead of subjects, we have bearers-of-subjectivity, personifications of economic categories.[28] Why must they bear this burden and personify these categories? Because of the prevailing economic social formation. Ergo, these socio-economic forms—not particular subjects within them—are the object of analysis in Marxian natural history.

---

26 Wertform, a concept developed in *Capital*, chapter 1, section 3.
27 Kōjin Karatani, *Transcritique: On Kant and Marx* (Boston: MIT Press, 2004), 18.
28 Marx's terms are "Träger" and "Personifikation ökonomischer Kategorien."

Marx is not saying that everyone in a capitalist society is either a worker or a capitalist: He was perfectly aware that this was not the case in the 1860s. Neither is he suggesting that, since we personify economic categories, these categories are the only things that exist or that matter. Marx writes that "individual persons play a role" in *Capital* "only insofar as they are the personifications of economic categories, or the bearers of particular class relations and interests."[29] He is announcing a simplification so that we can grasp his object of analysis. A concise illustration: Wealthy individuals in capitalist society often harbor strong feelings about the natural environment. I do not doubt the genuineness with which some very wealthy people speak of their love for nature. But those individuals live their lives caught up in capital, where accumulation remains imperative. Hence, they extract themselves from their wealth, not their wealth from extraction.

With these propositions, Marx's standpoint in *Capital* opens an aperture through which we can analyze capitalist society as a socio-economic form. His emphasis on natural history is not only a nod to Darwin but also an indication that he treats society's economic formation as entailing nature and society at once: While the analysis in *Capital* generally concerns the social relations of different classes (especially proletariat-bourgeoisie), the analysis points to the ways these social relations necessarily shape and reflect relationships between humans and nature (HH×HN). In chapter 5, we will see that Marx contends that the capitalist form of society "reflects back at people the social characteristics of their own labor as objective characteristics of their labor products, as socio-natural properties of those things."[30] Marx stakes this position *not* because the human individual is reduced to nature, as in social Darwinism, but because the individual remains a creature of social conditions. For Marx, the individual as such is not the object of analysis in *Capital*, nor in natural history. The object of analysis is the social-economic formation in which those individuals live.

---

29 Marx, *Capital*, 8.
30 Marx, *Capital*, 48–9.

Marx's standpoint and his critique of Hegel

Let me anticipate a criticism. It would be reasonable if my readers thought: "Well, perhaps Marx wrote like that in one paragraph in *Capital*—but perhaps he changed his mind? After all, he never wrote like that elsewhere." Perhaps I am blowing this out of proportion?

In part II, I elaborate upon how Marx's natural-historical perspective influences *Capital*. For now, I would like to emphasize that Marx did explicitly return to the natural-historical framing of his project in *Capital* via a comment from a critic.[31] In the afterword to *Capital*, volume I, Marx quotes at length from a review by an anonymous, contemporary German critic. The critic summarizes Marx's perspective:

> Marx is concerned with one thing alone: To prove, by way of a precise scientific investigation, the necessity of certain orders of social relations... Marx regards social movement as a natural-historical process that is governed by laws that are not only independent of the will, consciousness, and intentions of a person but that themselves determine their will, consciousness, and intentions... [For Marx], every major historical period has its own laws... as soon as it has outlived a given period of development,... it already begins to be governed by different laws. In short, *economic life presents to us in this case a phenomenon perfectly analogous to what we observe in other categories of biological phenomena.*[32]

This review of *Capital* shows that Marx's contemporaries recognized strong parallels between his project and *Origin*.

If my interpretation of *Capital* is wrong, then Marx have should have rejected this review out of hand (Marx was extremely sensitive to misrepresentations of his thought). Instead, after quoting from this review—I only know of it because Marx found it so valuable that he quoted it at length in the postscript to later editions of *Capital* (and this is the only review to make that cut)—Marx responds to it, favorably:

---

31 Marx makes five direct references to natural history in *Capital*. I discuss two in this chapter and the others in part II.

32 Marx, *Capital*, 707–8, my italics.

> Insofar as *the author* [of this review] *depicts my real method*, as he calls it, *with great accuracy*, and insofar as he proceeds with a great deal of good will where he is concerned with how I apply that method, he can't depict anything but the dialectical method.[33]

Marx says that the reviewer has correctly described his standpoint, which is nothing but "the dialectical method." Yet, Marx emphasizes, his dialectic is not the same as Hegel's. This is where Marx writes his famous statement about flipping Hegel around:

> my dialectical method doesn't just differ from the Hegelian one: They are utter opposites. Hegel goes so far as to transform the process of thinking into an independent subject, doing so under the name "the idea." For him, that process is the demiurge of the real, while the real merely constitutes its external appearance. For me, conversely, the ideal is nothing but the material as it is transposed and translated inside human heads . . . Hegel mystified dialectics, but that didn't stop him from being the first to consciously and comprehensively represent its general forms of movement. Here the dialectical method stands on its head. You have to flip it around in order to find the rational kernel encased in its mystical husk.[34]

Marx returns here to his critique of Hegel from 1844–46, when he wrote that Hegel's philosophy lacks grounding in nature and history (see chapter 1). The "rational kernel" in Hegel's philosophy is his dialectical conception of being and becoming; the "mystical husk" is the teleological metaphysics of Spirit and Reason. After you "flip it around," the result is to understand social and economic forms as results of "a natural-historical process that is governed by laws," in the words of the reviewer Marx compliments for depicting his method with "great accuracy."

To recapitulate: In his Paris manuscripts, Marx criticized Hegel for reifying nature and treating history as teleological. The solution Marx found was to treat nature as historical and to recognize humanity's place in natural history. In the words of Yusuke Akimoto, "Marx

---

33 Marx, *Capital*, 709.
34 Marx, *Capital*, 709.

reached his own concept of natural history by distancing himself from Hegelian teleology, which reduces nature to one step in the realization of the absolute."[35] The result is a complex conception of the human relationship to nature. On one hand, humans are distinguished from other natural beings by the unique character of our consciousness; but this distinction is in turn treated as historical, since humanity generally and our consciousness in particular are results of evolution, that is, products of natural history.[36] Marx also goes beyond Hegel by insisting upon the analysis of how these relations are transformed by the adoption of a capitalist form of society.[37] Capitalism must be "denaturalized" (shown to be a historical product of social relations) to be able to recognize capital's nature as a peculiar social formation that, since emerging through class struggle, has transformed what it means to be human on Earth. This is the meaning of what Marx wrote immediately after reading Darwin: His *Origin of Species* "is most important and . . . provides a basis in natural science for the historical class struggle."[38]

To simplify my argument and put it in a form preferred by some Hegelians, we might explain the emergence of Marxian natural history as the result of negation-of-negation:

1. Positing: Hegel posits a teleological philosophy of history that is absorbed and endorsed by Marx as a student.
2. Negation: Marx criticizes Hegel's philosophy of history on two grounds: First, it was teleological; second, Hegel's philosophy of history and reason is not grounded in practical life and humanity's

---

35 Yusuke Akimoto, "Marx's philosophy on natural history," in *Natural Born Monads: On the Metaphysics of Organisms and Human Individuals*, eds. Andrea Altobrando and Pierfrancesco Biasetti (Berlin: De Gruyter, 2020), 137–52, 138.

36 Hence, to quibble with Wendy Brown's (generally excellent) foreword to the 2024 Princeton University Press translation of *Capital*: When Brown asks, "What light might *Capital* shed on the planetary ecological catastrophe . . . since Marx joined his contemporaries in differentiating humans from 'nature' and followed Aristotle and Hegel in casting us as bound to incessantly transform nature for our own comfort and benefit?" (xxix), her question is based upon false premises. For, in truth, Marx broke with most of "his contemporaries" by locating humanity within natural history and rejected Aristotle and Hegel for their ahistorical conceptions of human relations with nature.

37 Marx's key statements on this theme are in *Capital*: See part II.

38 Marx, Letter to Lassalle, January 16, 1861, MECW 41: 246–7.

relations with nature. Marx was correct, but in 1844–46, he could provide no scientific alternative to Hegel, only a philosophical critique. So, Marx set the matter aside.[39]
3. Negation-of-negation: After reading Darwin, Marx could *realize* his critique of Hegel. Darwin provided Marx with a basis for achieving a nonteleological, natural-historical explanation of humanity and its present form.[40] Marx discovered a natural-historical foundation for his theory; he devoted the rest of his life to explicating this standpoint, what I call Marxian natural history.

How might Marx's natural-historical standpoint help us to understand the emergence of capitalism and our planetary crisis? I take up this question in part III. Before doing so, as a preliminary step, we should ask: How does clarifying this standpoint help us read *Capital*? I answer this question in part II.

But, before turning to *Capital*, let me briefly address two confusing episodes that have complicated prior studies of Darwin and Marx and provide three brief clarifications about Marx and *Capital*.

---

39 Marx jokes of *The German Ideology*: "We were informed that owing to changed circumstances it could not be printed. We abandoned the manuscript to the gnawing criticism of the mice all the more willingly since we had achieved our main purpose—self-clarification": Marx, Preface to *A Contribution to the Critique of Political Economy* (New York: International Publishers, 1977 [1859]), 22.

40 After reaching this conclusion in my research, I discovered two sources which independently arrived at a similar finding: [1] "Although in his early work Marx tried to develop a consistent natural science perspective, he does not appear to have achieved this objective until Darwin's evolutionary research was published in 1859": Paul Heyer, *Nature, Human Nature, and Society: Marx, Darwin, Biology, and the Human Sciences* (Westport: Greenwood, 1982), 59; [2] "Instead of interpreting the course of human history as a series of stages of exteriorization of the Idea [as with Hegel—JDW], Marx shows us earthly social forces in conflict, from which results the characteristics of mundane reality . . . However, until the mid-nineteenth century, an alternative substantive approach for addressing . . . the origin of species, was not available. Before Darwin . . . there was a dissatisfaction with the theological mode of approaching questions related to the human species . . . but not yet a substantive theory that offered a real alternative": Mauricio Vieira Martins, *Marx, Spinoza and Darwin*, 179–80. Though I have my quibbles with these sources (which differ considerably in aims and method), I am pleased that we converge on this finding, and I recommend these studies to future scholars of the Marx-Darwin relationship.

Two Confusing Episodes Concerning Darwin and Marx

1. *The dedication myth and exchange of letters*

Discussions about Marx and Darwin were long distorted by a myth that Marx offered to dedicate a volume of *Capital* to Darwin, who declined the offer. This dedication myth started in 1931, when a letter from Darwin to Edward Aveling—a promoter of Darwinism, socialist, and romantic partner of one of Marx's daughters, Eleanor—was discovered in Moscow and mistakenly published as a letter from Darwin to Marx.[41] (Aveling had sought Darwin's endorsement of his student's guide to Darwinism; Darwin gently declined.)[42] From the 1930s until the late 1970s, most debates surrounding Darwin's influence upon Marx were muddled by this error.[43]

While Marx never sought Darwin's public approval of his work, *Capital* did form a connection between them, and this tie, however slight, remains worthy of consideration. On June 16, 1873, Marx mailed a copy of *Capital* to Darwin. He inscribed his book to "Mr Charles Darwin / on the part of

---

[41] On the debunking of the *Capital*-dedication myth, see Ralph Colp, "The contacts between Karl Marx and Charles Darwin," *Journal of the History of Ideas* 35, no. 2 (1974): 329–38; Louis Feuer, "The 'Darwin-Marx correspondence': A correction and revision," *Annals of Science* 33, no. 4 (1976): 383–94; Margaret Fay, "Did Marx offer to dedicate *Capital* to Darwin?: A reassessment of the evidence," *Journal of the History of Ideas* 39, no. 1 (1978): 133–46; Margaret Fay, "Marx and Darwin: A literary detective story," *Monthly Review* 31, no. 10 (1980): 40–57; Joel Barnes, "Revisiting the 'Darwin-Marx correspondence': Multiple discovery and the rhetoric of priority," *History of the Human Sciences* 35, no. 2 (2022): 29–54. David Stack argues that the myth would never have taken root had the ground not already been cleared for its belief; hence, "Much of the blame for this myth must also rest with Engels ... [who] explicitly encouraged socialists to regard Marx and Darwin as complementary": David Stack, *The First Darwinian Left: Socialism and Darwinism, 1859–1914* (Cheltenham: New Clarion, 2003), 1.

[42] Edward Aveling, *The Students' Darwin* (London: Freethought Library, 1881). Apropos Darwin, Marx, and Aveling, see Joel Barnes, "'This great principle of the continuity of phenomena': Edward Aveling on the evolutionism of Darwin and Marx," in *Imagining the Darwinian Revolution: Historical Narratives of Evolution from the Nineteenth Century to the Present*, ed. Ian Hesketh (Pittsburgh: University of Pittsburgh Press, 2022), 121–36.

[43] The works of the second and third phases in the literature on Marx and Darwin (see note 8 of this chapter) are distinguished in part by this debunking of the dedication myth: Thanks to certain works from the second phase, the third is (almost entirely) released from its spell.

his sincere admirer / [signed] Karl Marx." Darwin did not read Marx's book. He could not do so; his German was far from fluent. Moreover, Darwin spent very little time reading works in political economy. Nevertheless, Darwin replied to Marx with modesty:

> Dear Sir:
>
> I thank you for the honour which you have done me by sending me your great work on Capital; & I heartily wish that I was more worthy to receive it, by understanding more of the deep and important subject of political Economy. Though our studies have been so different, I believe that we both earnestly desire the extension of Knowledge, & that this is in the long run sure to add to the happiness of Mankind.
>
> I remain, Dear Sir
> Yours faithfully,
> Charles Darwin[44]

In 2024, Darwin's copy of *Capital* returned to public view at the Darwin Museum (Down House) after some preservation work. The museum's curator was quoted by liberal media outlets stating that Darwin's copy of *Capital* brings "amusing insight into the dynamics between these two prominent intellectuals."[45] If the relations between Darwin and Marx seem merely "amusing" to us today, the joke is on us. Marx saw that *Capital* should be gifted to Darwin, for he knew the importance of Darwin's thought for his standpoint. I suspect that Marx was delighted to receive Darwin's warm reply and felt that he had settled his debt to Darwin.

In *Capital*, Marx repaid Hegel as well. For, if *Capital* flirts with Hegelian language, it was because the pupil could now look his former teacher in the eye. Darwin undermined teleology in nature and history. Darwin is therefore the ultimate critic of Hegel. Marx thought with these two in counterpoint. The result is Marxian natural history.

---

44 Darwin, Letter to Karl Marx, October 1, 1873, Darwin Correspondence Project.

45 Jack Guy, "'Amusing insight' revealed in Marx's *Das Kapital* gift to Darwin," CNN, February 22, 2024; Ez Roberts, "Charles Darwin's copy of Karl Marx's book back on display," BBC, February 23, 2024.

2. Engels's speech at Marx's graveside

Marx died on March 14, 1883. Three days later, a small group gathered to pay their respects at his gravesite in Highgate Cemetery, London. In his eulogy, Engels made a remarkable statement about his old friend:

> Just as Darwin discovered the law of development of organic nature, so Marx discovered the law of development of human history: The simple fact, hitherto concealed by an overgrowth of ideology, that mankind must first of all eat, drink, have shelter and clothing, before it can pursue politics, science, art, religion, etc.; that therefore the production of the immediate material means, and consequently the degree of economic development attained by a given people or during a given epoch, form the foundation upon which the state institutions, the legal conceptions, art, and even the ideas on religion, of the people concerned have been evolved, and in the light of which they must, therefore, be explained, instead of vice versa, as had hitherto been the case.[46]

Engels's remarks by Marx's graveside were soon published in *Der Sozialdemokrat*, a leading newspaper of the Left in Germany.[47]

There are several problems with Engels's statement about Marx with Darwin in the graveside speech. It does not make sense to call Darwin's theory of natural selection a "law of development of organic nature"; nor does Marx's critique of political economy provide us with a "law of development of human history." Engels distorted both theories to fit them into a false equivalence.

There is no shame in a little hyperbole at a loved one's funeral. Engels wanted to praise his friend highly and Darwin was the most famous scientist in Britain. But the publication and circulation of Engels's remarks contributed to a popular view that Marx had fashioned a theory akin to Darwin's which explains the evolution of human societies, where fitness is based on socio-technological development and class struggle.[48]

---

46 Fredrick Engels, Speech at the Grave of Karl Marx, Highgate Cemetery, London, as published in *Der Sozialdemokrat*, March 22, 1883. MECW 24: 467–71.

47 Reproduced at MECW 24: 465.

48 Bober's influential study *Marx's Interpretation of History*, for instance, begins with the claim that "Twelve years before the *Origin of Species* was given to the world

This distorted a basic lesson of *Capital*. Nevertheless, a version of just such ideas became widespread in the twentieth century, by which time it was converted into something called "dialectical materialism"—a term that Marx never used to describe himself and provides a poor handle for his thought.

The persistent Engelsian framing has also created problems for appreciating Darwin's influence on Marx's thinking, by generating a countertendency of denying that Darwin influenced Marx at all. Delivering the 2016 Ralph Miliband Lecture, Gareth Stedman Jones argued that Marx *opposed* Darwin because Marx sought to hold onto historical teleology.[49] Since the argument here is rather convoluted, I will interpolate my views in bracketed comments:

> In 1883, at Marx's graveside, Engels did his best to associate Marx's work with that of Darwin [true]. "Just as Darwin discovered the law of development of organic nature, so Marx discovered the law of development of human history." ... But this argument was fundamentally forced [true]. Marx's objection to Darwin was that he regarded progress as "purely accidental" [debatable; here Stedman Jones cites the aforementioned letter to Engels]. Darwin himself did not believe that history possessed any unilinear meaning or direction [true]. As he wrote, "I believe in no fixed law of development." Marx, on the other hand, maintained that man was not simply a creature of his environment [no citation is provided for this claim, making this claim difficult to evaluate] ... There is no evidence that Marx ever

---

a book appeared in which was formulated an evolutionary theory of history at once comprehensive and challenging ... The book was Karl Marx's *Poverty of Philosophy*." M. M. Bober, *Karl Marx's Interpretation of History* (Cambridge: Norton Library, 1965 [1927]), 4. Bober's book is subtitled *A Study of ... The Marx-Engels Doctrine of Social Evolution*." Few today would classify Marx as a "social evolutionist."

49 For a stronger interpretation of Marx as a teleological thinker, see Allen Wood, *Karl Marx* (London: Routledge, 1981), chapter VII, passim. Part of the issue is that thinkers like Stedman Jones and Wood accepted the base-superstructure model of Marxism. Yet, as Kōjin Karatani observes: "While the theories of Marx or Freud ... are often described as discoveries of a kind of substratum or base, what they actually accomplished was a dismantling of precisely that teleological ... perspectival configuration that produces the concepts of substratum and stratification": Karatani, *The Origins of Modern Japanese Literature* (Durham: Duke University Press, 1993), quoted in Fredric Jameson, "In the mirror of alternate modernities," *South Atlantic Quarterly* 92, no. 2 (1993): 295–310, 304–5.

abandoned his 1844 idea that history was the process of the humanization of nature through man's "conscious life activity" [this is, at best, an oversimplification]. While later admirers thought that Marx started precisely where Darwin left off, Marx did not accept the fundamental continuity between natural and human history, as argued by the Darwinists [ditto].[50]

Notwithstanding my quibbles, Jones clarifies the essential question: Did Marx accept the thesis of continuity between natural and human history?

The evidence shows that, by the time of Marx's writing of *Capital*, this question must be answered affirmatively—albeit with critical qualifications. From Marx's standpoint, the continuity of natural and human history does not imply perfect *equivalence*. As natural beings, Marx distinguished humans by consciousness. It follows that the continuity of human and natural history takes the form of a *differentiated* totality, where the differentiation concerns human consciousness within, and as a consequence of, natural history. Moreover, Marx argues that the capitalist form of society historically generates a sort of separation of nature and humans. One of the curious qualities of life under capital is that it exhibits *both* continuity between nature and humanity (since science demonstrates that everything humans do changes nature in some fashion, and that environmental conditions influence our lives and bodies) *but also* a fundamental break (which we experience as a growing separation from the natural world). Continuity is maintained, but differentially; it is never absolute. *Capital* shows that to grasp this complexity we must study the changing economic forms of society as processes of natural history.

Engels's graveside speech continues to complicate commentaries on Darwin and Marx. In his final years (1872–1883), Marx took detailed notes on the history and conditions of precapitalist, non-Western societies. Among the works he studied was Lewis Henry Morgan's 1877 ethnographic study, *Ancient Society*.[51] After Marx's passing,

---

50 Gareth Stedman Jones, "History and Nature in Karl Marx: Marx's Debt to German Idealism," *History Workshop Journal* 83, no. 1 (2017): 98–117, 109–10.

51 Lewis Henry Morgan, *Ancient Society, or Researches in the Lines of Human Progress from Savagery, Through Barbarism, to Civilization* (New York: Macmillan, 1877).

Engels drew upon Marx's notes while writing his *Origin of the Family, Private Property, and the State*. Engels refers to Marx frequently in that work and claims in the preface—without providing any evidence—that Marx had "planned to present the results of Morgan's researches in connection with the conclusions arrived at by his own . . . materialist investigation of history."[52] Engels stresses the parallel between Morgan and Marx, noting, for example, that Morgan "rediscovered in America, in his own way, the materialist conception of history that had been discovered by Marx forty years ago."[53] Concerning this remark, Kevin Anderson perspicaciously observes that this "was not the first time Engels had drawn too easy a parallel between Marx and another thinker. At [Marx's] graveside in 1883, Engels had famously done so with regard to Charles Darwin, ignoring Marx's strictures concerning the English biologist in the first volume of *Capital*."[54] While I concur with Anderson apropos Engels's graveside speech, in clarifying one issue, Anderson threatens to create a new point of confusion: In fact, there are no "strictures concerning [Darwin] in the first volume of *Capital*."[55] Before turning to it, three preliminary clarifications are in order.

---

52 Engels, *Origin of the Family, Private Property, and the State*, in MECW 26: 131. Although Marx undoubtedly shared Engels's enthusiasm for Morgan's research, *Origin of the Family, Private Property, and the State* is Engels's book, not Marx's.

53 Engels, *Origin of the Family*, 131.

54 Kevin Anderson, *Marx at the Margins: On Nationalism, Ethnicity, and Non-Western Societies* (Chicago: University of Chicago Press, 2020), 278, note 10.

55 Similarly, Anderson, *The Late Marx's Revolutionary Roads: Colonialism, Gender, and Indigenous Communism* (New York: Verso Books, 2025)—an invaluable reading of the late Marx—downplays Marx's dedication to the natural sciences (including Darwin) even where Anderson's research documents the late Marx's passion for them. Anderson writes, for example, "Marx himself was *more critical of the perspectives of natural science* than has often been supposed . . . as seen in an often overlooked footnote to *Capital* on the British biologist [Darwin]: 'The weaknesses of the abstract materialism of natural science, *a materialism which excludes the historical process*, are immediately evident from the abstract and ideological conceptions expressed by its spokesmen whenever they venture beyond the bounds of their own specialty.'" (Anderson, 74, my italics): Anderson's assertion that Marx was referring to Darwin in this sentence is implausible. For although Marx refers (positively) to Darwin earlier in this footnote, it cannot be claimed that Darwin's materialism "*excludes the historical process*," since he was precisely the thinker who placed species-formation within natural history—as Marx recognized.

Three clarifications about Marx and Capital

[1] Marx has been seriously misunderstood, including by some of his fiercest advocates. When people say that Marx was wrong or that his ideas failed, they are often criticizing ideas that are not really Marx's, but someone else's. The widely criticized labor theory of value, for instance, comes from Locke, Smith, and Ricardo, not Marx. It *is* simplistic—which is why Marx elaborated a more subtle theory of value (see chapter 5). Similarly, the political strategy that many people associate today with Marx—communists form a party, seize the state, and smash capitalism—would be better attributed to Lenin, if not Blanqui.[56] Marx wrote that *if* capitalism was to be overcome, it would happen in a simultaneous worldwide revolution. He never said that such a thing was guaranteed, only that it was possible and desirable (I return to this in chapter 8). One more example: By the time he wrote *Capital*, Marx rejected the view that European colonialism was a progressive factor of history. On the contrary, Marx became a vigorous critic of colonialism.[57]

Marx generated important insights that are not widely appreciated today, and a great deal of supposedly Marxist thinking has nothing to do with him. I suspect he saw this coming. Late in life, informed of the views of a group that called itself "Marxist," he replied laconically, "all I know is that I am not a Marxist."[58]

One of these underappreciated insights is that the forms of human societies evolve and that the presently dominant form, capitalist society, is governed by a logic that can be studied scientifically. This is the aim of *Capital*, in which Marx shows that capital organizes our lives around definite class relations defined by competition, accumulation, exploitation, and the separation of humans from nature. Hence our world is organized by something amoral—the value relation of capital—which must be explained and somehow overcome. The standpoint from which

---

56 A simplification, obviously. Moreover, Lenin's political writings—*State and Revolution* (1917), for instance—reveals his debts to Engels, so one must also consider Engels's influence on this conception of communism (not to mention Blanqui).

57 See Anderson, *Marx at the Margins*; Marcello Musto, *The Last Years of Karl Marx: An Intellectual Biography* (Stanford: Stanford University Press, 2020).

58 He spoke these words in French ("Tout ce que je sais, c'est que je ne suis pas marxiste"), the language of the group in question: Marx, quoted by Engels in a letter on August 5, 1890: MECW 49: 7.

Marx frames his explanation is indebted to Darwin. And Marx says he treats the evolution of economic life—the *form* of capitalist society—as a process of natural history. That idea is uniquely Marx's and particularly valuable today. Since this claim is only introduced in *Capital*—an odd and tricky book—it requires clarification and elaboration.

[2] *Capital* is a profoundly incomplete work. Marx never finished volumes II and III; neither did he make much headway into anticipated volumes on the state and the world market. Moreover, Marx never stopped revising the book that was published in his lifetime—*Capital*, volume I. Through multiple editions and translations, he fiddled with the work until his end; judging by letters he wrote shortly before he died, had he lived for a few more years, he might have produced a new edition with substantive changes.[59] It was left for Engels to cobble together *Capital* II and III from Marx's copious manuscripts. Partly for this reason, Engels has played a major role in debates surrounding ecological Marxism.

Marx did not fail to finish the project for lack of effort. He worked feverishly in his final years, filling notebooks with excerpts and commentaries on diverse topics.[60] Indeed, the very diversity of these topics had for some time led Marxist scholars to genuflect about their contents. From what I can ascertain from the published literature on them, three themes are central to these topics: World history; natural science and nature; and non-European social and economic formations. If we consider these three themes through the lens of Marxian natural history—understood as an incomplete project defined by an almost universal theme—then the diversity is complemented by a sense of unity.

[3] Marx's thinking never stopped evolving. *Pace* Althusser, who sought to divide the "early," Hegelian Marx from the "mature," scientific

---

59 "Marx's lifelong project of a *Critique of Political Economy* has never been completed ... Marx was not content with the result of his many efforts to improve the text of the first volume of *Capital* ... After nearly forty years of work—from the spring of 1844 until March 1883—the biggest project of his life remained unfinished, although he kept working on it until his very last days [with] intensive studies on a wide range of topics," including natural science and world history: Michael Krätke, "An unfinished project: Marx's last words on *Capital*," in *Marx and Le Capital: Evaluation, History, Reception* (New York: Routledge, 2022), 144–71, 144.

60 Musto, *The Last Years of Karl Marx*.

Marx, it is wrong to suggest that there are two Marxes.[61] For the purposes of this book, all of the following developments could be considered decisive in the formation of Marxian natural history (dates approximate):

Paris manuscripts, German Ideology (1844–45)	formation of philosophy of history and nature
political writings of 1848–55	elaboration of philosophy of history and politics
Grundrisse (1857–58)	distinction between living labor and C_{LP}
study of Darwin (1860)	scientific basis of philosophy of natural history
studies of theory of value (1857–72)	development of theory of value form, fetishism
studies of colonialism (1857–70)	analysis of natural history of capital in, for example, Ireland
late ecological notebooks	deepening analysis of energy, biology, geology
late ethnology and world history notebooks	capacious analysis of world history

I emphatically do not claim, therefore, that Marx's reading of Darwin was decisive for his entire analysis of capitalism. Nevertheless, without considering his study of Darwin, it is impossible to understand his shift in emphasis toward natural history. To appreciate *why* he spent his final years studying natural sciences, world history, and indigenous societies, we must consider that he read Darwin in 1860.

So let us finally open *Capital* and see how Marx explains the natural history of capitalism, which is nothing but a peculiar social formation: a way in which our species recently became reorganized—with fateful consequences for every species on Earth.

---

61 Althusser's position developed from a misleading starting point, viz.: "We must admit that *Capital* (and 'mature Marxism' in general) is *either an expression of the Young Marx's philosophy, or its betrayal*": Louis Althusser, "On the young Marx," in *For Marx* (New York: Vintage, 1970 [1960]), 49–85, 52. *Capital*'s relationship to the early writings is more complex than this bifurcation allows.

PART II
Reading *Capital* as Natural History

Preface to Part II

Marx read Darwin in late 1860 after writing his initial research notebooks on capitalist political economy (*Grundrisse*) and before starting composition of *Capital*.[1] One way to gauge the consequences of reading Darwin, therefore, is to compare how Marx discusses nature and history in *Grundrisse* and in *Capital*.[2] That is my aim in part II. More specifically, I examine Marx's treatment of labor, technology, population, value theory, and fetishism.

With the exception of technology, Marx does not cite Darwin in these sections of *Capital*. Nevertheless, I claim that reading *Origin* had significant consequences for Marx's critique of capital's political economy. Darwin's antiteleological natural history inspired Marx to look anew at

---

1 His 1857–58 notebooks came to be called the *Grundrisse*, or "ground-plan," but Marx did not name them anything in particular: They were simply his research notebooks. Though elements of these texts were published as early as 1904, the notebooks were not published until 1939–41 and remained largely unknown (except by specialists) until the 1960s. Their influence upon anglophone Marxism really only began with the 1973 publication of Martin Nicolaus's translation by New Left Books/Penguin. See Geoff Mann and Joel Wainwright, "Marx Without Guardrails: Geographies of the *Grundrisse*," *Antipode* 40, no. 5 (2008), 848–56.

2 An exhaustive analysis would consider all of writings of Marx in this period, such as his *Contribution to the Critique of Political Economy*, draft economic manuscripts, political writings, journalism, letters, and so on. These are all important. Yet because my aim is to clarify the implications of Darwin for *Capital*, the contrast with *Grundrisse* is especially useful.

human society in its capitalist form. The result was a distinctive ecological approach in *Capital* without parallel in Marx's earlier writings.

Before proceeding, some preliminaries. These chapters presuppose some familiarity with *Capital*, so I will not summarize the book.[3] I further presume that my readers are aware that capitalism is destroying our planet and does need a restatement of the brutal facts about our planetary crisis.[4]

Though I argue that Darwin influenced Marx, I do not claim that these thinkers held the same philosophy, politics, or method—they did not.[5] Still, what they shared was significant. Darwin and Marx saw themselves as scientists concerned with discovering the natural-historical processes that shaped the forms of living beings. They both lived in England at the same time and participated in an overlapping intellectual-cultural milieu. Whereas Darwin's *Origin* provides a general natural-historical hypothesis to explain how and why all living beings evolve into forms for which they are selected, Marx was concerned with one species and the implications of its adoption of a particular social form. By recognizing these commonalities between Darwin and Marx, we can appreciate why certain themes fundamental to *Origin*—competition, material exchange, technical change, adaptation, and surplus population—appear in *Capital* at critical points.

---

3   Others have written excellent guides to *Capital*. I particularly recommend Kōjin Karatani, *Transcritique: On Kant and Marx*, (Boston: MIT Books, 2004); Michael Heinrich, *An Introduction to the Three Volumes of Karl Marx's Capital* (New York: Monthly Review Press, 2012); Fredric Jameson, *Representing Capital: A Reading of Volume One* (New York: Verso Books 2014); Teinosuke Otani, *A Guide to Marxian Political Economy* (New York: Springer International, 2018); David Harvey, *A Companion to Marx's Capital: The Complete Edition* (New York: Verso Books, 2018); and Ryuji Sasaki, *A New Introduction to Karl Marx: New Materialism, Critique of Political Economy, and the Concept of Metabolism* (New York: Springer Nature, 2021).

4   Readers seeking to understand the links between capitalism and ecological crisis may consult the writings of ecological Marxists such as Paul Burkett, Mike Davis, Andreas Malm, Kōhei Saitō, and others.

5   It should not be presumed that Darwin was a hypothetico-deductive scientist, Marx a dialectical materialist. Neither is the case.

4

Labor, Nature, and Technology

Capital as a Distinctive Social Form: The Labor Power Commodity

On the Origin of Species enabled a new conception of humanity. Breaking with all previous theories, Darwin allowed us to conceive of humans as animals—primates, specifically—which evolved within and adapted to particular environmental conditions. While previous thinkers had hypothesized that species change, the real novelty of Darwin's book was to provide a natural-historical, causal explanation for the forms adapted by species (see chapter 2). By 1844, Marx had embraced humanity's historical naturalness and insisted that the "first fact to be established" about this condition "is the physical organization of these individuals and their consequent relationship to the rest of nature"[1] (see chapter 1). This is how I propose to read *Capital*—as a book that analyses humanity's present-day organization and relationship to the rest of nature. Marx shows that the distinctiveness of capitalism does not lie in anything essential about money, commodities, technology, markets, trade, or the state. Rather, capitalism is a form that combines these things (money, commodities, technology, etc.) into definite, purposeful relationships. Capitalism is thus the most recent natural-historical form to shape human societies.

---

1 Marx and Engels, *German Ideology* (1845–6), MECW 5: 31.

The fundamental relationship within this form is the generalization of the sale of the commodity labor power. Labor power is a unique commodity: It is inseparable from living human beings and it alone is the source of value. To say that the commodity labor power produces value means that labor may be exploited. Such a claim had already been made by other political economists, as noted by Marx in 1844. At that point, Marx characterized the problem with capitalism in terms of a labor theory of value, emphasizing alienation. By producing commodities for sale (for someone else's profit), the laborer becomes alienated from the product of their labor and their essence as a laboring being. By contrast, in *Capital*, Marx strips away these Ricardian and Hegelian arguments to begin again from a natural-historical standpoint. He links humanity's naturalness with the central object of his critique of capital—the laboring human body. To appreciate Marx's conception of capital as a distinctive social form, therefore, we should begin there.

In chapter 5, "The Labor Process and the Valorization Process,"[2] Marx writes:

> Labor is a process involving human beings and nature; in it, their own activity mediates, regulates, and controls their metabolizing of nature. When human beings work with materials found in nature, they are acting as natural forces. They set in motion the natural powers that belong to their bodies—arms and legs, head and hands—in order to appropriate natural materials in forms in which such materials serve human life. In applying this movement to the natural world around them, human beings alter it and at the same time alter their own nature.[3]

Note that this definition of labor does not say a word about money, capital, or commodities. Human labor did not begin with the capitalist form of society—humans have labored for as long as we have existed.

Since the activity of laboring long predates the capitalist form of society, it may seem like working for a living under capitalism is only

---

[2] In Fowkes's translation, this was chapter 7: "The Labour-Process and the Process of Producing Surplus-Value."

[3] Marx, *Capital*, 153.

natural. Indeed, the way people tend to speak about the necessity of work has the effect of making it seem normal that we sell our labor power for a living: The labor-capital relationship is naturalized. But as a consequence of the adoption of a capitalist social form, the character of human labor has changed fundamentally.

From reading earlier political economists like Adam Smith and David Ricardo, Marx recognized that to understand the human condition in a way that affirms both the general fact of human labor and the specificity of labor's commodification under capitalism we need to grasp labor historically (as apparent in *Grundrisse*). Reading Darwin provided Marx with a natural-historical standpoint that encompasses more than the capitalist epoch while clarifying how and why human labor under capital is distinct from previous epochs. Before elaborating, let me briefly summarize Marx's presuppositions about labor and labor power under capital.

Labor is inherently social and natural. Labor is natural because human beings are part of nature and our labor always acts with and upon nature; through labor, we reproduce our lives and change our environment. Labor is a means by which humans mediate our relations with other beings and the world around us. We do this socially—the individual at work is always already connected to many other people in a web of relations of mutual reciprocity and exchange. From the standpoint of natural history, labor under capital takes a different form: The form of a commodity—the labor power commodity. A capitalist society appears to us as a place where wealth "appears in the form of an enormous accumulation of commodities."[4] A commodity is anything people produce not to meet their immediate needs, but to be sold. These commodities could be tangible things like cell phones or the book in your hands, or services like music lessons or provision of health care (assuming the latter is commoditized, as is standard in the USA, where I live). The commodity that a worker brings to the market is labor power. This is what most people in capitalist societies have to sell.

The social relationship that distinguishes capitalism from all previous forms of economic life is the regular, generalized sale of labor power as a commodity. The everyday activity of human labor—what Marx calls

---

4   Marx, *Capital*, 13 (the book's first sentence).

"living labor"—is analytically and existentially distinct from the commodity labor power.[5] Formulaically,[6]

living labor $\neq C_{LP}$

It is important to note in passing that, for the vision of human society expressed in *Capital*,

human life $> C_{LP}$

meaning that the lives of human beings should not be defined by the act of their selling labor power as a commodity (or failing to do so).

For a capitalist society to function, workers must reproduce their labor power and sell it. If they stop doing so—either because they go on strike (temporarily refuse to sell C_{LP}) or because they are prevented from working due to an environmental or health emergency (as occurred for some due to Covid in 2020–21)—the result is a decline in production. Commodities cannot be produced without the use of labor power. Moreover, it is the purchase of commodities by workers with the money they earn selling their labor power that drives global capitalism today. So, if workers refuse to spend their money on something—whether because of a change of taste or because of a boycott—the result is harmful for the producer of that commodity.

Workers are human beings living on Earth. The reproduction of the labor-power commodity (C_{LP}) requires a livable environment for human beings. Nevertheless, we cannot claim that C_{LP} is "natural," for it is the result of capital's reformation of human life. Nature only furnishes the conditions of possibility for surplus labor and surplus value."[7] The emergence of a large group of workers who are "free" to sell C_{LP} is a significant natural-historical event:

---

5 Marx first distinguished *living labor* and the commodity *labor power* before reading Darwin, in the *Grundrisse*. On the timing and significance of this distinction, see Martin Nicolaus (1973) "Foreword," *Grundrisse*, 44–52.

6 Throughout this book, I use the following notation from Marx's *Capital*: Commodity is C, money is M, and "–" signifies exchange. The general formula for capital is M–C–M'. To facilitate clarity about the unique commodity labor power, I designate it by C_{LP}. The purchase and sale of the commodity labor power = M–C_{LP}.

7 Marx, *Capital*, 471.

Labor, Nature, and Technology

A money owner has to find a free worker in the commodity market—free in two senses—in order to turn money into capital. As a free person, the worker can do whatever he wants with his labor-power: He can sell it as his own commodity. Furthermore, he is otherwise commodity-free: He has none of the things he needs to realize his labor-power. The money owner ... isn't interested in why he encounters this free worker in the circulation sphere ... But one thing is clear: Nature doesn't produce money owners or commodity owners on the one side, and people who own only their labor-power on the other. This relation comes from natural history just as little as it is a social relation that we find in all historical periods. Clearly ... it was produced by many economic revolutions in the past, or a whole series of older formations of social production going under.[8]

In short, "free labor" is not provided naturally. Human evolution did not produce workers who could do whatever they wanted with their labor-power and were yet otherwise commodity-free. This condition is the result of economic revolutions.

Why do workers sell their labor power? To earn a wage, which is, in turn, used to pay for commodities: Shelter, food, and the other necessities of life. Capitalism could be described as a form of society in which most people sell their labor power to earn money to buy commodities produced by other people. This arrangement was established not for the well-being of the worker but because it suited the interests of the owners of the means of production: The capitalist class, also known as the bourgeoisie, which legally accumulates the surplus value, or unpaid labor, contributed by the workers.[9] Nonetheless, once such an arrangement is established, we are compelled to act accordingly: A member of the proletariat who comes to agree with Marx will still raise their children as proletarians. By providing the young person with food, shelter,

---

8 Marx, *Capital*, 142. For reasons elaborated below, *Capital* would be clearer if Marx had noted here that "natural history" did not end with the emergence of capitalism (as could perhaps be implied from his statement that the capitalist labor relation "comes from natural history just as little as it is a social relation that we find in all historical periods").

9 On the production of surplus value, see Marx, *Capital*, §§III–V, 153–487.

education, and health care, the child is being prepared to earn a living: Social reproduction creates new C_{LP}.[10]

Some critics of Marx have argued that his schema in *Capital* precludes the complexity of everyday economic life. Not everyone is either an employer or worker. Of course, Marx knew this from direct experience: After all, he was a man of nineteenth-century Europe, when there were still many peasants (to name only one prominent non-proletarian social category). *Capital* provides an *abstraction* of the capitalist form of class relations;[11] this enables us to see that capital is beset with deep

---

10   On social reproduction as a process inherent to capital, see Nancy Fraser, *Cannibal Capitalism: How Our System Is Devouring Democracy, Care, and the Planet and What We Can Do About It* (New York: Verso Books, 2023). The education system in capitalist society is oriented toward the production of worker-consumers. The student is a proletarian-in-training. A degree—from high school, university, or graduate school—certifies the quality of labor power. Roughly speaking, the fancier the name of the school and the higher the level of the degree, the greater the exchange-value of the C_{LP} sold by the degree-bearer.

11   "Microscopes and chemical reagents are of no help to us when we analyze economic forms. Our power of abstraction must do the work of both things, for in bourgeois society, the commodity-form of labor products, or the value-form of commodities, is the economic cell-form" (Marx, *Capital*, 6). Bob Jessop asks, Why does Marx begins *Capital* with the analysis of the commodity? This marks a clear shift from *Grundrisse*. By examining Marx's preparatory texts and statements on method, Jessop shows that Marx's study of cell theory between 1857 and 1867 led him to use concepts from cell biology to explain capitalism: Jessop, "'Every beginning is difficult, holds in all sciences': Marx on the economic cell form of *Capital* and the analysis of capitalist social formations," in Marcello Musto, ed. *Marx's Capital After 150 Years* (New York: Routledge, 2020), 54–82. Just as all bodies are comprised of cells, the social body of capitalist society is constituted from complex arrangements of the value-form, of which the elementary form is the commodity; the "cell form" (*Zellenform*) of capitalist society. These commodities are at once independent entities yet shaped by the whole of capitalist society, which presents itself as an "enormous accumulation of commodities" (*Capital*, 13). Just as cells arise from other cells, commodities are produced from other commodities. Cells persist through a lifetime of metabolic exchanges, which can be modified through cellular differentiation and form organs and organisms; under capitalism, the whole social metabolism that defines human existence on Earth comes to be reorganized by the production and circulation of value through the use of the labor-power commodity. Yet Marx's use of cell theory to explain capitalism was essentially metaphorical, for "whereas cells are the universal basis of organic life and operate through known universal chemical, physiological, and metabolic processes, the value form of the commodity as the economic cell-form of the capital relation is historically specific and its laws and tendencies are doubly tendential, in the sense that, they exist only to the extent that the contradiction-rife and crisis-prone capital relation is reproduced in and through social practices that are historically contingent and contested" (Jessop, 75).

contradictions—not from anything "outside" of capital but from "within" capital itself. Marx titrated the purest possible elements of capital to exclude from his analysis the prospect of capital being "fixed" by some impure element from outside.[12]

The general formula for capital

Let us briefly consider the logic of a society prior to the emergence of capitalism. Suppose that we live in a place that has the things often associated with capitalist society, such as money, a state, commodity production, and market exchange. Is this capitalism? Not necessarily. In themselves, none of these things produce capitalist society. What distinguishes capitalism is the reorganization of these elements into a definite social form, which is to say, their organization through a specific set of social relations.

Marx makes this point in *Capital* in three different ways. In chapter 1, Marx provides a brief *abstract typology* of types of economic forms (discussed below). At the end of *Capital*, he provides a *historical illustration* by showing that European colonialism could not create capitalist societies without adjusting the markets for land and labor to produce dependent wage laborers.[13] Between these two instances, in chapters 4 and 5, Marx provides a *formulaic expression* of the point in his discussion of the shift from an economic form of type C–M–C to M–C–M'. Where economic life is organized around simple commodity exchange, producers create commodities (C) which they bring to market to exchange for money (M) to buy commodities (C). Since the things that they sell and purchase are different, we can describe the process with the formula:

---

12 The power of this approach is to illuminate the nature of capital. One limitation is that it contributes to a common misunderstanding that "Marx thought the capital-labor relation was the sum of capitalist society, the only relevant social relation." Since this claim is obviously false, Marxists sound ridiculous. Ergo we must confront the common misunderstanding. Relatedly, Marx's "discovery that historical relations are objectified in the form of the commodity can be misinterpreted so as to produce the idealist conclusion that, since Marx reduces all economic categories to relationships between human beings, the world is composed of relations and processes and not of bodily material things . . . [In truth, all] social relations are mediated through natural things, and vice versa. They are always relations of [humans] 'to each other and to nature'": Alfred Schmidt, *The Concept of Nature in Marx* (London: New Left Books/Verso, 1971), 69.

13 I discuss this process in part III, chapter 6.

$$C_x - M - C_y$$

where C_x might be language lessons or a computer code and C_y a Bible or a pint of beer.[14] These are only examples. Of course, for the person selling C_x and C_y, it matters a great deal what exactly they have to sell (hence all the anxiety about studying for a given career) and what they want to buy (as a Bible ≠ beer). Regardless of the specific desire, it is the general social process that concerns us: Someone with something produced for sale (a commodity) sells it to someone, acquires money as a result, and buys something else for consumption. This simple form of commodity exchange includes all the elements we expect from a capitalist marketplace (production, consumption, and commodity exchange via money), but it does not define capitalism as a form of society. It merely describes a simple exchange relationship, one that could even do away with the mediation of money if the owners of C_x and C_y are fortunate enough to meet and barter their goods directly ("I will buy you a beer if you teach me some Mandarin"). Such simple exchanges certainly occur in capitalist society, thankfully; but, as a socio-natural formation, capital works otherwise.

To grasp this, it is best to begin from a different vantage point: Not that of someone with something to sell (C_{LP} or a computer code) but someone with a lot of money (M). What does a person with a lot of money do with it? Imagine that you win the lottery. Congratulations! What next? You might throw a party for your friends (who for once drink beer without having to provide language lessons), then upgrade your lifestyle by buying a house (or a second one, if you were fortunate to own one already), a car (perchance a private jet), some spiffy artwork (you have style, you know), and—do not forget this one, or the M will go fast—the specialized labor power of a clever financial advisor. Once this clever advisor takes charge of your finances, they will offer some predictable advice: "Invest your capital wisely." What does this mean? You should buy assets—stocks, for example (fractions of ownership in a firm).[15] These firms, in turn, hire workers and buy means of production

---

14 "The form of the direct exchange of products is x use-object A = y use-object B. Things A and B aren't commodities before they are exchanged; only through exchange do they become commodities": Marx, *Capital*, 63.

15 A variation: Buying assets from which you can charge rent. On the centrality of asset investment to contemporary capitalism, see Brett Christophers, *Our Lives in Their Portfolios: Why Asset Managers Own the World* (London: Verso Books, 2024). I thank Christophers for his stalwart support of this project.

Labor, Nature, and Technology 121

to produce commodities; then they sell those commodities to make money. The making money is the point: Investing your money (M) facilitates commodity (C) production, for the sake of sale (C–M'), thereby obtaining more money at the end of the process (where M' > M):

$$M - C - M'$$

Marx calls this the "general formula for capital," a double entendre.[16] It is the formula for capital because, in Marx's logic, capital is not money but the entire movement M–C–M'. If capital = M–C–M', the stuff that "moves" through this series of exchanges is value, which first takes the form of money, then the form of a commodity, then the money form, and so on ad infinitum. But M–C–M' is also the general formula for capital because it describes the *general form of a capitalist society*. This is the general form of society because it expresses the fundamental relationships determined by that form.

So understood, we can use these two formulas to express one of the essential insights of *Capital*. The shift from C–M–C to M–C–M' describes a break in the natural-historical form of human society. We are dealing in each case with the same human beings and the same earth, but the natural-historical form of human society has changed. But how so, and to what end?

In capitalism (M–C–M'), commodities are not produced for the sake of production or simply to meet people's needs (use-value). They are produced to be sold to generate profit. The underlying aim is the accumulation of value in the form of money: For M' to be greater than M. In Marx's terms, the commodity's sale is necessary for the realization of the commodity's value. The completion of the circuit of capital requires not only production, but sale (consumption).

Note that, for proletarians, capitalism is experienced as C–M–C. What the worker sells is the commodity labor power (C_{LP}); what they receive is the wage (M_{wage}); they spend their wage buying commodities needed to survive, to reproduce themselves and their labor. Naturally, working people today (like everyone else in history) tend to believe that the way they experience things reflects the world as it truly is. Yet, as Marx is at pains to show in *Capital*, this is not the case. C–M–C is not

---

16 Marx, *Capital*, 121–30.

the essence of capital, nor its form. Capitalist society is defined by the *inversion* of this form = M–C–M'. *Capital* teaches that labor's commonsensical understanding of life under capital actually presents an inverted view of reality; for the ordinary laborer, the situation is worse that it appears.[17] Like *Origin*, therefore, *Capital* can be seriously disorienting and, for some readers, distressing.

Material Exchange, Means of Labor, and Natural History

Human beings labor and, in so doing, change the physical environment. Technology mediates this relationship between human labor and the environment. These truths were apparent long before Marx. The novelty of *Capital* is that Marx explains why and how the change in *form of society*—from various precapitalist economic forms to capitalism—constitutes a fundamental change in social (class) relations and also in our social relations with nature.[18] This is distinct from saying that human societies had significant environmental consequences only after they

---

17 "C-M-C and M-C-M' seem like the front and back of the same cycle, but are completely different because the initiative of the circulation is seized and controlled by the possessor of money": Karatani, *Transcritique*, 208. This helps explain the centrality of ideology to governing capitalist society. If laborers genuinely believe that their commonsense view of the economic form of society is the correct one, there will be less need to repress their demands, apart from occasional wage adjustments; their conception of a better life within capitalism will remain an extension of the only life they have known: C-M-C, with a little more M. To a bonded or enslaved person, by contrast, the exploitation of their labor is direct. Ideology may play a relatively minor role in maintaining such a relationship: Force is necessary. By contrast in a capitalist society, the character of labor's exploitation is not obvious, particularly where proletarians sell their labor power for a wage that is formally contracted and allows for basic needs to be met. Indeed, many economists teach that, so long as no one forces a worker to sign a labor contract (and the agreement to sell C_{LP} was based on informed consent), the deal must be fair. So long as the use of the labor power stays within the terms of the contract, the relationship is not exploitative. By such reasoning, the concept of labor exploitation has been effectively removed from most contemporary economics. Herein lies, I suspect, one reason that psychologists find that studying microeconomics stimulates antisocial behavior and moral debasement (for a review: Amitai Etzioni, "The moral effects of economic teaching," *Sociological Forum* 30, no. 1 [2015]: 228–33).

18 Marx outlines some forms of society that preceded capitalism in his preparatory notebooks on the critique of political economy (*Grundrisse*): MECW 28: 399–439 (see chapter 6).

became capitalist, which is emphatically not Marx's claim.[19] Rather, Marx was the first to explain that the emergence of capitalism changes both the human relationship with nature and the relations between humans—and to explain the underlying, causal connections between these changes. HH×HN long predates capitalism, but capitalism changes it decisively.

Once some part of Earth has been somehow isolated by human labor and put on the market as a commodity—once this fragment of nature has become a means of production—it exists in a curious position.[20] While still an element of nature, it has been cut away from its preexisting position; while it exists as a commodity, it is not exactly a freestanding thing, for it exists as one element of an ensemble of a class process—the expansion of production and accumulation—which give it exchange-value and other definite qualities (legal rights, for instance).[21] In *Capital*, volume II, Marx makes the following observation concerning the treatment of the earth as a means of production (by "the act M—*mp*," Marx means the sale of some part of the earth for use as means of production):

> As soon as act M—*mp* is completed, the commodities (*mp*) cease to be commodities and become one of the modes of existence of

---

19 Marx was well aware of large-scale anthropogenic environmental changes caused by precapitalist forms of society. Discussing Marx's writings on the botanist Karl Fraas—who examined the degradation of soil in different historical empires—Kōhei Saitō writes: "Capitalism alone does not create the problem of desertification" (*Karl Marx's Ecosocialism*, 250). None of the cases Marx cites from Fraas were capitalist societies. They were precapitalist societies in which the state was the predominant form of power.

20 Marx notes that what is difficult to grasp about *money* in capitalism is not "recognizing that money is a commodity. It is figuring out how and why it got to be one" (67). The same could be said for *nature*. The commodification of Earth is necessary for capital, yet Earth cannot be produced and its commodification always generates complications. The difficulty for us lies not in seeing that the human relationship with nature (that is, material exchange) is commodified, but in grasping how and why a commodity is a social relationship with nature. This explains why environmental problems—and human-nature relations more generally—are inherently ideological in a capitalist form of society, and inseparable from the power of fetishism (see next chapter).

21 A piece of fee simple property is neither simple nor a thing with absolute properties: Nick Blomley, "The ties that blind: Making fee simple in the British Columbia treaty process," *Transactions of the Institute of British Geographers* 40, no. 2 (2015): 168–79.

industrial capital in its functional form of P, productive capital. Their provenance is therefore obliterated; they now exist simply as forms of existence of industrial capital, and are incorporated into it.[22]

Once capitalism takes hold someplace, the fate of the earth there is to become a means of production, incorporated into the bodies of commodities. But that is not the end of the story. The means of production, having been consumed by the production process, must be replaced: "Their replacement requires their reproduction, and to this extent the capitalist mode of production is conditioned by modes of production lying outside of its own stage of development." Capital must go out again into the world—particularly into areas "lying outside of its own stage of development"—to obtain more means of production. The tendency of capitalism is therefore

> *to transform all production as much as possible into commodity production*; the main means by which it does this is precisely by drawing this production into its circulation process; and developed commodity production itself is capitalist commodity production. The intervention of industrial capital everywhere promotes this transformation, and with it too the transformation of all [direct] producers into wage-labourers.[23]

Capital's drive transforms Earth into commodities and human beings into proletarians. Herein lies the fundamental cause of our planetary crisis.

The remains of means of labor

A terminological clarification: In *Capital*, not all natural things that humans use constitute means of *production*. Marx uses the term "means of labor" to describe any "thing or group of things that a worker puts between himself and the object of his labor. These things serve as

---

22 Marx, *Capital* II, MECW 36: 115. This translation is lightly modified based upon Lasker's translation (113).
23 Marx, *Capital* II, MECW 36: 115–6, my italics. This line of thinking was further elaborated by Rosa Luxemburg, *The Accumulation of Capital* (New York: Routledge, 2015 [1913]).

Labor, Nature, and Technology

conduits for his activity, conveying his labor to its object."[24] Marx's definition of means of labor includes important remarks on HN relations:

> With the exception of a ready-made means of subsistence that a person gathers using only his own body as his means of labor, such as fruits, the first thing a worker takes hold of is a means of labor, not an object of labor. Thus the natural world itself comes to function as an organ in the worker's activity, an organ with which he supplements the organs of his own body . . . Just as the land is the worker's original pantry, so it is also his first toolbox. It supplies the stones, for example, that he throws and uses to grind, press, cut, and so on. The land itself is a means of labor, yet a whole series of other means of labor have to be invented, and labor-power has to reach a relatively advanced stage, before the land can serve as a means of agricultural labor.[25]

For humans, Earth is a direct source of "fruits," the "original pantry," and our "first toolbox"; the land is at once *means* of labor and *object* of labor. Human labor upon Earth is a natural force at work on nature, mediated by means of labor, an ensemble evolving historically:

> The moment the labor process starts to develop beyond its initial form, it requires means of labor that have been crafted by labor. We find tools and weapons made from stones in the oldest human dwellings. When human history was in its earliest stages, domesticated animals counted among the primary means of labor—i.e., animals that had been acted upon by labor or bred for particular purposes. So did stones, wood, bones, and shells that had been modified by purposeful human activity. Although some animals create and use means of labor, albeit in very rudimentary ways, these activities are characteristic of a labor process that only human beings can carry out. Hence Franklin defines the human being as "a toolmaking animal."[26] *The remains of means of labor are as important for understanding past*

---

24 Marx, *Capital*, 154.
25 Marx, *Capital*, 155.
26 Marx does not provide a reference to this pithy statement by Benjamin Franklin. MEGA2 notes that Marx took it from Thomas Bentley, *Letters on the Utility and Policy of Employing Machines to Shorten Labour* (London, 1780), 2–3, which cites Franklin as saying that man is a tool-making animal.

economic formations of society as the remains of bones are for understanding extinct species of animals. The distinguishing feature of an economic epoch isn't which things are made, but rather how things are made: Which means of labor are used. Means of labor aren't simply yardsticks that tell us how far human labor-power has advanced; they also reflect the social conditions under which labor is performed.[27]

We see here an imprint of *Origin* upon Marx's conception of technology. I have found nothing in Marx's oeuvre from before 1860 to suggest he arrived at this thought—"The remains of means of labor are as important for understanding past economic formations of society as the remains of bones are for understanding extinct species of animals"—independent of Darwin. In making this claim, Marx is adopting a strategy popular in the years after *Origin*, explaining his argument by using Darwin's evidence from skeletal remains as a metaphor.[28] Yet, for Marx, this is no mere metaphor: Since humans are the tool-making species, he reasons, the discarded tools lying around the earth's surface are evidence "for understanding past economic formations of society"—the very purpose Marx defined for *Capital*.[29]

---

27 Marx, *Capital*, 155–6, my italics.

28 It is metaphorical since Darwin applied natural selection to *species* and Marx implies that natural selection works upon *means of labor*—a concept that sits in *Capital* somewhere between metabolism (qua HN mediation) and machinery (as competitive compulsion of M–C–M' form).

29 As specified in *Capital*'s preface (see chapter 3). Similar ideas were developed in the 1970s in the subdiscipline of Marxist anthropology, which peaked circa 1979: See Marshall Sahlins, *Culture and Practical Reason* (Chicago: University of Chicago, 1976); Maurice Godelier, *Perspectives in Marxist Anthropology* (Cambridge, UK: Cambridge University Press, 1978); Marvin Harris, *Cultural Materialism: The Struggle for a Science of Culture* (Walnut Creek, CA: Altamira Press, [1979] 2001); Maurice Bloch, *Marxism and Anthropology: The History of a Relationship* (Milton Park: Routledge, 2013 [1983]). Curiously, Marxist anthropology went into rapid senescence in the 1980s, for reasons that I have been unable to discern (notwithstanding fruitful conversations with several scholars of the field, whom I thank). I find Mick Taussig's assessment ("Marxist Anthropology adopted the reifying optic of positivist social science and plodded diligently into the swamp-world of a different sort of make-believe") simplistic and unconvincing: Taussig, "The rise and fall of Marxist anthropology," *Social Analysis: The International Journal of Anthropology* 21 (1987): 101–13. Of the four fields, Marxist anthropology was best preserved among archaeologists: See, for example, Thomas Patterson, *Marx's Ghost: Conversations with Archaeologists* (London: Routledge, 2020); cf. Marcus Bajema, *Throwing the Dice of History with Marx: The Plurality of Historical Worlds from Epicurus to Modern Science* (Leiden: Brill,

Human labor mediates the metabolizing of nature

Earlier, I emphasized that Marx distinguishes living labor from C_{LP}. This could make it seem as if Marx is splitting an arbitrary philosophical distinction, or stating something banal, that two things are different (like saying, "cats and dogs are different species"). But Marx's distinction is neither arbitrary nor absolute—and the distinction is necessary for understanding human life today. *Capital* shows that these two phenomena, living labor and C_{LP}, distinct forms of human labor, emerge *as* distinct through historical processes that require explanation. Simply put, we need to explain how people come to be dependent upon selling C_{LP} as a means of producing a livelihood. More fundamentally, from Marx's natural-historical standpoint, tracking the emergence of this distinction requires analysis of the material exchanges between humans and the environment. The emergence of a proletariat transforms HH×HN.

Herein lies the essential meaning of Marx's use of the term Stoffwechsel in *Capital*, a word commonly translated as "metabolism" but better understood as material exchange.[30] Marx first used this term in the 1850s to describe exchanges that occur between living beings and their environment.[31] In *Grundrisse*, Marx came to distinguish material exchange from the changes for the form of these material exchanges: "A system of exchanges, changes of material, from the standpoint of use-value. Changes of form, from the standpoint of value as such."[32] But

---

2023), chapters 5–6. Regardless, the decline of Marxism has left anthropology poorly situated to analyze the climate crisis as an effect of capitalist social relations. Were Marxist anthropology to revive, I anticipate that it would be in response to the climate crisis and adopt some form of natural-historical critique of social formations.

30 See preface, note 20. Marx uses of Stoffwechsel in *Capital* at 20, 79, 88, 103, 111, 116, 153, 159, 359, and 460 (Reitter trans.); 133, 198, 207, 210, 217, 228, 283, 290, 637–38 (Fowkes trans.). For readings of *Capital* that emphasize metabolism, see Kōhei Saitō, *Karl Marx's Ecosocialism: Capital, Nature, and the Unfinished Critique of Political Economy* (New York: Monthly Review Press, 2017), chapter 3; Ryuji Sasaki, *A New Introduction to Karl Marx: New Materialism, Critique of Political Economy, and the Concept of Metabolism* (New York: Palgrave, 2021), chapter 3.

31 On the history of Marx's use of Stoffwechsel, see Saitō, *Karl Marx's Ecosocialism*, chapter 2, passim. Material *exchange* with nature implies the production and recycling of *waste*: See Vinay Gidwani, "Six theses on waste, value, and commons," *Social and Cultural Geography* 14, no. 7 (2013): 773–83. I thank Gidwani for his generous insights.

32 Marx, *Grundrisse*, quoted in Saitō, *Karl Marx's Ecosocialism*, 75. Marx carries this distinction between material exchange (Stoffwechsel) and changes of form (Formwechsel) over into *Capital*, but the significance of the former is amplified by its place in the text's argument.

Marx's genuinely novel use of the concept only occurs in *Capital*—after his reading of Darwin—through an analysis that links the *processes specific to capitalism* as a social formation to changes in *human relations with nature*.

In *Capital*, human labor and nature are conceived as *material processes*, intertwined *natural-historical processes*. The laborer is a human being, a person on Earth, a product of natural history. Their laboring is a natural-historical action insofar as it is inseparable from the whole evolution of the human species and the recent emergence of a capitalist form of society.[33] The practical activity of laboring always entails material exchange between the human body and its natural environment. In *Capital*, Marx first invokes this material exchange to observe that human labor—understood here not as a commodity but as a practical activity whereby humans produce use-value—mediates human exchanges with nature: "As useful labor, labor is a condition of human existence independent of all forms of society. It is an eternal natural necessity, needed to mediate the human metabolizing of nature, and, thus, to mediate human life itself."[34] I take this as *Capital*'s definition of *living labor*. Against this, C_{LP} is a unique commodity sold by one human to another—purchased by the buyer to obtain a unique quality of this commodity, its capacity to produce surplus value. Since C_{LP} is also a form of organizing material exchanges between humans and nature, we are dealing with two distinct forms of one social activity.

I emphasize *human relations with nature* and *social activity*. These expressions may appear hopelessly anthropocentric, so let me be clear on this point. I recognize (as did Marx) that we humans live on Earth with myriad other species and that the boundaries between the life-forms of human beings and non-human others are complex and

---

33   Marx pinpoints labor as fundamental to the analysis of natural history of socio-economic forms of society. On one hand, labor is material exchange (Stoffwechsel), the mediation of humans and nature. On the other hand, labor is compelled to act as the commodity labor power, exchanging labor time for a wage. These are two distinct phenomena. They are unified practically and existentially in our bodies. This unity of difference—living labor/labor power—constitutes an embodied antinomy. No solution to this antinomy can be obtained under capitalism, because it is fundamental to capital. If there is to be an end to it, it must be created via the transcendence of capital as a social form.

34   Marx, *Capital*, 20.

porous.³⁵ Still, *Capital* presupposes a distinctly human form of labor. Marx does not deny that other species engage in acts akin to human labor, nor that other species contribute to commodity production. Certainly, they do. Yet what distinguishes *human* labor in capitalism is not the specific physiological capacities of our bodies, but the fact that *humans alone buy commodities*, including the unique commodity (C_{LP}). Despite the involvement of other species and beings in our lives, under capitalist commodity production, C_{LP} is sold by humans to other humans. If cats sold their labor to earn a wage and buy commodities like cat food, we would say that they create value.³⁶ Only humans have formed our society in this peculiar fashion (which cats in turn observe with their sly judgment). However much they may provide humans with comfort, companionship, or other acts akin to human labor, non-human animals do not produce value. Value is a uniquely human social relation—peculiar to its capitalist form—which drives and organizes our material exchanges with nature.³⁷

In *Capital*, Marx shows that the human relationship with nature underwent a profound change with the emergence of capitalist social relations. Humans still have metabolic relationships with Earth—we cannot cease to exchange energy and matter—but our relationships have been thoroughly redirected and re-formed. By separating producers from the means of labor, reorganizing how people obtain food and shelter, encouraging the endless production and consumption of commodities, and so on, the material exchanges between humans and the rest of nature have been transformed in ways that systematically degrade our natural environment.

Some of our material exchange relations have been so thoroughly deformed that we sometimes say that they are broken, or that humans are so effectively separated from nature that a fundamental gap has

---

35 Spare a thought for the trillions of microbes (of 10,000+ species) comprising a healthy human microbiome: Walaa Mousa, Fadia Chehadeh, and Shannon Husband, "Recent advances in understanding the structure and function of the human microbiome," *Frontiers in Microbiology* 13 (2022): 825338.

36 This is not to deny the *axiological* claim that the life of a cat has value—nor the fact that when a cat is sold as a *commodity*, it bears exchange-value. On cats and capitalism, see Leigh Claire La Berge, *Marx for Cats: A Radical Bestiary* (Durham: Duke University Press, 2023).

37 What is distinctive about human labor under capitalism is not specific to human consciousness: See chapter 7.

opened up between our lives and nature's cycles. This is the intuition behind the so-called "metabolic rift": An expression never used by Marx,[38] but inspired from this passage of *Capital*:

> Capitalist production draws people together into the great urban centers whose inhabitants make up an ever-larger majority of the overall population ... [This] disrupts the metabolizing [Stoffwechsel, material exchange] that goes on between human beings and the earth. The natural elements that people consume as food and clothing can no longer return to the land.[39]

Marx claims here that, by concentrating people in urban areas and failing to recycle waste nutrients into the soil, agriculture in capitalist society degrades the environment.[40] We might ask whether it is still true that the "natural elements that people consume" in urban spaces fail to "return to the land": In fact, material recycling is more efficient at scale, e.g., when people are concentrated together in cities (although the relative efficiency gains vary by type of material).[41] Regardless, this simple metaphor ("metabolic rift") has led to a regrettable tendency among ecological Marxists to overgeneralize this pattern to all environmental problems.

At any rate, Marx's contribution with his conception of material exchange does not lie with the so-called metabolic rift. Rather, the real

---

38   It was coined by John Bellamy Foster in "Marx's theory of metabolic rift: Classical foundations for environmental sociology," *American Journal of Sociology* 105, no. 2 (1999): 366–405.

39   Marx, *Capital*, 460. In making this claim, Marx was influenced by studying natural scientists writing on the biochemistry of plants and soil, particularly Justus von Liebig: Kōhei Saitō, "Marx's ecological notebooks," *Monthly Review* 67, no. 9 (2016): 25–42. If the paradigmatic condition of metabolic rift is the spatial division between the town and the country, leading to declining soil fertility in rural areas and despoiled urban environments, the paradigmatic discipline for the analysis of this problem was soil chemistry (for which von Liebig was exemplary).

40   As Liebig and Marx both knew, this problem—urbanization-induced waste of soil nutrients—predated the capitalist form of society by thousands of years. Indeed, this problem is best grasped via analysis of the natural history of the state (empire): Sing Chew, *World Ecological Degradation: Accumulation, Urbanization, and Deforestation, 3000 BC–AD 2000* (Lanham: Rowman Altamira, 2001).

41   Mingzhen Lu, Chuanbin Zhou, Chenghao Wang, Robert Jackson, and Christopher Kempes, "Worldwide scaling of waste generation in urban systems," *Nature Cities* 1, no. 2 (2024): 126–35.

novelty of Marx's conception of material exchange in *Capital* lies in his claim that *labor mediates and regulates the human relation with Earth in ways that are determined by social form*. None of the natural scientists said this before: This is Marx's original formulation.

There are strong grounds for seeing Darwin's influence here. Recall that *On the Origin of Species* begins not with natural selection, but with *artificial* selection, human-breeding of plants and animals. Darwin begins his argument by showing how human labor modifies species like pigeons and sheep. This served as Darwin's grand metaphor for *natural* selection. In *Capital*, Marx generalizes Darwin's point about artificial selection to make a critical natural-historical observation: All human labor entails human-environmental exchange. Just as Darwin emphasizes that human labor changes the form of certain species, Marx defines human labor as the mediation of our dynamic relations with nature. That is the meaning of Marx's claim that human labor is "a process involving human beings and nature; in it, their own activity mediates, regulates, and controls their metabolizing of nature. When human beings work with materials found in nature," they "alter [the natural world] and at the same time alter their own nature."[42] This observation is fundamental to Marx's critique of capitalism, for capitalism is a form of human social relations with definite consequences for how human labor organizes our metabolic relationship with Earth.

Appreciating the influence of Darwin on Marx here can also help to explain the fate of a pair of concepts from Marx's early writings in his later economic works. After Marx's early works were published in the twentieth century, many readers noticed the centrality of the concepts *alienation* and *praxis*, neither of which appear to play any role in *Capital*. In essence, Marx's use of alienation explains a divided relationship between a subject and its object. A subject is alienated from an object when the latter is divorced or separated from the subject in a nontrivial sense. Such subject-centered concepts are basically foreign to the tone of *Capital*, where subjectivity is bracketed: Individuals, as Marx states plainly in the preface, are treated merely as the bearers of class positions. Rather, the divided

---

42   Marx, *Capital*, 153. Darwin's *Origin* "roused Marx's interest in a natural scientific theory—nor for the sake of its practical application but for its theoretical implications": Sven-Eric Liedman, *The Game of Contradictions: The Philosophy of Friedrich Engels and Nineteenth-Century Science* (Chicago: Haymarket, 2022 [1977]), 107.

relationship between subjects and objects appears in a new guise in *Capital* as a natural-historical account of changes in material exchange. Marx claims that the history of capitalism has entailed separating people from the land and means of labor: Capitalism moves things out of place, changes material and energetic flows, re-forms the fundamental processes of generation and reproduction, and so on. The critique of alienation in Marx's early notebooks is thereby revised in *Capital* as a critique of the natural history of the capitalist form of society—a form that is predicated upon these separations and reinforces them.

Accordingly, Marx's concepts of species-essence and praxis also drop out. Recall that in 1844–45, Marx's critique of Hegel's teleology lacked a nonteleological, scientific explanation of the natural history of human society. Only after reading Darwin could Marx find a novel solution to the problem that initially inspired his use of concepts like sensibility, alienation, species-essence, and praxis. The role played by these ontological concepts in 1844–45 would come to be filled by his analysis of labor as practical activity through which humans regulate our material exchanges with nature. The chief philosophical difference lies in the conception of unity and telos. Alienation presupposed a potential reunification of subject and object, species-being a transhistorical human essence, and praxis a name for the means to achieve unification and realization of our essence. In *Capital* these concepts are unnecessary and effectively replaced by human labor mediating material exchange through evolving socioeconomic forms of society. This standpoint promises neither unity nor telos: "no vestige of beginning, no prospect of an end."[43]

Technology

Marx was a keen observer of technology. His early writings sometimes adopt a technophilic tone; these works have been accused of "Promethianism," a faith in human ingenuity to overcome its worldly challenges by dominating nature with technology.[44] Although that

---

[43] James Hutton, *Theory of the Earth* (1899): see chapter 2.
[44] See Paul Burkett, "Was Marx a Promethean?" *Nature, Society, and Thought* 12, no. 1 (1999): 7–42; *Marx and Nature: A Red and Green Perspective* (Chicago: Haymarket, 2014 [1999]), chapter 11, passim.

charge is often overblown, Marx's early writings do contain passages in which he implies that technological changes drive social change in an historically progressive fashion. And in the *Manifesto of the Communist Party* of 1848, Marx and Engels celebrate the "revolutionary" dynamism of bourgeois society for its relentless development of science and technology:

> The bourgeoisie cannot exist without constantly revolutionising the instruments of production, and thereby the relations of production, and with them the whole relations of society. Conservation of the old modes of production in unaltered form, was, on the contrary, the first condition of existence for all earlier industrial classes. Constant revolutionising of production, uninterrupted disturbance of all social conditions, everlasting uncertainty and agitation distinguish the bourgeois epoch from all earlier ones ... In place of the old wants, satisfied by the production of the country, we find new wants, requiring for their satisfaction the products of distant lands and climes. In place of the old local and national seclusion and self-sufficiency, we have intercourse in every direction, universal interdependence of nations.[45]

At this stage in his thinking, Marx associates the bourgeois drive for technological innovation with the globalization of capital, "revolutionizing ... instruments of production" and undermining archaic social relationships with cheap commodities. From such passages, it is not difficult to appreciate why Marx would come to be treated as the patron saint of industrial state socialism in the twentieth century: One only needs to imagine that the revolutionary dynamism of technology could be transferred to proletarian goals via the state. The result will be a metaphysics of communism qua technological development, production planning, and state regulation of socio-natural relations.

Marx and his generation of communists suffered a decisive political defeat in 1848. The revolutionary enthusiasm of the 1840s which suffuses Marx's early writings culminated in the great "springtime of the peoples" of 1848, crushed by the ascendent bourgeoisie and scattered by their

---

45 Marx and Engels, *Manifesto of the Communist Party* (London: Penguin, 2002 [1848]), 222–3.

emerging interstate system for harassing and policing dissent.[46] This experience forced Marx into exile to London and compelled him, after a dark period of reflection, to restart his studies of political economy in 1857. There is a marked shift in tone between Marx's writings on technology in these notebooks (*Grundrisse*). In particular, his notebook from February 1858 contains some of his most brilliant criticisms of the logical consequences of the dynamics of technological development in bourgeois society.[47] Consider this passage in which Marx describes a condition where capital has advanced to the point where the activity of labor "is determined and regulated on all sides by the movement of the machinery":

> The appropriation of living labour by objectified labour—of the power or activity which creates value by value existing for-itself—which lies in the concept of capital, is posited, in production resting on machinery, as the character of the production process itself, including its material elements and its material motion. The production process has ceased to be a labour process in the sense of a process dominated by labour as its governing unity. Labour appears, rather, merely as a conscious organ, scattered among the individual living workers at numerous points of the mechanical system; subsumed under the total process of the machinery itself, as itself only a link of the system, whose unity exists not in the living workers, but rather in the living (active) machinery, which confronts his individual, insignificant doings as a mighty organism.[48]

Note the Hegelian tone. Machinery figures here as an abstract phenomenon. The active agent—akin to *spirit* in Hegel—is *capital*, which posits a relation between living labor (human) and objectified labor (machine) through production. Marx recognizes this as a *political* relationship, mediating between ruler and ruled:

> In machinery, objectified labour confronts living labour within the labour process itself as the power which rules it; a power which, as the

---

46 On 1848 as the culmination of a revolutionary era, see Eric Hobsbawm, *The Age of Revolution: 1789–1848* (New York: Vintage, 1996); on 1848 and its reactionary aftermath, see Eric Hobsbawm, *The Age of Capital: 1848–1875* (New York: Vintage,1996).
47 Marx (1973) *Grundrisse*, Nicolaus trans., 692–711 and 819–33.
48 Marx (1973) *Grundrisse*, Nicolaus trans., 693.

appropriation of living labour, is the form of capital. The transformation of the means of labour into machinery, and of living labour into a mere living accessory of this machinery, as the means of its action, also posits the absorption of the labour process in its material character as a mere moment of the realization process of capital. The increase of the productive force of labour and the greatest possible negation of necessary labour is the necessary tendency of capital ... The transformation of the means of labour into machinery is the realization of this tendency. In machinery, objectified labour materially confronts living labour as a ruling power and as an active subsumption of the latter under itself, not only by appropriating it, but in the real production process itself ... In machinery, objectified labour itself appears not only in the form of product or of the product employed as means of labour, but in the form of the force of production itself. The development of the means of labour into machinery is not an accidental moment of capital, but is rather the historical reshaping of the traditional, inherited means of labour into a form adequate to capital.[49]

Marx's notebooks (*Grundrisse*) are replete with such dizzying passages which, in their unsparing description of the worker's fate under the logic of capital, anticipate the arguments on technology in *Capital*'s celebrated chapter 13, "Machinery and Large-Scale Industry," in which Marx brings his anti-Promethean theory of technology onto the historical terrain of capital's history. As this is well-trodden terrain, I will only briefly recapitulate Marx's core claims.[50]

In a capitalist society, producers (or firms) compete with one another to produce and sell commodities for profit. Producing these commodities requires two things: Labor power (which Marx will call variable capital or v) and means of production (constant capital or c). Spurred by competition, producers will look for opportunities to reduce their costs by replacing laborers (v) with machines (c). Marx calls the ratio between these two forms of capital, c/v, the organic composition of capital.

---

49   Marx, *Grundrisse*, Nicolaus trans., 693.
50   Andy Merrifield describes the chapter as "a book in itself, a staggeringly dense and expansive discussion that could easily stand alone—not only as a brilliant exegesis of capitalist machinery, but also as a sweeping social history of technology": Merrifield, "Marx on technology," *Monthly Review* blog, May 7, 2021.

Competition between producers leads to the replacement over time of labor by technology: An increasing organic composition of capital. At the level of capitalist society as a whole, this generates contradictory pressures. For value to be produced, there must be workers; for capital as a whole to be realized, there must be a class of consumers with means to buy commodities: For these reasons, the expansion of capital requires an increasing number of worker-consumers (proletarians). However, the market compulsion to reduce labor costs to a minimum will lead individual firms to replace labor with technology (in Marx's terms, "machinery"). These two tendencies are contradictory: Fewer paid workers, fewer consumers. If every employer replaces workers with technology, insufficient wages will be earned by workers to sustain the levels of consumption needed for growth. What is rational for an individual capitalist generates a crisis for the bourgeoisie as a whole: A crisis of value realization.

Capital shows that this contradiction cannot be completely resolved under capital, but only managed in some fashion (through measures to regulate the wage rate or state-driven consumption, for instance). The recurrence of this contradiction and the ability of such measures to address it oscillate with a rather predictable historical periodicity. During periods of relatively steady accumulation by capital—usually associated with economic growth and deepening of capitalist social relations—demand for labor power increases, leading to a rise in wages and consumption. Competition between firms leads to investment replacing high-wage-earning workers with machines (as well as hoarding of money by the bourgeoisie if the rate of profit on investment declines below a reasonable level), leading to stagnation. Accumulation slows; demand for labor-power declines; recession sets in. After some time—and, typically, war—the cycle begins again with expanded investment and a new cycle of accumulation.[51]

Returning to his description of capitalist industry as "revolutionary" from 1848, Marx extends the interpretation in new directions in *Capital*:

Modern industry never views or treats the existing form of a production process as definitive. Its technological foundation is therefore

---

51 On capital's cycles of accumulation, see Giovanni Arrighi, *The Long Twentieth Century: Money, Power, and the Origins of Our Times* (New York: Verso Books, 1994).

revolutionary, whereas that of all earlier modes of production was essentially conservative. Using machinery, chemical processes, and other methods, modern industry continuously transforms the functions of workers and the social combinations of the labor process as it improves the technology on which production is based ... The nature of large-scale industry is thus such that it requires labor to be variable, labor's functions to be fluid, and workers to be generally mobile. On the other hand, the capitalist form of large-scale industry reproduces the old division of labor and the petrified specializations that go with it. Readers have seen how this absolute contradiction strips the worker's life circumstances of all calm, stability, and security, and how it constantly threatens to tear his means of labor—and with them his means of subsistence—from his hands, making his specialized function, and thus him, superfluous. They have also seen how the violent force of this contradiction is channeled into the nonstop festival of sacrificial slaughter inflicted on the working class, the heedless squandering of the bearers of labor-power, and the devastation caused by social anarchy ... It becomes a matter of life or death, too, to replace the horror of an impoverished reserve population of workers—a population kept always at the ready as capital's exploitation needs change, with a human being's absolute readiness to respond to labor's changing demands. In other words, the specialized individual who is merely the bearer of one narrow social function must be replaced with a fully developed individual who treats his different social functions, each of which is supplanted by the next, as the different modes of activity he engages in one after the other.[52]

In this passage, a figure appears which will take on profound importance. That of the human being driven to become adequate to capital—capable of becoming a universally capable worker ("a fully developed individual who treats his different social functions ... as the different modes of activity he engages in"). Meet the forerunner to the neoliberal entrepreneur of the self. What compels this figure to develop these capacities is the fear that, upon pain of failure, they will fall into the ranks of the "impoverished reserve population of

---

52 Marx, *Capital*, 448–9.

workers—a population kept always at the ready as capital's exploitation needs change."[53]

For an individual capitalist, the ideal situation would be the total replacement of labor (variable capital) with machinery (constant capital): A production site where machines program or make machines which program or make machines which program or make machines . . .

Such a world is no longer so difficult to imagine. The advance of robotics, computing, biotechnology, and artificial intelligence (AI) have reached a stage that many contemporary speculations on the future sound a good deal like Marx in chapter 13 of *Capital*, and not without reason. AI and robotics have generated a widespread fear of the potential elimination of many entire categories of labor. What sort of future can be imagined for taxi drivers in cities filled with driverless cars directed by AI? If AI can produce computer code more efficiently than human programmers, even computer engineers could be displaced—except, perhaps, to monitor the AI and advise the owners of those programs on potential threats and opportunities.

We see on our screens recurrent images of a raft of billionaires investing in cutting-edge technologies, competing with others to capture early-mover advantages and monopoly rents by dominating one field or another. It is not entirely clear, however, that any of these dynamics are fundamentally new.[54] Marx's standpoint helps to confront such dizzying combinations of change and continuity. One virtue of reading *Capital* today is that it reminds us that the basic logic of capital which drives the creation of labor-displacing technologies has been shaping society for hundreds of years. But this insight does not come with a sense of relief, for *Capital* shows that this contradiction of capitalist society is generated by the very capitalist form of society. So long as we live under capital, we cannot overcome its crisis tendencies.

The climate crisis brings this home. Many liberals today dream that, with the proper technology—perhaps solar-powered, AI-directed geoengineering—we may engineer our way out of the planetary crisis.

---

53 Marx, *Capital*, 448–9.

54 For an illuminating discussion on this question, see Aaron Benanav, *Automation and the Future of Work* (New York: Verso Books, 2020); for a broader critique of technology's history, see Alf Hornborg, *The Magic of Technology: The Machine as a Transformation of Slavery* (New York: Routledge, 2022).

This is false;⁵⁵ nevertheless, the most common counter-response—the romantic claim that "modern technology is turning us into machines; we need to get back to what is real"—provides a weak basis for criticism. It presumes that there is some quasi-universal, organic quality to being human that was deformed by technology. There is no such thing. Humanity came into being through relationships (HH×HN) that were already technological: Think of our ancestors' manipulations of stone, wood, and fire. *Capital* teaches that, if we remain at the level of romanticism, we miss the history of the mediation of all social relations by technology; we also risk missing what is decisive about how this process works under capitalism—for there has been a qualitative break from earlier history. As Adorno explained in a characteristically sharp passage of *Minima Moralia*:

> The pat phrase about the "mechanization" of human beings is deceptive because it thinks of them as something static which, through an "influence" from outside, an adaptation to external conditions of production, suffers certain deformations. But there is no substratum beneath such "deformations," no ontic interior on which social mechanisms merely act externally: The deformation is not a sickness in human beings but in the society . . . Only when the process that begins with the metamorphosis of labour-power into a commodity has permeated human beings through and through and objectified each of their impulses as formally commensurate variations of the exchange relationship, is it possible for life to reproduce itself under the prevailing relations of production. Its organizational follow-through demands the amalgamation of what is dead. The will to live sees itself referred to the repudiation of the will to live: Self-preservation annuls life in subjectivity. It follows that all the achievements of adaptation, all the acts of conforming described by social psychology and cultural anthropology, are mere epiphenomena.⁵⁶

---

55 See Andreas Malm and Wim Carton, *Overshoot: How the World Surrendered to Climate Breakdown* (London: Verso Books, 2024).

56 Theodor Adorno, "Novissimum Organum," in *Minima Moralia* (New York: Verso Books, [1947] 2021), 243–4, translation modified—with thanks to Dennis Redmond.

Under the rule of capital, "self-preservation annuls life in subjectivity": The will to live is transmogrified into subjection, specifically of the proletarian-consumer type. If we feel that technology is turning us into machines, it is not because of some magical agency on the part of any specific technology, but because of the incorporation of death into life. Under the shadow of dead labor, society collectively repudiates its will to live. In such conditions, what many blithely call "adaptation" today is nothing to celebrate. It is the signature of collective adjustment to a deadening social formation.

Darwin has drawn our Attention to the Natural History of Technology

Marx worked out his critique of technology in the 1850s before he read *Origin*. Nevertheless, his presentation of these ideas adopts a different tone in *Capital*—less mechanical and Hegelian—for reasons that we can attribute to his reading of Darwin. Darwin's explanation of the evolution of the mechanisms species use to move, obtain resources, and reproduce inspired Marx's sense of the processes of technological competition among capitalists. This helps us to appreciate why, comparing *Grundrisse* with *Capital*, Marx came to give greater emphasis to the evolution of machinery as a central form of intra-capitalist competition.[57]

Marx cites Darwin twice in *Capital*. Here is first instance:

---

57 The shift may also be discerned in chapter twelve on the *division of labor*—a theme central to political economy since Adam Smith sought to explain the productivity of market society through that concept. In *Capital*, Marx writes that the "division of labor within society places opposite one another independent commodity producers who recognize no authority except that of competition, i.e., the coercive force exerted by the pressure of their competing interests, just as in the animal kingdom the *bellum omnium contra omnes* preserves every species' conditions of existence to a greater or lesser degree" (327). Here Marx locates Hobbes's formula in the animal world (much as Marx accuses Darwin of doing in a famous letter to Engels on June 8, 1862: MECW 41: 381) to criticize the formation of division of labor through competition. A few pages earlier, Marx emphasizes the social formation of division of labor using language that invokes natural law and the formation of species:

the transformation of a specialized activity into a person's lifelong occupation resembles the older practice of making occupations hereditary. Trades calcified into castes, or, where historical conditions caused individuals to vary in ways that weren't

Even before Wyatt, machines, albeit rudimentary ones, were used for spinning, a practice that likely began in Italy. A critical history of technology would show how little any eighteenth-century invention should be attributed to a single individual. Currently, no such work exists. Darwin has drawn our attention to the natural history of technology—i.e., the development of plant and animal organs as the instruments for producing their respective lives. Shouldn't we devote just as much attention to how the productive organs of human beings in society developed historically—that is, how the material basis of every organization of society developed historically? And wouldn't this history be easier to write, since, as Vico says, what distinguishes human history from natural history is that we make the one but not the other? Technology reveals the active relation of human beings to nature, or the process whereby their lives are directly produced. In doing so, it also reveals the process through which the social relations of their lives—and the intellectual creations that arise from those relations—are brought about.[58]

---

compatible with the caste system, into guilds. *Castes and guilds result from the same natural law that regulates the division of plants and animals into species and subspecies*, although once a certain stage of development has been reached, people enshrine as social laws the tenets that caste status is inherited and guilds are exclusive.

Marx gestures here toward a natural history of social division of labor, employing metaphors from the natural world in an imprecise fashion. Recall that Marx criticized Darwin for doing the inverse, drawing metaphors from political economy to explain the animal kingdom. (For a careful assessment of Darwin's debts to political economy, see Greg Priest, "Charles Darwin's theory of moral sentiments: What Darwin's ethics really owes to Adam Smith," *Journal of the History of Ideas* 78, no. 4 [2017]: 571–93.)

58   Marx, *Capital*, 342. Note how Marx emphasizes the entanglement of "the active relation of human beings to nature" with "the social relations of their lives"—this sentence is as close to the Paris manuscripts as anything in *Capital*. For a thoughtful commentary on this note, see David Harvey, *A Companion to Marx's Capital* (New York: Verso Books, 2010), 189–201. Some scholars (Kevin Anderson, for example, in *The Late Marx's Revolutionary Roads: Colonialism, Gender, and Indigenous Communism* [New York: Verso Books, 2025], 73–4) interpret Marx's criticism of "the shortcomings of the abstract materialism of natural science" at the end of this note as a dig at Charles Darwin. I find this implausible. Marx's critique in this note is squarely directed toward those natural scientists who "fail to begin with the actual, existing relations of life [and] exclude the historical process"; surely Darwin, of all people, does not meet these criteria—as Marx would have known.

Marx's argument that science and technology reveal "the active relation of human beings to nature, or the process whereby their lives are directly produced" is an elaboration of this argument in a direction inspired by Darwin. It is employed here to support his more general criticism of the logic of capital, specifically to show how its drive to accumulate surplus value leads inevitably to the replacement of labor with machines and hence crisis. Chapter 13 says little about class struggle as a means to overcome this tendency.[59]

While this crisis is manifest ordinarily in the form of unemployment or underemployment, in certain periods it adopts a more serious form of general economic crisis. Marx's argument in *Capital* is that such crises are inevitable under capital. With the growth of a population of proletarians, periods of crisis become increasingly threatening to the social order and must be managed. In Marx's analysis of this challenge, we can see traces of Darwin's influence, with its emphasis on the complications that arise when a species's success leads to changes in a physical environment that are contrary to that species's survival. Marx concludes the chapter on machinery in *Capital* by introducing an explicitly natural-historical dimension to his crisis theory:

> Every time the earth's fertility is successfully increased for a given period, this ruins some part of the earth's sources of long-lasting fertility ... Capitalist production thus advances the technological means of social production processes and combines those processes more and more only by damaging the very founts of all wealth: The earth and the worker.[60]

---

59 Andy Merrifield astutely observes that "Marx is surprisingly quiet in chapter fifteen [chapter 13 in the second German edition I am citing] about the role of class struggle. Toward the end, in part 9, over several pages, he projects the immanent possibilities for a technologically-driven society, one that functions around people's needs, varies work and even shortens the working day. But he hardly says anything about how we might reach that utopian point": Merrifield, "Marx on technology."

60 Marx, *Capital*, 461. Translating *die Erde* as "the earth," with a lower case "e" (as Reitter does), could lead English readers to think that Marx means that wealth comes from labor and *soil* (the word used in Fowkes's translation: *Capital* [1976], 638). Yet, by my reading, Marx is saying that what we usually call "nature" is a source of wealth: Die Erde equals planet Earth, all-inclusive of oceans, rocks, wind, incoming solar radiation, etc.—certainly not just earth qua land or soil: "Labor is *not the source* of all wealth. *Nature* is just as much the source of use-values ... as is labor, which itself is only the manifestation of a natural force, human labor power": Marx, *Critique*

Labor, Nature, and Technology

Technology changes labor and human relations with nature in ways that transform HH×HN on an ever-expanding scale: "When capital incorporates into itself the original cocreators of wealth, labour-power and the earth, it acquires elastic factors of reproduction on an ever larger scale, and thus also of accumulation, that don't depend on its material dimensions."[61]

Menaced by a slippage into technological determinism

Notwithstanding the critical nature of Marx's analysis of machinery in capitalist society, Fredric Jameson perceptively observes that, in *Capital*, Marx mainly uses the word "revolutionary" for economic transformations resulting from technological innovations.[62] It is as if Marx's appraisal of the dynamic consequences of human ingenuity in a capitalist form of society—all those machines revolutionizing our lives—is mirrored by his utopian hope that we may transform society into some as yet unknowable collective form. Building upon this insight, Jameson provides a remarkable image:

---

of the Gotha Program (New York: International Publishers, 1966 [1875]), 3; MECW 24: 81.

 The Physiocrats, defending precapitalist forms of rent, claimed that value derived from fertile land. Already by the 1850s, Marx firmly rejected this position, arguing that value comes from human labor. By elaborating his theory of value in *Capital*, Marx went further, providing a means to distinguish *wealth* (from Earth and labor) from *value* (what labor is *condemned to produce* in a society of a capitalist form). Ergo, in a capitalist form of society, Earth is a source of wealth but not value; labor creates wealth and value (including surplus value). Today a neo-Physiocratic view rejects *Capital*'s distinction of "wealth" and "value," arguing for a return to the claim that land creates value. I reject this view, which, if endorsed, would require rejecting *Capital*'s value theory in toto.

 61 Marx, *Capital*, 553. Some important implications of this claim are elaborated by Hideto Akashi, "The elasticity of capital and ecological crisis," *Marx-Engels Jahrbuch* 2015, no. 1(2016): 45–58. I thank Akashi for his comradely encouragement.

 62 Jameson is correct: By my count, 63 percent of Marx's uses of "revolution" and "revolutionary" in *Capital* describe economic transformations, mainly attributed to technological change; 23 percent concern political revolutions and revolutionaries (particularly the cases of France and England); 10 percent appear in the expression "the industrial revolution"; 3 percent describe the act of turning or rotating something; and in one instance (the penultimate paragraph of the afterword), the dialectical method: Marx, *Capital*, 710.

> We may be tempted to take these two celebrations [by Marx]—collectivity and machinery—as the convex and concave of a single process (Hegel's subject or system) in which it is the technological that stands as the concrete realization of the collective at the same time that it reverses its human meaning: Far from constituting the allegory or reification of cooperation, it would seem as though it stands as the latter's fate or doom.[63]

Jameson thereby suggests that we read *Capital*'s affirmations of technological revolutions, on the one hand, and social cooperation, on the other, as coinciding fatefully in a telos (or, indeed, "doom"). By this reading, the development of technology and forms of production could be mirrored historically in the cultivation of new collectivities, resulting in a fateful omega point of rationally managed, collective abundance. But, as Jameson elaborates:

> Marxism is in that sense always menaced by a slippage into technological determinism. This [menace of determinism] is as it were the other face of its opposite number, the temptation of a sublimation into Hegelian categories, of a dialectical metaphysics into which the contingencies of history and production dissolve.[64]

I think Jameson is correct that *Marxism*, a socio-political phenomenon distinct from Karl Marx, has been "menaced" by "technological determinism." But I cannot find the same in *Capital*—nor a "dialectical metaphysics into which the contingencies of history and production dissolve." On the latter, much depends on the weight we give to what Marx claims to be doing in *Capital*—always a fraught point in literary criticism. If we emphasize Marx's statement in the preface concerning his standpoint, then we see how he avoids the risk of "slippage into technological determinism"—for, by defining his standpoint in terms of processes of the natural history of socio-economic formations, Marx cleared a view of capital that emphasizes competition-induced technological revolutions without ultimate direction nor end. Here again, Marx's textual strategies for a non-deterministic interpretation of the role of machinery in capitalist competition changed after reading *Origin*.

---

63 Fredric Jameson, *Representing Capital* (New York: Verso Books, 2014), 55.
64 Jameson, *Representing Capital*, 55.

Consider now the second passage from *Capital* in which Marx discusses Darwin:

> In his epoch-making work on the origin of species, Darwin observes about the natural organs of plants and animals, "As long as the same part has to perform diversified work, we can perhaps see why it should remain variable, that is, why natural selection should not have preserved or rejected each little deviation of form so carefully as when the part has to serve for some one special purpose. In the same way that a knife which has to cut all sorts of things may be of almost any shape; whilst a tool for some particular purpose must be some particular shape."[65]

This quotation was first written down by Marx in his economic manuscripts of 1861–63, in which he quotes the following passage from Darwin's *Origin*:

> I presume that lowness in this case means that the several parts of the organisation have been but little specialised for particular functions; and as long as the same part has to perform diversified work, we can perhaps see why it should remain variable, that is, why natural selection should have preserved or rejected each little deviation of form less carefully than when the part has to serve for one special purpose alone. In the same way that a knife which has to cut all sorts of things may be of almost any shape; whilst a tool for some particular object had better be of some particular shape.[66]

The editors of Marx's collected works note that Marx quotes this passage from Darwin "with minor alterations," albeit without specifying them.[67] The main alteration is that Marx leaves off Darwin's concluding thought, the final sentence of this paragraph: "Natural selection, it should never be forgotten, can act on each part of each being, solely through and for its advantage."[68] As I have emphasized, Darwin's natural selection acts

---

65 Marx, *Capital*, 313, n6.
66 Darwin, *Origin* (first and second edition), 149. To compare this text with other editions of *Origin*, see darwin-online.org.uk.
67 Marx, MECW 33: 387–8.
68 Darwin, *Origin* (first and second edition), 149.

upon individuals and manifests as changing proportions of genes within a population. Although Marx (like Darwin) did not have the benefit of understanding genetics, he would have known that Darwin's conception of evolutionary change could not be directly transposed to human social groups without modifying the mechanism of natural selection. Perhaps the most "Darwinian" argument we can find in *Capital* (as that word is usually understood) is that competition among individual producers leads spontaneously to differentiation of machinery and specialization of technology:

> This differentiation, specialization, and simplification of the means of labour therefore originates spontaneously with the division of labour itself, without any need for a prior insight into the laws of mechanics, etc. Darwin, [in *On the Origin of Species*], makes the same remark on specialization and differentiation in the organs of living beings.[69]

The implications of this "spontaneous" effect, however, take on greater weight when humans live within capitalist societies. To extend this logic while examining humans living within a social formation dominated by capital, Marx needed to modify Darwin's thinking. He sought to do so by expounding a theory of human population specific to capitalist society.

---

69 Marx, MECW 33: 388.

5
Population, Value, and Commodity Fetishism

Capital as a Theory of Surplus Population

Marx presents *Capital* as a study of the natural history of socio-economic forms that aims to elucidate the natural laws of capital. When Marx finally reveals the "general law" of capitalism in chapter 23, it concerns, of all things, the size of the human population.

Marx's arguments about human demography in *Capital* have proven tricky terrain for many Marxists, who seem to feel awkward about Marx's claims. But there is no hiding the fact that *Capital* culminates in a theory of "surplus population," a concept that represents, I will argue, the *scientific* conclusion of Marx's analysis of capitalism. Simply stated, Marx finds that capitalism is a form of society that condemns many humans to become "surplus population." Put so starkly, this sounds a lot like Malthus—but Marx wrote *Capital* to displace Malthus, not to affirm him. I believe that Marx's conclusion was scientifically correct and that—as our world confronts the specter of an ascendent fascist, neo-Malthusian social Darwinism—we must address the politics of human population more effectively and coherently. This means returning to Malthus with Marx and Darwin. For, as Marx wrote, in his "splendid" work, *Origin*, "Darwin *overthrew* Malthus's theory," providing "the detailed refutation, based on natural history, of the Malthusian theory."[1]

---

1 Marx, *Theories of Surplus Value* (1861-63), §g (Rodbertus); MECW 31: 350–1.

Although Marx offers acute observations on human demography and surplus population in his economic notebooks of 1857–58 (*Grundrisse*), these themes are not central. Concerning population, the key passage in these notebooks begins:

> It is a law of capital . . . to create surplus labour, disposable time; it can do this only by setting *necessary labour* in motion—i.e. entering into exchange with the worker. It is its tendency, therefore, to create as much labour as possible; just as it is equally its tendency to reduce necessary labour to a minimum. It is therefore equally a tendency of capital to increase the labouring population, as well as constantly to posit a part of it as surplus population—population which is useless until such time as capital can utilize it . . . It is equally a tendency of capital to make human labour (relatively) superfluous, so as to drive it, as human labour, towards infinity.[2]

In the Hegelian form in which Marx spins this argument, zero and infinity are almost equivalents.[3] This passage on the population dynamics of capital—with its emphasis on positing and not-positing, existence and non-existence, necessity and non-necessity, objectification and realization—might be among the most Hegelian passages in the entire Marxian corpus:

> But labour as such is and remains the presupposition, and surplus labour exists only in relation with the necessary, hence only in so far as the latter exists. Capital must therefore constantly posit necessary labour in order to posit surplus labour; it has to multiply it (namely the *simultaneous* working days) in order to multiply the surplus; but at the same time it must suspend them as necessary, in order to posit them as surplus labour . . . It is, on the other side, a tendency of capital—just as in the case of the single working day—to reduce the many

---

2   Marx, *Grundrisse*, trans. Martin Nicolaus (London: New Left Books/Penguin, 1973), 399.

3   On the Hegelianism of Marx's *Grundrisse*, see Hiroshi Uchida, *Marx's Grundrisse and Hegel's Logic* (New York: Routledge, 1988); Vinay Gidwani, "Capitalism's anxious whole: Fear, capture and escape in the *Grundrisse*," *Antipode* 40, no. 5 (2008): 857–78; Geoff Mann, "A negative geography of necessity," *Antipode* 40, no. 5 (2008): 921–34.

simultaneous necessary working days ... to the minimum, i.e. to posit as many as possible of them as *not necessary* ... At the same time, the newly created surplus capital can be realized as such only by being again exchanged for living labour. Hence the tendency of capital simultaneously to increase the *labouring population* as well as to reduce constantly its *necessary* part (constantly to posit a part of it as reserve). And the increase of population itself the chief means for reducing the necessary part.[4]

After this dialectical tour de force of positing, negation, transposition, necessity, and non-necessity, Marx plants his central claim:

Here already lie, then, all the contradictions which modern population theory expresses as such, but does not grasp. Capital, as the positing of surplus labour, is equally and in the same moment the positing and the not-positing of necessary labour; it exists only in so far as necessary labour both exists and does not exist.[5]

By "modern population theory," Marx means Malthus and his followers. Marx dismisses Malthus's supposed law—humans multiply exponentially; our food stocks grow geometrically; ergo overpopulation is natural and inevitable—as ahistorical and false.[6] We shall return to this. First, we need to see where Marx landed when he brought these Hegelian ideas down to Earth.

The absolute, general law of capitalist accumulation

Chapter 23 of *Capital*, "The general law of capitalist accumulation," provides the first of the three conclusions to volume I (see Table 5.1).[7] If the third conclusion has the virtue of being the last one, and

---

4 Marx, *Grundrisse*, Nicolaus translation, 399–401.
5 Marx, *Grundrisse*, Nicolaus translation, 401.
6 On Malthus, population, "overpopulation," and surplus population in *Grundrisse*, see Marx, *Grundrisse*, Nichlaus translation, 398–402, 413–20, and 595–610.
7 The chapter numbers here are those used in second German edition. In previous English editions of *Capital*, they are chapter 25, "The general law of capitalist accumulation" (first conclusion), chapter 32, "The historical tendency of capitalist accumulation" (second), and chapter 33, "The modern theory of colonialism" (third). To a reader of the second German edition—the last German edition published in Marx's lifetime—the

the second is the best known (for its prediction of the end of capitalism), the first is generally overlooked.[8] Marx begins his discussion with a remarkable assertion: "Every particular historic mode of production has its own laws of population, which hold only for individual historical moments."[9] Each distinct socio-economic formation generates particular population dynamics which in turn shape that socio-economic form. Population growth or decline is neither inherently a dependent nor independent variable, but historically and dialectically both cause and effect (as Darwin argued for all species in *On the Origin of Species*). Marx arrived at this claim after reading Darwin, and it is one of the most important arguments Marx made in a natural-historicist tenor. Nevertheless, Marx's argument remains little more than "a bald assertion which subsequent generations of Marxists have never elaborated upon or sought in any sustained study to substantiate."[10] At a time when many conversations on the climate crisis concern population, chapter 23 deserves rereading.

---

simple structure of the book's final section better represented the logic of *Capital*'s tripartite conclusion. The disaggregation of chapter 25 in later editions of *Capital* obscured this structure.

8   I discuss the second and third conclusions in part III.
9   Marx, *Capital*, 578.
10 Wally Seccombe, "Marxism and demography," *New Left Review* 137 (1983): 22–47, 32. In the voluminous literature on *Capital*, there is a major gap regarding Marx's claims about demography in chapter 23. Seccombe's generalization still holds:

> The primary form of Marxism's traditional address to demography ... has been through a virulent denunciation of its Malthusian versions. These polemics, however programmatically justified ... nevertheless have had an anesthetic effect ... placing the demographic realm itself beyond the pale of legitimate scrutiny and investigation. In the process of dismissing Malthus and his successors, Marxists have abandoned the terrain to our enemies (22).

The virulent denunciations, anesthetic effects, and abdication persist. If previously Marxists downplayed his demographic theory on political grounds—Leninists, who emphasized the revolutionary capacity of the proletariat, had little use for Marx's "law" of surplus population—today the greater factor is psychological: It is distressing to acknowledge that Marx's law may be right.

Table 5.1. *Capital*'s three conclusions

#	Chapter	Nature of conclusion	Object of analysis	Conclusion as expression of theme of social division	Key claim
I	Twenty-three. "The general law of capitalist accumulation."	Scientific and analytical.	Ratio of the population of surplus laborers relative to value.	Human society becomes increasingly divided as the expansion of capital generates a growing body of surplus population.	"The greater society's wealth, the greater the functioning capital, the extent and energy of that capital's growth, and thus also the absolute size of the working population and labor's productive power, the larger the surplus population or industrial reserve army will be" (589).
II	Twenty-four. "The so-called original accumulation."	Political and speculative.	Contradictions of capital relative to the capacity to regulate them within the prevailing social formation.	The becoming of capital entails the separation of people from Earth (means of labor).	"The concentration of the means of production and the socialization of labor reaches the point where neither process is compatible with its capitalist shell. This bursts, and now the bell tolls for capitalist private property. The expropriators are expropriated" (691).
III	Twenty-five. "The modern theory of colonialism."	Literal and geographical.	Extension and consolidation of capitalist social relations in the colonies.	Capital's drive for accumulation expands via the state's externalization of domestic class division on a world scale.	"Capital, rather than being a thing, is a social relation between persons that is mediated by things" (694).

Recall that, in *Capital*'s preface, Marx writes that his "ultimate aim" is to "lay bare the economic law of motion of modern society." So, what is this law of motion? The answer in chapter 23 concerns the dynamic between the *expansion of capital* and the *population size of the proletariat*. There can be no accumulation without more worker-consumers, who produce value as they labor and realize that value as they consume. Hence the spread of capitalist social relations are strongly correlated with a rapid increase in the number of proletarians in the world. Yet capital, Marx reasons, also produces a contrary demographic response. The expansion of capital results in greater investments in labor-saving technology, and a rising organic composition of capital: This generates, over time, a growing "relative surplus population," a term Marx uses for the fraction of the proletariat who are surplus to capital. Whereas earlier writers on human demography described surplus population in terms of *available resources*—meaning that a population is "surplus" if there is insufficient food, water, or shelter, for instance—Marx defines surplus population in terms of *value*.[11] This distinction is fundamental.

Let us sharpen the contradiction that generates the "law." Capital can only expand by increasing the number of worker-consumers (the proletariat)—but capital's drive undoes the proletariat:

> The nature of wage labor is such that the worker must always supply a certain quantity of unpaid labor ... [A] rise in wages indicates at best a merely quantitative decrease in the amount of unpaid labor the worker has to perform. The decrease [in unpaid labor] can never

---

11   Marx's argument that workers' struggles merely modify "the length and weight of the golden chain" they wear (since rising wages do not change the fundamental dynamics nor the drive for further accumulation) is proffered just before he states the "absolute general law of capitalism"—that the expansion of capital is not about the satisfaction of human needs, but the accumulation of surplus value. Once a society is organized in the form of M–C–M', workers will be pulled in—proletarianization—from regions that were previously marginal to these dynamics. Historically, this has generated three key dynamics: [a] migration of people from the rural areas to the city: This is the underlying cause of the world-historical urbanization that we have seen over the past century; [b] demographic increase: Improvements in health care, particularly in relatively affluent societies, mean that workers can work longer, ergo, produce more surplus labor; [c] expansion of capitalist social relations geographically: The story of the world since 1945 is that every society has become capitalist—even China.

reach the point where it seriously threatens the capitalist character of the production process and the reproduction of its basic conditions, namely, the means of production and subsistence existing as capital on one side and labor-power as a commodity on the other side—or, the capitalist existing on one side of the capital relation and the wage laborer on the other.[12]

Hence there are "now too few, now too many" workers—an oscillation peculiar to the capitalist form of society. The desired number of workers changes constantly, independent of the resource needs of the working class:

> The greater society's wealth, the greater the functioning capital, the extent and energy of that capital's growth, and thus also the absolute size of the working population and labor's productive power, the larger the surplus population or industrial reserve army will be. The same things that increase capital's power to expand also cause the disposable labor-power to increase. The proportional magnitude of the industrial reserve army thus grows as the potency of wealth does. But the greater this reserve army is in proportion to the army of active workers, the more massive the consolidated surplus population whose misery stands in inverse relation to the amount of labor its members have to suffer through. Finally, the greater the immiserated sections of the working class, and the greater the industrial reserve army, the greater the amount of official pauperism will be. *This is the absolute, general law of capitalist accumulation.*[13]

Marx presents this, the first of *Capital*'s three conclusions, as a natural law, but I think it would be fair to call it a provocative hypothesis, since the "absolute, general law of capitalist accumulation" has not been proven true.

---

12 Marx, *Capital*, 568. Regrettably, the history of labor struggles under capital bears out Marx's argument on this point. The organization and struggle of workers—essential for building class solidarity, winning substantive gains, and maintaining collective dignity—has never yet "seriously threaten[ed] the capitalist character of the production process and the reproduction of its basic conditions." At most it has created parties and states that reorganize capitalism on terms more favorable for some workers, but even these gains are at risk in the face of the present planetary crisis.

13 Marx, *Capital*, 589, my italics.

Indeed, it has hardly had a hearing.[14] Nevertheless, I admit that—although this is not the place for a full-fledged demonstration—I find Marx's hypothesis likely correct, at least if we consider capital at a world-wide scale (which we should) and we focus on the concatenation of surplus population and the transformation of the natural environment. Looking at the growing numbers of would-be proletarians around the world who are effectively locked out of any prospective path to a secure livelihood, coupled with the climate crisis and degradation of natural environments everywhere, Marx's hypothesis is looking stronger than ever.[15]

What is less certain is Marx's claim—implicit to *Capital*'s second conclusion—that the global growth of surplus population will become an absolute barrier to capital's expanded reproduction. Simply stated, Marx expected the growing numbers of people who were surplus to capital's valorization to unify around a communist political project. This has yet to occur, in part because of divisions within and between subaltern social groups: Nationalism, racism, xenophobia, and other forms of discrimination, coordinated with state power, have checked the unification of those made surplus to capital under a radical program.

---

14 Few contemporary demographers adopt a Marxist framework and I suspect that most economists would reject this "law" on both empirical and epistemological grounds. To the extent that these themes are taken up in recent Marxist research, it has been to discuss the *racialization* and *policing* of surplus population: See, for example, Sara Farris, "Social reproduction and racialized surplus populations," in P. Osborne, E. Alliez, and E-J. Russell, eds., *Capitalism: Concept, Idea, Image: Aspects of Marx's Capital Today* (London: CRMEP Books, 2019), 121–34; Ranabir Samaddar, "Is there a theory of population in *Capital*?," in eds. Achin Chakraborty et al., *Capital in the East* (Singapore: Springer, 2019), 115–36; Cedric Johnson, "Trumpism, policing, and the problem of surplus population," in *Labor in the Time of Trump*, eds. J. Kerrissey, E. Weinbaum, C. Hammonds, T. Juravich, and D. Clawson (Ithaca, NY: ILR Press, 2019), 169–88; and Prem Kumar Rajaram, "Refugees as surplus population: Race, migration and capitalist value regimes," in *Raced Markets* (New York: Routledge, 2021), 97–109. Much as I appreciate this literature, it tends to presuppose the meaning of "relative surplus population" and this concept's relation to recent demography. Such presuppositions are not warranted. We are overdue for renewed debate on population, capitalism, and natural history. See Aaron Benanav and John Clegg, "Misery and debt: On the logic and history of surplus populations and surplus capital," *Endnotes* 2 (2010): 20–51.

15 This notwithstanding the UN Universal Declaration of Human Rights, which (as noted in chapter 1) includes "the right to security in the event of unemployment" as an essential element of "recognition of the inherent dignity and of the equal and inalienable rights of all members of the human family." Accessed October 22, 2024, un.org.

While there is no reason to presume that this challenge could not be overcome, and it is entirely possible to imagine difficulties for individual states in a world with abundant surplus workers organizing around a renewed Left, it is unclear when and how this could become a problem for capital's expanded reproduction.

The Detailed Refutation, Based on Natural History, of the Malthusian Theory

The great strength of the Marxist tradition of writing on population has been to develop a critique of Malthus and his descendants. The basic point of the critique is that Malthus posits as absolute and natural the tendency toward overpopulation, whereas Marx argues that it is relative and historical. Marx disagrees with Malthus about the reason for the existence of "surplus population": Malthus says it is the result of nature (the plants we eat do not multiply as fast as humans), behind which stands God's will; for Marx, surplus population is the result of the history and logic of capital.[16] Both thinkers acknowledge the centrality of surplus population to what we might call the governance of capitalist society. Whereas Malthus aims to eliminate the problem by checking population growth and criminalizing the poor, Marx aims to educate and organize the proletarian to re-form society on another basis. These positions are clear enough that many Marxists have missed the fact that Marx did not reject population as a factor from his theory. Quite the contrary: He gave it pride of place in *Capital* as an explanation of capital's contradiction. Although the nature-versus-history dyad often leads Marxists to say that Marx historicizes relative surplus population while

---

16 See Engels's (1844) "Outlines of a Critique of Political Economy" (MECW III: 418–43), a blistering critique of Malthus's population theory, which Engels calls "the crudest, most barbarous theory that ever existed, a system of despair"; "since it is precisely the poor who are the surplus," according to Malthus, "nothing should be done for them except to make their dying of starvation as easy as possible, and to convince them that it cannot be helped and that there is no other salvation for their whole class than keeping propagation down to the absolute minimum . . . Charity is to be considered a crime, since it supports the augmentation of the surplus population" (437). Engels concludes that "Malthusian theory is but the economic expression of the religious dogma of the contradiction of spirit and nature and the resulting corruption of both" (439).

Malthus naturalizes overpopulation, it would be more correct to say that Marx sought a form of historicism characterized by the unity of nature and history:

> The conservative interests that Malthus was a slave to prevented him from seeing that the heedless extension of the workday, combined with extraordinary advances in machinery and the exploitation of female and child labor, would make a large part of the working class "superfluous," especially once the demand created by war had ceased, and England had lost its monopoly over the world market. It was naturally much more convenient, and much more in line with the interests of the ruling class, a group Malthus idolized in a downright sacerdotal manner, to explain "overpopulation" using the eternal laws of nature, than it was to do so using *the laws of capitalist production that are merely part of natural history*.[17]

This statement reflects a minor but important shift from Marx's earlier critique of Malthus. Recall that in *Grundrisse* Marx writes that capital, "as the positing of surplus labour, is equally and in the same moment the positing and the not-positing of necessary labour; it exists only in so far as necessary labour both exists and does not exist." After reading Darwin, Marx's Hegelian logic is brought down to Earth. Marx's analysis here runs in parallel with Darwin's emphasis in *On the Origin of Species* on population dynamics as both cause and consequence of natural selection.[18] Hence Marx's insistence upon locating the "laws of capitalist production" as "part of natural history."

Still, one might ask: What evidence exists that Marx was inspired specifically by Darwin? Let me offer three additional clues to Darwin's

---

17 Marx, *Capital*, 482, my italics.

18 David Harvey argues that Malthus, Ricardo, and Marx each represented distinct class positions—Malthus representing the landed elite, Ricardo the rising urban bourgeoisie, Marx the proletariat—resulting in specific ideological positions on the relationships between population and resources: David Harvey, "Population, resources, and the ideology of science," *Economic Geography* 50, no. 3 (1974): 256–77. While there is much in Harvey's paper that I admire, his explanation of the ideological character of the three thinkers is unconvincing. We have a mandate to repeat Harvey's (1974) analysis.

influence, drawing first from Marx's economic writings of 1861–63, the years immediately following his reading of *Origin*.[19]

[1] One of Marx's more powerful criticisms of Malthus (he wrote many) appears in his economic notes of 1861–63 under the heading "Notes on the history of the discovery of the Ricardian law."[20] The bulk of these notes is comprised of a critical commentary upon James Anderson's *Enquiry into the Nature of the Corn-Laws*, a defense of duties on imported corn and a work that proposed *inter alia* a theory of ground rent that, though weak, would prove influential in political economy.[21] Marx notes that "Malthus used the Andersonian theory of rent to give his population law . . . an economic and a real (natural-historical) basis, while [Malthus's] nonsense about geometrical and arithmetical progression borrowed from earlier writers, was a purely imaginary hypothesis."[22] (I note, in passing, Marx's affirmative reference to natural history.) Then Marx makes the startling observation concerning the fact that

> although at first the development of the capacities of the *human* species takes place at the cost of the majority of human individuals and even classes, in the end it breaks through this contradiction and coincides with the development of the individual; the higher development of individuality is thus only achieved by a historical process during which individuals are sacrificed, for the interests of the species in the human kingdom, as in the animal and plant kingdoms, always assert themselves at the cost of the interests of individuals.[23]

This is a sentence that has, so far as I am aware, no analog in Marx's pre-Darwinian corpus. It is as direct a statement as one could find in the immediate wake of *On the Origin of Species* of an attempt to link the development of the capacities of human individuals with the evolution of the human species. It is therefore no surprise that, just two pages later,

---

19 These correspond to MECW volumes 30–4.
20 Marx, MECW 31: 344–51.
21 James Anderson, *Enquiry into the Nature of the Corn-Laws*, 1777. Accessed June 3, 2024, at archive.org.
22 Marx, *Theories of Surplus Value*, MECW 31: 345.
23 Marx, *Theories of Surplus Value*, MECW 31: 351. *Theories of Surplus Value*, sometimes called the "fourth volume of *Capital*," was written between *Grundrisse* and *Capital*, shortly after Marx read Darwin.

Marx cites Darwin's *Origin* (at the point where Darwin credits Malthus's inspiration) and then comments:

> In his splendid work, Darwin did not realise that by discovering the "geometrical" progression in the animal and plant kingdom, he *overthrew* Malthus's theory. Malthus's theory is based on the fact that he set Wallace's geometrical progression of man against the chimerical "*arithmetical*" progression of animals and plants. In Darwin's work, for instance on the extinction of species, we also find (quite apart from his fundamental principle) the detailed refutation, based on natural history of the Malthusian theory.[24]

While Marx's praise of Darwin here is genuine and merited, his statement that Darwin "did not realise that by discovering the 'geometrical' progression in the animal and plant kingdom, he overthrew Malthus's theory" is unfair. While Darwin's early thinking was partly inspired by Malthus (see chapter 2), Darwin knew that his theory undermined Malthus's simplistic model about food availability and human population by demonstrating the *dynamic* character of populations and the biological resources they use to survive. As Paul Heyer observes, although Darwin "conceded the acuity of Malthus in observing how the struggle for existence could limit population numbers," his "observations of nature led him to conclude that under such circumstances favorable variations would be [more likely to be] preserved and unfavorable ones [more likely] destroyed; the character of the population could, through time, alter in response to the circumstances."[25] This is what Marx meant by refutation of the Malthusian theory based on natural history: Darwin undermined Malthus's assumptions of the static character of species and the inevitability of population decline with resource shortages.

[2] In his notes on theories of surplus value, Marx makes a revealing comment about Darwin within a commentary upon the English liberal Thomas Hodgskin (1787–1869). The comment appears in a passage where Marx is clarifying the role of living labor in capital accumulation: "What is really 'stored up,' not however as a dead mass but as something living, is the

---

24 Marx, *Theories of Surplus Value*, MECW 31: 351. Marx cites Darwin, *Origin*, second edition (1860), in English.

25 Paul Heyer, *Nature, Human Nature, and Society: Marx, Darwin, Biology, and the Human Sciences* (London: Greenwood, 1982), 37. My interpolations in brackets.

skill of the worker, the level of development of the worker."[26] Accumulation of skilled, living labor, explains Marx, "means assimilation, continual preservation and at the same time transformation of what has already been handed over and realized." That is, accumulation is not just a matter of capturing skilled labor—like capturing specific motions of the laboring body on film—but of wholescale assimilation of those skills into the production process. The process of developing efficient production processes requires more than abstract or static knowledge, but an entire "process of development . . . preservation and at the same time transformation."[27] To flesh out his metaphor, Marx returns to Darwin's *Origin*:

> In this way Darwin makes "accumulation" through inheritance the driving principle of the formation of all organic things, of plants and animals; thus the various organisms themselves are formed as a result of "accumulation" and are only "inventions," gradually accumulated inventions of living beings.[28]

Marx recapitulates here the central thesis of Darwin's antiteleological theory. The so-called inventions of natural beings are nothing but the gradually accumulated modifications of species, products of natural selection at work upon inherently varied populations. Marx extends Darwin's insight to the dynamics of production in human society:

> Such a prerequisite in the case of animals and plants is external nature, that is both inorganic nature and their relationship with other animals and plants. Man, who produces in society, likewise faces an already modified nature (and in particular natural factors which have been transformed into means of his own activity) and definite relations existing between the producers. This accumulation is in part the result of the historical process.[29]

Marx emphasizes the conjoining of two sorts of processes: HN, the human relation with nature, always already modified by the presence

---

26 Marx, *Theories of Surplus Value*, MECW 32: 427.
27 Marx, *Theories of Surplus Value*, MECW 32: 427.
28 Marx, *Theories of Surplus Value*, MECW 32: 427–8.
29 Marx, *Theories of Surplus Value*, MECW 32: 428.

of humans ("Man ... faces an already modified nature") and HH, social relations ("definite relations existing between the producers"). As we saw in chapter 1, the entwined, historical character of these processes (HH×HN) animated Marx's thought since around 1844.[30] But they did not lead Marx to develop a specifically ecological critique of capital until the 1860s. After Marx read Darwin, his thinking developed natural-historical hypotheses on capital's ecological consequences.

[3] Marx uses the best-known concept from Darwin's *Origins*—natural selection—once in *Capital*. The species in question is *Homo sapiens*:

> What experience tends to show the capitalist is that there is chronic overpopulation: I.e., at any given moment *the population exceeds what capital requires for its valorization*, although the source of this excess is generations of worn-out, rapidly replaced people who die young—in a phrase, people plucked from the vine before they were ripe. On the other hand, experience shows the intelligent observer that even if capitalist production began just yesterday, historically speaking, it has quickly and firmly grabbed the nation's vital forces by their very roots ... The only thing slowing the degeneration of urban workers is the fresher elements from the country continuously being absorbed by the urban population. Yet despite the healthy rural air that these workers once took in and the principle of natural selection that reigns among them, letting only the strongest individuals survive, the intelligent observer sees that they, too, have already begun to die off.[31]

Marx does not use "natural selection" ironically here. His point is that capitalism despoils the natural environment so thoroughly that even when fresh workers are brought to the city—people who, per natural selection, are well adapted to hard work—they get sick and die. Bracketing for the moment the validity of his claim, note that in this passage Marx aligns three elements: "chronic *overpopulation*," "the

---

30 Discussing forms of production which have been "conditioned by the low level of development reached by labor's productive forces and the correspondingly limited relations of people within the process of creating and maintaining material life—that is, *their relations both to one another and nature*" (*Capital*, 55–6, my italics), Marx provides another illustration of what I have called HH×HN.

31 Marx, *Capital*, 239, my italics. Marx writes the words "natural selection" in English, underscoring its source: Darwin.

principle of *natural selection*," and capital's novelty within *natural history* ("even if capitalist production began just yesterday, historically speaking").[32] This trinity is distinctly Darwinian—further evidence of *Origin*'s impression upon *Capital*.[33]

Let us return briefly to Marx's expression "*what capital requires for its valorization*." This important qualification provides the definite separation of Marx's conception of overpopulation from Malthus's. For Malthus, overpopulation = insufficient food to feed the population. For Marx, such a formula (however true abstractly) is ahistorical, because human capacities and needs *change*. For instance, Marx describes how our use of language changes through time as human needs, and the means of satisfying them, evolve. People, Marx writes,

> do not by any means begin by "finding themselves in this theoretical relationship to the *things of the outside world*" [as implied by Wagner]. They begin, like every animal, by eating, drinking, ... [by] actively behaving ... and thus satisfying their needs ... At a certain stage of evolution after their needs, and the activities by which they are satisfied, have, in the meanwhile, increased and further developed, they will linguistically christen entire classes of these things which they distinguished by experience from the rest of the outside world.[34]

---

32   Marx cites an 1863 Public Health report documenting the rapid degeneration of the inhabitants of the area around Sutherland (*Capital*, 239–40, n77). Marx also cites Edward Wakefield—to whom we shall return in chapter 6—reporting that "the overworked die off with strange rapidity; but the places of those who perish are instantly filled" (Wakefield cited in *Capital*, 239, n76). These were contemporary texts from conservative sources. While he wrote *Capital*, Marx constantly read the work of sources he vehemently disagreed with and wove hundreds of such references into *Capital*—a methodological procedure analogous to his analytical procedure of demonstrating capital's self-contradictory character.

33   Cf. Marx's Darwinian remark apropos capital encountering a "natural obstacle" (Naturschranken) when the population of the "exploitable working population" is insufficient for profitable employment (*Capital*, 579). In some previous editions, this term was translated as "natural check," bringing Marx's language closer to Malthus. Marx's use of Naturschranken suits his style of using material language, rather than the "limits" and "checks" customary in Malthusian-inflected political economy. On the strange earthiness of Marx's language, see *Capital*, 791–2, note xli.

34   Marx, "Marginal notes on Adolph Wagner," MECW 24 (1975 [1881]): 531–59, 538–9.

There are no transhistorical laws concerning human population.

To study the natural history of human demography, Marx reasons, we must consider the population dynamics specific to distinct forms of human society. Marx proposes that there is a population "law" specific to the capitalist form. This is the tendency for the relative surplus population to increase, that is, an ever-larger number of people who are not "required" for expanded accumulation—for capital's *valorization*.

Further discussion of Marx's law requires discussion of *Capital's* theory of value. But, before doing so, let us briefly look back and take stock of the argument to this point. I claim that Darwin influenced Marx's critique of capital and conception of the natural history of capitalism (see Table 5.2). It was not a matter of inserting missing pieces into an almost complete puzzle. Marx's thought changed decisively in the 1860s: It became more natural-historical. His characterization of labor, technology, and population in capitalist society shifted, revealing Darwin's influence and bringing Marx toward his ecological critique of capitalism.

Table 5.2. Précis of argument

Theme	Shift between *Grundrisse* and *Capital*	Evident influence of Darwin
Labor	Marx distinguishes between first- and second-order material exchange ("metabolism"): (1) material exchange and (2) human mediation of exchanges with nature. Only (2) is original to Marx.	The shift from (1) to (1+2) partly inspired by Darwin's use of *artificial* selection to explain *natural* selection.
Technology	Marx gives greater emphasis in *Capital* to machinery as a mechanism of competition that links accumulation of surplus value with competition and centralization.	Darwin's explanation of evolution of "nature's technology" sharpens Marx's historical sense of technological change and his critique of technology as a factor in capital's crisis.
Population	Marx moves beyond a critique of Malthus to provide an alternative historical theory of population.	Marx, seeing that Darwin overthrew Malthus, is inspired to attempt a more ambitious explanation of the demographic "law" of capitalism as a social form.

Natural History, Value Form, and Commodity Fetishism

Marx's theory of value is the conceptual backbone of *Capital* and one of the most contentious topics of discussion in Marxism. I make no claim that Darwin's ideas directly informed Marx's value theory. Nevertheless, it is noteworthy that Marx's value theory changed decisively after reading Darwin, with greater emphasis on the ecological dimension of the theory. In the early 1860s, Marx made his final break with the Ricardian "labor theory of value" and elaborated a set of concepts—value form, labor as material exchange, the distinction between value and wealth, commodity fetishism—that are fundamental to *Capital*. This change cannot be solely attributed to Darwin. Still, the revisions of the theory of value coincide with Marx's reconceptualization of history, nature, and capital's emergence in the 1860s (after reading Darwin).

To appreciate Marx's value theory, it is necessary to say a few words about the ideas of those political economists who preceded him. During the 1700s, as capitalist social relations deepened on both sides of the Atlantic, a body of scholarship developed to explain the ensuing changes: *Political economy*, a discipline that encompassed fields that are now separated: Economics, political science, and philosophy. The early political economists were fascinated by the divergence in wealth between different social classes and nations (hence Adam Smith's famous book is entitled *The Wealth of Nations*). To explain these changes, they looked to the marketplace, where commodities produced by human labor were exchanged and consumed. Examining these processes—production, exchange, consumption—demonstrated that *commodities bear value*, expressed in two distinct ways. On one hand, commodities are useful. The book in your hand, for instance, can be used to learn about Marx and Darwin (or, in a pinch, to start a fire, but this is not recommended). Adam Smith calls this value in use; Marx calls it use-value; most economists today call it utility. On the other hand, each commodity can be exchanged for another: x commodity $A = y$ commodity B. You could exchange this book for another that you wanted to read even more, for instance (I recommend you hold on to it for future rereading, but I trust your judgment). Smith calls this value in exchange; Marx calls it exchange-value; most economists today call it price—whereas for Marx, a commodity's price is its exchange-value expressed in the form of money. At any rate, all commodities exhibit both qualities, use-value and exchange-value.

Naturally, the early political economists asked, From where does this value arise? The answer provided by thinkers like Ben Franklin, Adam Smith, and David Ricardo was that value comes from the investment of human labor.[35] Their intuition is simple. People obviously value their own labor time; we apply it thoughtfully and, if an activity that requires our labor does not seem worthwhile, we avoid doing it. And if we are thinking of hiring someone else to do something, we will ask ourselves: Will we get more value from that person's labor time than we will have to give to them in the form of the wage? Whichever way you look at it, value originates in labor time. This is the essence of classical political economy's labor theory of value.

Marx studied classical political economy in the 1840s and his early critique of capitalist society reflects the labor theory of value. However, when Marx was working on his critique in 1857–58, he found that this approach exhibited limitations. Then he studied the works of a (now littleknown) critic of Ricardo's labor theory of value, Samuel Bailey. In his 1825 *Critical Dissertation on the Nature, Measures, and Causes of Value*,[36] Bailey contended that earlier political economists overstated the importance of labor and, by organizing their value theory around labor exclusively, generated a vague and incoherent basis for political economy.[37] Against those (like Smith and Ricardo) who said that value came from labor, Bailey insisted that value was purely *relational*: "value denotes ... nothing positive or intrinsic, but *merely the relation in which two objects stand to each other as exchangeable commodities*."[38] Bailey saw value as akin to distance: "A thing cannot be valuable in itself without reference to another thing, any more than a thing can be distant in itself without reference to another thing."[39] He argued that there was no such thing as value: There is

---

35   Franklin, Smith, and Ricardo conceived of the labor-value relation differently—and the physiocrats had another view altogether (see chapter 4, note 60)—but for present purposes we may ignore these differences.

36   Samuel Bailey, *A Critical Dissertation on the Nature, Measures, and Causes of Value; Chiefly in Reference to the Writings of Mr. Ricardo and His Followers* (London: St. Paul's Churchyard, 1825).

37   "Writers on political economy have generally contented themselves with a short definition of the term value, and have then proceeded to employ the word with various degrees of laxity": Bailey, *A Critical Dissertation on the Nature*, 1.

38   Bailey, *A Critical Dissertation on the Nature*, 4–5, my italics.

39   Bailey, *A Critical Dissertation on the Nature*, 4–5. Marx's contribution may be appreciated by considering that the OED says that value means the "adequate equivalent"

only what Adam Smith calls value in exchange; the value of commodity x can *only* be expressed as a quantity of commodity y (and vice versa). Hence, Bailey concludes, there is no need to explain *why* commodities have value—and no need to recognize labor as its source.

Marx was fascinated with Bailey's arguments and recognized the legitimacy of his critique of Smith and Ricardo. Yet, notwithstanding Bailey's criticisms of Smith and Ricardo, there was also no denying the rational kernel of truth in the old labor theory of value. Bailey was right to claim that we must conceive of value as relational, but he mistook this relationality for value *as such*.[40] In Marx's reasoning, value *is* relational, but is not *only* relational, because value has a substantial basis: It comes from the actions of people, laboring. The things that humans produce have value because of the labor that goes into making them. Bailey repressed this truth. Having posited the purely relational character of value, it was a small step for Bailey to claim that the value of any commodity is the same as its *price*. In taking this step Bailey was combining the repression of value's origin in labor with the affirmation of *money* as the true measure of all values. Yet Bailey had no theory of money, nor any explanation for why exchange-value must be expressed in the money form. In sum, Bailey did not offer a coherent value theory—indeed, his claims amount to an argument against any need for a theory of value, only of price determination. (Regrettably, conventional economics largely followed Bailey—albeit without recognizing his role in the discipline's history.)

Marx's value theory in *Capital* is positioned between Ricardo and Bailey, transcending their mutually opposed positions.[41] Marx affirms

---

between two things, but may again also refer to the very "standard of estimation of exchange." Bailey emphasized the former (value as relation, or adequate equivalent), but he begins his essay by treating value as *esteem*—his implied standard of estimation of exchange. For Marx, this will not do—Bailey's reference to "esteem" smuggles in an inexplicable factor and cannot answer the question *why commodities are held in esteem* (by such-and-so person). Marx's critique of Bailey's "esteem" undermines any transhistorical notion of value's "mere relationism" (also not explained by Bailey) by introducing the concepts of the *labor time socially necessary for production* and *value form*.

40 Bailey "confused value with the value form": Marx, *Capital*, 27. On Marx's critique of Bailey, see Marx, *Capital*, 27, 39, and 59.

41 Whereas Marx saw that Ricardo's value theory was organized to account for *labor*, Bailey's centers upon the *relationality of commodity exchange*. Since neither system was coordinated by the two factors, they could not grasp the form of value: See Marx, *Capital*, 27, n17, and chapter 1, section 3, passim; Karatani, *Transcritique*, part II.

that value is produced by human labor while also confirming value's inherent relationality within a capitalist form of society. In *Capital,* value is *produced* by human labor and *realized* through sale. The value of a commodity is defined by the labor time socially necessary for its production. Value is relational since a commodity's value is only realized through exchange with something else. A collection of commodities represents congealed value, but unless they are sold, the value is not realized: The value is lost and it is as if it never existed. By implication, through commodity production and exchange, the value of every person's labor time is perpetually put into comparison with everything else that is on the market.

Some critics of Marx have proposed that his perspective is anthropocentric, arguing that non-human animals also produce value.[42] Two points deserve clarification here. First: Marx consistently distinguishes *value* (a central object of analysis in *Capital*) from *wealth*. These two terms are often conflated, but for Marx's theory of value they must be carefully distinguished. There is a myth that Marx believed that all wealth comes from human labor. This is wrong. In chapter 13 of *Capital,* Marx writes that "capitalist production ... advances the technological means of social production processes ... by damaging the very founts of all wealth: The earth and the worker."[43] Or consider the opening lines of his 1875 critique of the draft program of the United Workers' Party of Germany: "Labor is *not the source* of all wealth. *Nature* is just as much the source of use-values (and it is surely of such that material wealth consists!) as labour, which itself is only the manifestation of a force of nature, human labour power."[44]

Second clarification: Marx recognized the myriad contributions of non-human animals in commodity production. Without bees, no honey; without cats, no cute cat videos. But the truly distinctive relationship between humans and Earth under capitalism is the production and accumulation of value, which is a human social relation. Let us

---

42 For criticism of Marx's ostensible anthropocentrism, see Ted Benton, *Natural Relations: Ecology, Animal Rights and Social Justice* (London, Verso, 1993); Katherine Perlo, "Marxism and the Underdog." *Society and Animals* 10, no. 3 (2002): 303–18.

43 Marx, *Capital,* 461.

44 Marx (1875) "Marginal notes on the Programme of the German Workers' Party," MECW 24: 75–99, 81.

return to our friend, the cat, from the previous chapter. If she is fortunate enough to have a human that buys cat food, then, obviously, she lives within a capitalist society.[45] This lucky cat does not sell *her* labor for a wage to buy food produced by other cats (still less to hoard money—behavior our cat might find ridiculous). Then, while both the cat and her proletarian companion live in a capitalist world—which explains why both cat and human live so differently from their natural-historical ancestors—only the human life is defined by value. This is emphatically not to say that the life of the cat "has no value."[46] It is to recognize that, although the cat's world is organized by capitalism—and the cat's biology and behavior have changed accordingly—our cat does not live a life of a capitalist sort. Unlike the human she lives with, she bears no class relation. Whether playing, hunting, cuddling, or sleeping, the cat's activities are naturally directed toward immediate ends. The labor of the human she lives with, by contrast, is organized by the form of value. As the sticker on the human's laptop says: "I work hard so that my cat can have a better life."

Human-nature material exchanges and the form of value

After reading Darwin and Bailey, Marx came to recognize that classical political economists had never accounted for what he calls in *Capital* the *form of value*. Marx argues that we must account for the form of value

---

45 I presume that the cat food is purchased from a profit-seeking firm with money earned by selling C_{LP}.

46 The widespread confusion surrounding these points originates partly from the generally positive valence associated with the term *value*. In Marx's thought, by contrast, "value" bears multiple valences: For a commodity to have use-value is good, but humans bear the burden of living under the law of value. Consider a kitten for sale in a pet store: The cat's exchange-value (expressed in money) is its price; its use-value is as a cute, sly companion; its value reflects the human labor time socially necessary for bringing the cat into its form as a commodity (a simplification for present purposes). If we then say that "the cat's life has intrinsic value," we are making an axiological claim distinct from (and not mutually exclusive to) these analytical descriptions concerning value and its expressions under capitalism. This perspective is anthropocentric to the extent that capitalism is a socio-economic form that places everything into value form, including non-human beings—an anthropocentric operation. Alas, we cannot overcome this quality of capitalism simply by criticizing anthropocentric language and treating cats respectfully. To live non-anthropocentric lives, our socio-economic form must change.

that makes value into exchange-value:[47] "One of the fundamental shortcomings of classical political economy is that it has never managed to analyze the commodity and, more specifically, commodity value, to the point where it could discover *the form of value that makes value into exchange-value*."[48] But why does this form have to be "discovered"? Because the form that value assumes today is not transhistorical, but only emerged when society adopted a specifically capitalist form. This generates a specific challenge for social analysis:

> The value-form of a product of labor is the most abstract but also the most universal form of the bourgeois mode of production, and it therefore marks that mode of production as a specific kind of social production, and, thus, as historically specific. So if one misperceives it by taking it to be the eternal natural form of social production, then one will necessarily fail to see what is specific about the value-form and, in turn, about the commodity-form and the forms that develop from it: The money-form, capital-form, and so on.[49]

Marx's value theory compels us to ask how it is that things come to be understood as bearing value—how, for example, we come to believe that "nature provides us with valuable services" that can be expressed by a money price. Such thinking is necessary in a society in which the relations between humans and nature have been re-formed for the expanded accumulation of capital through acts of commodity production and consumption that cannot but change humans and their environment. In a word, in any capitalist society, exchanges between humans and nature will be determined by value.[50]

Prior to *Capital*, nature played a minor role in the debates over money, price, and value.[51] Marx brought the question of nature to value theory in political economy in the 1860s, that is, after reading Bailey and Darwin.

---

47 "The exchange process gives the commodity it transforms into money not its value, but its specific value-form": Marx, *Capital*, 66.
48 Marx, *Capital*, 57, my italics.
49 Marx, *Capital*, 57.
50 "The metabolic interaction between humans and nature must be organized despite the private character of labor, and this metabolism can only be mediated by the pure social value": Saitō, *Karl Marx's Ecosocialism*, 118.
51 Except in the literature on rent, which I am unable to discuss here.

His analysis in *Capital* shows that the Smithian-Ricardian labor theory of value, and Bailey's relational critique thereof, were both symptomatic of the separation of laborers from the means of labor. By historicizing the emergence of the value form as a dimension of capitalism as a social formation, Marx provides a way out of their inadequate conceptions of value that naturalize capital. As Paul Burkett writes, in Marx's conception, all human societies "must engage in a reproductive allocation of their labor time, and the social relations regulating this allocation always imbue labor with a specific social form." Prior to the emergence of the capitalist form of society, this reproductive allocation of social labor

> occurred through relations of direct personal interdependence and/or hierarchical dependence mutually constituted with the laborers' social ties (including spiritual ties) to the natural conditions of production. ... [P]roduction was mostly for use; and even when some products became commodities, a general regulation of production by socially necessary labor time in the form of exchange-values did not take place. In short, although precapitalist people-nature relationships were socially mediated [HH×HN], compared to capitalism they were not as socially autonomous from the use-value's natural basis and substance insofar as the precapitalist human producers were not as socially separated from natural conditions.[52]

By contrast, under capitalism, where "value-formed production" is the rule, "use-values are only produced as means of obtaining exchange-value, not of satisfying human needs, including the need for a sustainable and fulfilling coevolution with nature. Capitalism only validates human and extra-human nature as necessary ... insofar as they can be profitably objectified in vendible use-values," that is, as commodities intended for profitable sale.[53] By explaining how and why human-nature material exchanges (metabolism) are made subject to capital's form-determinations, *Capital* dispels liberal notions about capitalism's environmental consequences, which tend to treat

---

52   Paul Burkett, *Marx and Nature: A Red and Green Perspective* (Chicago: Haymarket, 2014), 82. The paper upon which this chapter is based (a landmark text in ecological Marxism) was written circa 1995.
53   Burkett, *Marx and Nature*, 83.

pollution and degradation as accidental or moral failings to be corrected by green policies.

Political economists before Marx had reflected upon nature, labor, use-value and exchange-value without asking *why* humans living in capitalist society are compelled to organize their lives around value:

> Political economy has in fact analyzed value and magnitude of value, although not at all exhaustively, and uncovered the content hidden in these forms. But *it has never even posed the question of why this content takes that form*, why labor is represented in value and the amount of labor in terms of duration represents the labor product's magnitude of value. Formulas clearly marked as *belonging to a social formation whose production process controls people*—and isn't yet under their control—are as much a self-evident natural necessity in the bourgeois consciousness of political economists as productive labor itself. [54]

Marx contributes to explaining how this "social formation . . . controls people" in section 4 of chapter 1, where he introduces the concept of fetishism (in its inaugural appearance in social analysis).

---

54 Marx, *Capital*, 56–7, my italics. Why did political economists fail to pose Marx's question? To be sure, thinkers like Adam Smith and David Ricardo shared his desire to explain capitalism as a socio-economic formation; they also saw "the content hidden in" exchange-value, namely, labor power. Yet fetishism led them to stop there and naturalize capital via an abstraction—the individual person in the marketplace, looking to buy or sell a commodity—which allowed them to see the individual's decisions apropos commodity exchange as rational expressions of social relationships in general. Commodities, meanwhile, seemed to be merely things supplied to meet demand, crystallizations of simple relationships that need no explanation; the exchange of commodities and the natural course of commerce therefore provide their own alibi. Hence their labor theory of value reflects their treatment of commodity relations as the real substance of human life; they thought that by honoring the labor behind commodity production, they could do justice to life under capital. For all its limitations, this marked a step forward. Earlier classical political economists avoided this problem by invoking God; Edmund Burke, for instance: "The laws of commerce, which are the laws of Nature, and consequently, the laws of God." (Burke, cited in *Capital*, 689). Later, neoclassical economists would address these issues by appeal to supply and demand—concepts used in *Capital* for describing price movements and market transactions, but which in themselves cannot explain why our lives are organized by capital.

Fetishism

Because "fetish" later came to have erotic connotations (decades after *Capital*), the word sometimes makes people giggle.[55] That is understandable, but fetishism is no laughing matter. It describes the uniquely human practice whereby we subject ourselves to things of our own making. When people create religious objects of worship that they "fetishize"—treating the sacred object as if it were imbued with spiritual power, instead of just another material thing someone made—they not only place themselves under the fetish's power, but strengthen whatever social relations surround it (whether patriarchal family, church, what have you). A fetish is therefore a human-made object that obscures the very social relations reinforced through its veneration. Marx observes that, in capitalist societies, rather than engage one another simply and directly as social and natural beings, people behave as if our social relations are defined by our relationships with *commodities*. Whereas in the religious world, "things produced by the human mind seem endowed with lives of their own: They seem to be autonomous figures interacting with one another and human beings . . . in the world of commodities . . . we are dealing with things made by human hands."[56] Commodity fetishism thus describes, in the first instance, a distorted self-conception specific to the capitalist form of society—one where we take our measure of ourselves and others from commodities and the value represented by them.

Such is the standard account of Marx's theory of fetishism. But there is much more to *Capital*, chapter 1, section 4. Without going too deeply into it, for sake of my broader argument, I will extend two observations concerning nature, history, and the fetish.

First, fetishism is not simply a mental activity, for it describes an operation of *power* with material consequences for human life.[57] One

---

55 On the complex genealogy of "fetish," see William Pietz, *The Problem of the Fetish*, eds. Francesco Pellizzi, Stefanos Geroulanos, and Ben Kafka (Chicago: University of Chicago Press, 2022). The first edition of *Capital* I (1867) did not include the discussion of fetishism. Marx added it to the second German edition, written between 1870 and 1873. Marx had used the term previously (once in *Grundrisse*, twice in the 1867 edition of *Capital*) but without any explanation of the phenomena.

56 Marx, *Capital*, 49. In this sense, the analysis of fetishism in *Capital* is Marx's critique of capital as a magical power or religion (of money).

57 As a mode of power, it is derived from the way human life is organized, a mode that we participate in *unconsciously*: "People don't put their labor products into relation

might hope that we simply need to remind ourselves of history—to say, "I need not bow down before this thing, it is just something humans made"—to abolish fetishism. But it is not so simple as this, because fetishism is an effect of the prevailing socio-economic formation, not the mentality of the specific individuals living in it. Unfortunately, we cannot escape the power of fetishism just by studying the first chapter of *Capital*. Still, let us stay with the text.

Marx's analysis of fetishism starts by observing that commodities are made out of *natural materials*: "Anyone can see that human activity modifies natural materials so as to make them useful for people."[58] But if his analysis of fetishism *begins* with our material exchanges with nature, fetishism is not *caused* by these exchanges, nor from our obtaining use-value from natural products. Fetishism does not emerge simply from our interactions with the natural world; yet, as we will see, it has consequences for it.[59] Fetishism arises, rather, from the social character of labor under capitalism. It results from things taking the form of a commodity when "the equality existing among different kinds of human labor takes on a thingly form of labor products' equal value-objecthood. Duration as the measure of expended human labor-power takes on the form of labor products' magnitude of value"; hence "the mystery of the commodity form" is that "the form reflects back at people the social characteristics of their own labor as objective characteristics of their labor products."[60]

---

with one another as values because they regard these things as mere thingly husks that encase homogeneous human labor. It is the other way around. *When people exchange their different kinds of products, they are equating them as values, and they are thus equating their diverse instances of labor as human labor. They know not what they do*": Marx, *Capital*, 50, my italics. While the fetishism of money is specific to capitalism, previous socio-economic forms organized around religion exhibited fetishism: See Kōjin Karatani, *Power and Modes of Exchange* (Durham: Duke University Press, forthcoming).

58   Marx, *Capital*, 47.

59   Marx emphasizes that, in the activity of useful labor, human bodies act naturally as earthy organisms: "The varieties of useful labor are functions of the human organism, and that every such function, whatever its purpose and form, is essentially an exertion of a human brain, nerves, muscle, sensory organs, and so on": *Capital*, 48. Fetishism does not arise from the mere action of such bodies nor the activity of useful labor, but from the social organization of labor under capital.

60   Marx, *Capital*, 48–9. Marx explains that people do not "put their labor products into relation with one another as values because they regard these things as mere thingly

The object of analysis in *Capital* is capitalism as a socio-economic form. Since Marx begins this analysis with the commodity, his emphasis in chapter 1, section 4, is specifically with *commodity* fetishism. However, by chapter 3, it becomes clear that the central fetish of capitalist society is, in fact, *money*. Capitalism is a socio-economic formation that organizes our lives in a way that generates a universal compulsion to accumulate money: This is the *sine qua non* of lives under capital.[61] We might imagine that our desire to accumulate money reflects something rational: the need to buy commodities to acquire use-values and meet our immediate needs. That is to view our social relationship from the view of a proletarian, whereas our natural-historical standpoint encourages us to examine these relations from the vantage of capital.

Let us consider a rational capitalist who has been carefully hoarding up surplus value in the form of money. What is it all for? Our capitalist wants a lot of money, of course, because money is the universal equivalent: It is a unique commodity which can be exchanged for any other. But if he spends his money, it is gone; better, then, he thinks, to invest the money (M) to earn more (M'). He exchanges M for the unique commodity that generates surplus value, C_{LP}, and realizes the surplus value as money. Now that he has more money, what does he need? More, again: M' ... M'' ... Since money is an end in itself in a capitalist society, this

---

husks that encase homogenous human labor. It is the other way around. When people exchange their different kinds of products, they equate them as values, and when they do that, they equate their diverse instances of labor as human labor": *Capital*, 50.

61 Even just to earn a living wage. Everyone who has had to try to sell C_{LP}—everyone who has been on a labor market in a capitalist society—will be familiar with the awkward rituals required by the hunt for a decent job. If we ask ourselves why this process can feel strange and dehumanizing, we find two answers. On one hand, the compulsion to sell labor power means that we must make ourselves into a saleable thing (C_{LP}) and thus to relate to others not directly as ordinary people, that is, as the social and natural beings that we are, but as potential value for them. And they cannot but see us this way, for what kind of firm would hire someone who would produce less value for the firm than the firm would pay out in the form of the wage? (Answer: An unprofitable one, put out of business by competition.) On the other hand, applying for any job means placing oneself into competition with others—perhaps even living in another part of the world—where the competition is to sell labor power: We say to the would-be buyer of our C_{LP}, "take part of *my* life, not of the other's." If the labor market reveals something dehumanizing about relating to other people as and through commodities, this estrangement is a real abstraction of social life under capital. It is not a power imposed upon us from outside, so to speak. *Capital* teaches us that these qualities are aspects of one phenomenon—fetishism—through which we unconsciously reproduce our own social domination.

process has no telos; it will continue long after his needs are met. Indeed, to encourage his money to grow faster, our capitalist will even forgo his basic needs.[62] As Karatani writes,

> What is the perversion that motivates the economic activity of capitalism? It is the fetishism of money ... Marx discovered the miser (money hoarder), who lives the fetishism of money in reality. Owning money amounts to owning "social prerogative," by means of which one can exchange anything, anytime, anywhere. A money hoarder is a person who gives up the actual use-value in exchange for this "right." Treating money not as a medium but as an end in itself ... is not motivated by material need ... In a miser there is a quality akin to religious perversion. In fact, both money saving (hoarding) and world religion appeared at the same time, that is, when circulation—which was first formed "in between" communities and gradually interiorized within them—achieved a certain global nature. Therefore, if one sees the sublime in religious perversion, one should see the same in a miser's perversion; or if one sees a certain vulgar sentiment in the miser, one should see the same in the religious perversion. It is the same sublime perversion.[63]

The capitalist form of society is unique in making the accumulation of money an end in itself. No one living in such a society can completely escape this sublime perversion.

This brings us to the second observation. Many people today believe that their relationship to the natural world is free of the sort of religious metaphysics we associate with pre-modern society. Unlike our ancestors, we do not burn offerings to a god to bring rain or guarantee a good hunt. But *Capital*'s analysis of fetishism reveals an operation of power specific to contemporary human social relations arising from the form governing human relations with nature.[64] Earlier I noted that fetishism

---

62 Think of the billionaire investor who, needing more money (one billion is not enough, two billion is not enough, three billion . . .), invests in fossil fuel exploitation—even as the planet upon which they live becomes less livable.

63 Kōjin Karatani, *Transcritique: On Kant and Marx* (Boston: MIT Books, 2005), 208.

64 Karatani proposes that the practice of making ourselves subject to things we produce began with attempts to use magic to mediate the human relationship with

does not emerge directly from our interactions with the natural world. Nevertheless, fetishism is inseparable from the HH×HN specific to the capitalist form of society and bears important consequences for human relations with nature. As Marx emphasizes, this perversion is naturalized by making this social formation appear in the socio-natural characteristics of the things around us:

> The mystery of the commodity-form amounts, then, simply to this: *The form reflects back at people the social characteristics of their own labor as objective characteristics of their labor products, as socio-natural properties of those things.* And so the commodity-form also reflects back at people the producers' relation to the totality of labor as a social relation among objects that exists apart from and outside the producers themselves.[65]

Let us pause on Marx's claim that the commodity form reflects the social character of our own labor back to us as the socio-natural properties of commodities. (Marx's expression "socio-natural properties" [gesellschaftliche Natureigenschaft] is his neologism.)[66] The socio-natural properties of commodities act like mirrors that reflect back at us our relation to the totality of labor as a socio-natural relation.[67] Living in capitalist society, we do not recognize ourselves simply and directly as the social beings that we are, beings on Earth organized in one peculiar, recent socio-economic form; rather, the organization of our lives under capitalism is naturalized, and we see ourselves as particular individuals

---

nature: Kōjin Karatani, *The Structure of World History: From Modes of Production to Modes of Exchange* (Durham: Duke University Press, 2014).

65 Marx, *Capital*, 48–9, my italics.

66 According to Google's ngrams, this expression only entered the German lexicon in 1867, the year the first edition of *Capital* was published. Similarly, the English adjective "socio-natural" first appears around 1976—right after the publication of Fowkes's translation of *Capital*. The contemporary uses of "socio-natural" in geography and anthropology therefore recapitulate a post-Darwinian concept of Marx's.

67 Marx, *Capital*, 49. Thus, the fetishism inherent to capitalist society cannot be understood without considering nature—a point made sharply by Alf Hornborg and Andreas Malm, "Yes, it is all about fetishism: A response to Daniel Cunha," *Anthropocene Review* 3, no. 3 (2016): 205–7; see also Alf Hornborg, "Machine fetishism, value, and the image of unlimited good: Towards a thermodynamics of imperialism," *Man* (1992): 1–18.

competing to survive within it. In the socio-natural properties of the commodities circulating around us, a whirl driven by accumulation of surplus value as money, we catch glimpses of ourselves, as if through a cracked mirror. This provides us with a vague sense that we lack a "natural species-connection with fellow humans."[68] Commodity fetishism is thus one of the names for a more general phenomenon that Marx analyzes in *Capital*, namely, the historical divorce of human social relations from the physical environment in which these relations remain embedded. Fetishism reveals not so much an objective separation from nature but an estranged relation to ourselves as beings somehow cut out from natural history.[69]

Capital is a form of organizing human society and our relationship with nature whereby through our labor, that is, our material exchanges with the earth, we bind ourselves to capital. All of nature comes to be treated as a means to an end: We live in a world of commodities, organized for the accumulation of money. Fetishism thus contributes to the objectification of Earth as means of production and as a mass of commodities. By adopting this socio-economic form, we humans produce these relations that dominate us: "In the realm of religion, people are ruled by a product of their own heads, and it is just so in capitalist production, except that here *people are ruled by the products of their own hands*."[70] How powerfully are we ruled by this fetish? Enough that we are witnessing the undermining of the very environmental conditions conducive to human life on Earth—and panicked reactions to our planetary crisis.[71]

---

68   Marx, *Capital*, 55.

69   Without a critique of fetishism, economics treats nature as if it were a resource *within* capital (which is a human social relation). Complaining about the "fatuous debate about the role of nature in creating exchange-value," Marx observes that it reveals "how deeply some political economists are deluded by the fetishism of the commodity world" (*Capital*, 57).

70   Marx, *Capital*, 569, my italics.

71   Right-wing ideology promises people an authentic relationship with nature. What compels capitalist society toward fascism today is not only the bourgeois reaction to the threat of communism, but a romantic desire to overcome estrangement (fetishism) through an unmediated relationship with blood and soil, nation and nature. Of course, fascism does not satisfy this desire.

The conditions for transparent, rational relations among people and between people and nature

Long before capitalism, land and labor existed; people fought over land and plundered one another to accumulate goods. None of that equals capitalism. Human societies were not fully capitalist until the compulsion to produce and reinvest surplus value had been generalized. This is the meaning of the compulsion toward expanded accumulation, which Marx shows is fundamental to capital. Unlike all previous social formations, in which cycles of production were completed by the consumption of what was produced, the drive of a capitalist form of society begins and ends with money—for which there are no limits to accumulation. There is, however, a way in which we can conceive a limit to capital:

> The religious mirroring of the real world [that is, fetishism] won't vanish *until the workaday world's practical relations become consistently transparent, rational relations among people and between people and nature.* The form of the social life-process, i.e., the material production process, will not shed its foggy shroud of mystery until it becomes the product of freely associated people, consciously planned and controlled by them. But for this to happen, a society must attain a certain material basis or multiple material conditions of existence, which will arise spontaneously from a long and painful history of development.[72]

This "long and painful history of development" could be read politically to describe the long history of *class struggle*; by my reading, it also refers to the natural-historical processes through which humanity would transform the very "form of the social life-process" on Earth. Taken together, we have a description of the end of rule by capital.

What would it take to re-pair humans and nature? Recall that "it is not the *unity* of ... humanity with the natural ... conditions of their metabolic exchange with nature ... which requires explanation or is the

---

72   Marx, *Capital*, 56, my italics.

result of a historic process, but rather the *separation*."⁷³ The unity of humanity with the natural conditions of their metabolic exchange with nature is the result of a historic process we now call evolution: Humans evolved historically within natural conditions. What needs to be explained historically is how humans ever came to see ourselves as separate from nature—and how humans became destructive of the natural conditions that support our species (and many others). While many thinkers had examined such environmental issues before Marx, *Capital* was the first book to pose this question: Through what processes did HH×HN change such that we must speak of a "separation" between humanity and nature (to be repaired)?

Marx answers: Capitalism is a socio-economic formation through which humans organize themselves; through the course of its development, capital separates people from their means of labor. People who historically lived on the land and produced their own food by hunting or fishing, gathering, and farming, tend to become urban proletarians, people who sell the commodity labor power to earn a living. At the same time, the earth comes to be reorganized as a giant collection of commodities: "real estate," "environmental resources," "ecological services," and so on.⁷⁴

By inducing competition between producers, capitalism spurs relentless adaptations in commodity production, technological advances, and (often) cheaper products. These commodities can improve people's quality of life, but the body of each commodity bears some consequence for nature. When producers strive to expand their business by creating and selling more of their commodities—which they must do, given the competitive nature of the social formation—the result will be an increase in the material and energetic throughput. It has always been the hope that this sort of expansion in material and energetic throughput could be ameliorated by efficiency gains in technology or new sources of energy. To be sure, in many instances, producers do find such efficiency gains, and the growth of solar and wind power attests to the potential for

---

73   Marx, *Grundrisse*, Nicolaus translation, 489.
74   "Human beings often made other human beings, in the form of slaves, into the original money material, but they couldn't have imagined doing that with land. Such a thought could occur only in an advanced bourgeois society. The idea dates to the last third of the seventeenth century, and people first attempted to act on it on a national scale only a century later, during France's bourgeois revolution": Marx, *Capital*, 64.

low-carbon energy. But producers in capitalist society will only adopt more efficient forms of production or cleaner forms of energy if it is *profitable* to do so. It is not enough for there to *be* clean technology or fuel; it must be cleaner *and more profitable.* This explains why, although it is now cheaper to produce electricity from renewable sources (like solar and wind power) than fossil fuels (coal, gas, oil), the global shift in energy systems that we desperately need to avoid climate crisis has not really begun. From the vantage of capital, what is essential is not the cost of electricity production nor the long-term fate of humanity, but short-term *profitability*, and it remains more profitable to derive and sell electricity from fossil fuels than renewables.[75]

While it is sometimes implied that Marx failed to see the unity of humans and nature (as if he mistakenly split them apart), in truth, Marx proposed a natural-historical account of this division. The historical separation of humans from nature was an achievement of capital. What is difficult is not recognizing capitalism's ecological crisis but in figuring out how and why we natural beings could fatally degrade our habitat.[76] But what if capital's expansion reached some sort of natural limit? How could we work ourselves out from that end?

Capital's four socio-economic formations

These questions go some way to clarifying something odd about *Capital*: the section on fetishism provides—for the only time in *Capital*—something like a description of the history of the economic forms of society. Observing that the "categories of bourgeois political economy" are "socially valid for . . . the relations of production in this historically specific social mode of production," he proposes that to get away from "the mysticism of the commodity world, all the magic and phantoms enshrouding labor products made on the basis of commodity production," we must "escape" to other forms of society.[77] Recall that in the preface, Marx said that his standpoint was one that treats the development of

---

75 Brett Christophers, *The Price Is Wrong: Why Capitalism Won't Save the Planet* (London: Verso Books, 2024).
76 A play on Marx's claim that what is really difficult is *not* "recognizing that money is a commodity," but "figuring out how and why [money] got to be one": Marx, *Capital*, 67.
77 Marx, *Capital*, 52.

socio-economic forms as a process of natural history. Yet here, the only time in *Capital* that Marx describes such forms, there is no mention of natural history, only abstract sketches of four scenarios—including one that is fictional and one that has never yet existed. Briefly, they are:

[1] *Robinson Crusoe*: A fictional case. One laborer produces all he consumes. Here there are no commodities and no commodity fetishism: "While he engages in a wide range of productive functions, he recognizes that, being performed by one and the same Robinson Crusoe, they represent different forms of his own activity, and are thus merely different modes of human labor."[78] This is an imaginary world that has never existed. Marx begins with this case not only because it is simple but because it is an abstraction functional to (perhaps necessary for) capitalist ideology, which often portrays life as if individual producers are independent actors, acting transparently and naturally.

[2] Economic life in *medieval Europe*: Relations of dependence mediate the labor of every person. Serfs are dependent upon masters, for instance; but even the king is dependent upon the lowest strata of society because the subjects produce the goods that allow the king to exist. The social relations of production are therefore stamped by relations of personal dependence. Exploitation—the expectation, for example, that the peasants will give a tithe of their produce to the church, or that serfs must provide a certain amount of labor-power to their lord—is concrete and visible. In this economic arrangement, commodities exist, but most production is oriented towards use-value. Since "personal relations of dependence constitute the existing social foundation" of medieval Europe, Marx writes, "labor and its products have no need to take on a fantastic form at odds with their reality."[79] "Labor's natural or particular form is in this case its directly social form, whereas its general form is its directly social form under commodity production . . . Every serf knows that he expends a certain quantity of his own personal labor-power in the service of his lord." Whatever we might say about exploitation in medieval Europe, it was transparent to the oppressed. Peasants, serfs, and bondsmen are not separated from the means of production (means of labor). Social relations were not defined by commodity fetishism: "Whatever one thinks of the different actors' masks in which people

---

78   Marx, *Capital*, 53.
79   Quotes in this paragraph from Marx, *Capital*, 53–4.

interact with one another in such a society, at least the social relations among laboring people present themselves as their personal relations, and they aren't disguised as social relations among things."

[3] The *capitalist form* is the exception. In a capitalist social formation, we find the generalized sale and purchase of the labor power commodity. Labor produces value that is congealed in commodities. These commodities circulate; labor is reproduced by receiving wages and buying commodities. Social relations between people take on the appearance of relations between things, which people take as the reality of social existence (commodity fetishism). Since the end of the cycle of capital is M (exchange-value) not C (use-value), the cycle has no limit nor end. Capital is inherently expansionary.

[4] A *future communism*: Marx invites us to imagine an economic arrangement in which "an association of free people" use "communal means of production" to unify "their many individual instances of labor-power as one social labor-power."[80] In such a scenario, Marx contends, we gain the simplicity of the Robinson Crusoe scenario, but elevated to a higher level, made real by meeting everyone's needs:

> Everything Crusoe produced was nothing but his personal product and therefore served him directly as a use-object. *The total production of this association of free people is, in contrast, a social product.* Part of this product is used as new means of production. This part remains social. Members of the association consume another part as their means of subsistence, however, and so this latter part must be distributed among them. How it is distributed varies with the social organism of production itself, and also according to the producers' corresponding level of historical development.[81]

Now labor is for use-value production; the means of production are owned and managed collectively; consumption is organized to meet social needs. The power of fetishism fails. *How* this could be achieved is a historical question, but Marx suggests that the form of society to be taken: An association of associations (see chapter 8).

---

80   Marx, *Capital*, 54–5.
81   Marx, *Capital*, 54–5, my italics. By "social product," Marx means "produced and owned socially."

Conclusion: On Reading *Capital* as Natural History

Humans qua *Homo sapiens* have been growing in number and transforming the earth for a few hundred thousand years. What has changed during this time is not the biological character of individual humans, but the socio-economic form of our existence. As these forms have changed—as HH×HN relationships have been organized differently—the implicit aims of human life have evolved. Life in a capitalist world is organized for a specific purpose: Accumulation. "Accumulate, accumulate! This is Moses and the prophets! Accumulation for the sake of accumulation, production for the sake of production!"[82] Accumulation of what? Accumulation of surplus value = money = social power. That is the meaning of human life in a capitalist society. Here is the fundamental reason why there cannot be something like balanced and harmonious relationships between different social groups, or between humans and nature, under capitalism. Achieving such balance or harmony is not the purpose of our social-ecological formation. Capital's goal is expanded accumulation, without end.

If the lesson of *Capital* is that capitalism is a peculiar form of organizing human and ecological relations, it follows that the struggles described by Marx throughout the book—struggles between capitalists and workers, landowners and renters, colonists and native people, and so on—are all forms of socio-ecological class war. So long as humans have lived in state-dominated societies, human relations with nature have been mediated by the state as a socio-ecological relation. Capitalism has decisively modified this history by reorganizing the social actors and their aims, with deadly consequences for life on Earth. Such a perspective is implied by *Capital* but only makes sense if we subsume human history into natural history, and vice versa.

We have seen that Marx took up important concepts from Darwin and that his natural-historical standpoint in *Capital* was inspired by *The Origin of Species*. While Darwin did not directly influence Marx's value theory, by compelling Marx to write in a more natural-historical way, he indirectly contributed to the breakthroughs of the 1860s—value form and fetishism—and the reconfiguration of the place of nature and value in Marx's critique of capitalist political economy. Marx also

---

82 Marx, *Capital*, 545.

deepened his critique of Malthus and generated world-revolutionary conclusions from his hypothesis apropos human population; yet, Marx's prediction of the end of capital has not been borne out.[83] To recognize this is not to deny capitalism's role in driving the climate crisis (which is all too real), nor to suggest that *Capital* is wrong about capital (it is not). It is, rather, to peer into the depth of the present crisis. For there are few signs of simultaneous world revolution on the horizon.

We appear stuck in capital's natural history, the world getting warmer and weirder, with ever more "surplus population" ejected by capital into the wasteland of the dispossessed.[84] (Indeed, the very distinction between *relative* surplus population and *absolute* surplus population is becoming difficult to discern.) Today, we find everywhere an authoritarian mode of biopolitical governance that sustains all the contradictions of a world capitalist society in which large groups of people constitute "relative surplus population" and there are neither means to transcend capital nor eliminate poverty. This goes some way to explain the rise of fascism, which is nothing but a political reaction to such a conjuncture—one which promises to manage the human-nature relationship during a capitalist crisis while blaming the other for capital's problems (whether on the basis of patriarchy, racism, nationalism, antisemitism—it varies).

From a natural-historical standpoint, "relative surplus population" is *Capital*'s most important political concept. *Who will be made surplus?* Contemporary political economy revolves silently around this question.

This carries some ambiguous political implications. Generations of Marxist-Leninists were troubled by *Capital*'s apparent ambivalence about the prospects of revolution: Marx seems to suggest that capital will be undermined by its own contradictions, without clarifying the

---

83 "[C]apital won. Sometimes with armies, sometimes with persuasion, sometimes with money, and sometimes by accident, but it won. For at least the last thirty or forty years . . . capital has proven richer, more powerful, more expansive, more convincing, and more real than any other political economic force on the planet": Geoff Mann, *Disassembly Required*, 222–3.

84 I take "wasteland of the dispossessed" from Kalyan Sanyal, *Rethinking Capitalist Development: Primitive Accumulation, Governmentality and Postcolonial Capitalism* (Calcutta: Routledge, 2014). For a sympathetic critique of Sanyal's concept, see Vinay Gidwani and Joel Wainwright, "On capital, not-capital, and development: After Kalyan Sanyal," *Economic and Political Weekly* 49, no. 34 (2014): 40–7.

specific role of labor's organized revolt via communist parties.⁸⁵ Compare the writings of Marx during the 1840s—or even his economic writings in the *Grundrisse*—with *Capital* and the implied revolutionary agency of workers within capitalism is decidedly more limited. One could claim that, after the defeat of 1848 and the reactionary wave that swept Europe, Marx lost some of his youthful confidence in labor's revolutionary defeat of capital. In Marx's framing of *Capital* as natural history we can read another implication: Whereas in his writings from the 1840s, Marx anticipated that humanity would realize or achieve a set of revolutionary reversals—labor overthrowing capital; humans dominating nature; chance brought under rational regulation; the gods brought down to Earth—in *Capital*, the analysis concludes with a demonstration of the fundamentally contradictory character of the system as a whole, and therefore only a promise of further change to come.⁸⁶ Reading *Capital's* political denouement, the shedding of one social form for another, we should recall Darwin's slightly defensive conclusion to *Origin* ("There is grandeur in this view . . ."). I am not suggesting that Marx became more politically conservative after reading Darwin. Rather, I am suggesting that the complex totality analyzed in *Capital* is—notwithstanding all the abstractions at play in the text—akin to the tangled bank, that ensemble of beings, knotted together. The political uncertainty of Marx's three conclusions to *Capital* is the price paid for analytical rigor and a commitment to a standpoint that—befitting natural history—has eviscerated teleology.

Let me briefly address here two texts by Marx that complicate my general argument. Notably, both concern the concept of progress in relation to *Capital*.

---

85 Hence Gramsci could write, weeks after the Bolshevik seizure of power, that the Bolshevik revolution is a "revolution against Karl Marx's *Capital*. In Russia, Marx's *Capital* was more the book of the bourgeoisie than of the proletariat": Gramsci, "The revolution against *Capital*," *Avanti!*, December 24, 1917, in *Selections from the Political Writings 1910–1920* (New York: International Publishers, 1977 [1917]), 34–7.

86 Recall (from chapter 2) that Darwin's theory made no claim to prediction. The uncertainties surrounding Marx's qualified prediction of the end of capital in *Capital* reflect, I surmise, his intuitive recognition that Darwin had it right. If Darwin's natural history could explain processes without capably predicting specific outcomes, the same should be true of Marx's analysis of the natural history of socio-economic formations.

[1] Marx wrote a letter to Engels on August 7, 1866, claiming—bizarrely, and wrongly—that Darwin's theory had been superseded by the work of a now-forgotten French Orientalist and natural historian, Pierre Trémaux (1818–1895).[87] If Marx was really influenced by Darwin, one might ask, why would he write the following? (Since this passage contains several statements which are incorrect, I will insert my interpolations in brackets.)

> A very important work which I shall send on . . . is: P. Trémaux, *Origine et Transformations de l'Homme et des autres Êtres*, Paris 1865.[88] In spite of all the shortcomings that I have noted, it represents a *very significant* advance over Darwin. The two chief theses are: *Croisements* [crossings—JDW] do not produce, as is commonly thought, variety, but, on the contrary, a unity typical of the *espèces* [this is false]. The physical features of the earth, on the other hand, *differentiate* [true] (they are the chief . . . basis [of evolutionary changes] [false]). Progress, which Darwin regards as purely accidental, is essential here on the basis of the stages of the earth's development [false][;] *dégénérescence*, which Darwin cannot explain [because there is no need to do so], is straightforward [in Trémaux]; ditto the rapid extinction of merely transitional forms, compared with the slow development of the type of the *espèce*, so that the gaps in palaeontology, which Darwin finds disturbing [an exaggeration], are necessary here. Ditto the fixity of the *espèce* [false], once established, which is explained as a necessary law (apart from individual, etc., variations). Here hybridisation, which raises problems for Darwin [false], on the contrary supports the system, as it is shown that an *espèce* is in fact first established as soon as *croisement* with others ceases to produce offspring or to be possible, etc. In its historical and political applications far more significant and pregnant than Darwin.[89]

---

87   Marx, Letter to to Engels, August 7, 1866, MECW 42: 303–5.
88   Accessible at gallica.bnf.fr.
89   Marx, Letter to Engels, 1866 MECW 42: 304–5. Marx is wrong to claim that "progress, which Darwin regards as purely accidental, is essential here on the basis of the stages of the earth's development, *dégénérescence*." This bizarre argument runs directly contrary to the antiteleological quality which Marx initially—rightly—celebrated in Darwin's theory. If "Darwin cannot explain" the lack of progress in the stages of the earth's development, as Marx says in this letter, that is because there are no such progressive stages to be explained.

In reality, Trémaux's *Origine et Transformations de l'Homme* is just one of the raft of mediocre books on human evolution written in the twelve-year gap between Darwin's *Origin* and *Descent*. Trémaux was a racist with an implausible theory. It is bewildering that Marx would have been taken in by it, even if only briefly.[90] The specific quality that Marx finds most impressive about Trémaux's *Origine*—its explanation of "progress"—is one that Marx had previously criticized harshly. I have no explanation for Marx's errant comments. Engels must have felt similarly, for he replied to Marx cooly, remarking that Trémaux's work was not good. Probably, Engels set Marx right.[91] By the time Marx published *Capital*, these errant ideas of Trémaux were gone.

[2] On one occasion, Marx explicitly emphasized the parallels between *Origin* and *Capital*—but this instance may cause confusion. Shortly after *Capital* was published, Marx wrote Engels, encouraging him to publish a review of his book in a liberal German news journal, *Der Beobachter*: "it would be an amusing coup if we could hoodwink Vogt's friend," the editor, into publishing a review. "It would be easy to contrive the thing as follows."[92] Then Marx proposes that Engels write a review of *Capital* along the following lines:

> When [Marx] demonstrates that present society, economically considered, is pregnant with a new, higher form, *he is only showing in the social context the same gradual process of evolution that Darwin has demonstrated in natural history*. The liberal doctrine of "progress" ... embraces this idea, and it is to his credit that [Marx] himself shows

---

90 Marx and Engels's correspondence about Trémaux has generated a considerable literature, particularly for its insights into their views on race. See, e.g., Ralph Colp, "The contacts between Karl Marx and Charles Darwin," *Journal of the History of Ideas* 35, no. 2 (1974): 329–38; John Stanley and Ernest Zimmermann, "On the alleged differences between Marx and Engels," *Political Studies* 32 (1984): 226–48; Erik Van Ree, "Marx and Engels's theory of history: Making sense of the race factor," *Journal of Political Ideologies* 24, no. 1 (2019): 54–73; and Sven-Erik Liedman, *The Game of Contradictions: The Philosophy of Friedrich Engels and Nineteenth Century Science* (Leiden: Brill, 2022).

91 Liedman writes that "Marx seems to have capitulated unconditionally" to Engels on this point (Liedman, 113) and that "Trémaux's book quickly met with well-deserved oblivion; the racist overtones in it are what stand out most clearly today. Objectively speaking, [his book] has nothing to do with Marx's theory of society" (Liedman, 133). I concur. For an opposing view, see Van Ree (see previous note).

92 Marx, Letter to Engels, December 7, 1867; MECW 42: 493.

there is hidden progress even where modern economic relations are accompanied by frightening direct consequences. At the same time, owing to this critical approach of his, the author has, perhaps *malgré lui* [despite himself], sounded the death-knell to all socialism by the book, i.e. to utopianism, for evermore.[93]

Engels followed through on Marx's proposal.[94]

We should not give great hermeneutical weight to this letter, since (as Engels would have understood) Marx is ventriloquizing *Capital* here in a deliberately liberal style for the sake of drumming up publicity and sales in Germany.[95] Nevertheless, with this letter, Marx encouraged readers to link his argument in *Capital* and Darwin's natural history. This letter provided a mandate for Engels's problematical graveside speech (see chapter 3).

History itself is a real part of natural history

If my proposed natural-historical reading of *Capital* has merits, it would follow that Marx's late intellectual work should bear certain qualities: A magnified scale of historical analysis; a fuller embrace of natural scientific research and natural-historical study of human-nature

---

93 Marx, Letter to Engels, December 7, 1867; MECW 42: 493, my italics.
94 See Engels, Marx-Engels-Werke 16 (1868), S.226–28.
95 This letter still haunts Marx. When Paul Reitter's English translation of *Capital* was published in 2024, the *New York Times* published a (largely favorable) review framed by this 1867 letter to Engels. Astonishingly, when quoting Marx's letter, the reviewer removed the scare-quotes around "progress" (thus implying that Marx sincerely portrayed *Capital* as a book about progress) in order to castigate Reitter and his coeditor Paul North—for downplaying the Darwinian theme of progress in *Capital*: "It's a sign of the times that the editors and translator of an eagerly anticipated new English edition of the book . . . largely ignore both Darwin and the idea of progress": James Miller "Karl Marx, weirder than ever," *New York Times*, September 9, 2024. North and Reitter replied: "James Miller maintained that it was 'weird' of us to point out how Marx tries to evoke the strangeness of capitalism while saying little about his notion of progress. For, of the two things, the latter [progress] mattered more to Marx, particularly in *Capital*—or so Miller contended. Miller did not actually cite evidence from *Capital* to support this claim, relying instead on a letter in which Marx conspired to 'trick' a liberal newspaper into running a review of his book": Paul North and Paul Reitter, "Was Marx's *Capital* a revolutionary text? Yes and no," *Chronicle of Higher Education*, December 16, 2024. The truth is that, *pace* Miller, "the idea of progress" was undermined by Darwin and Marx. This episode reveals more about the ideology of *The New York Times* than anything to do with *Capital*.

relations; a keen interest in the emerging anthropological research; and a search for diverse paths and prospects for social change. Recent research has identified exactly these novel qualities in the late Marx's work.[96] First: The geographical breadth of his vision expanded. Studying anti-imperial movements in Ireland, India, Mexico, and elsewhere, Marx overcame his earlier Eurocentrism.[97] His journalistic writings and notebooks are not only filled with notes about people and events in every corner of the world; by the mid-1860s, Marx was clearly developing a *global* perspective on capital and struggles against it. Second: His perspective became more world-historical. Earlier, Marx had posited that capital created, for the first time, *world history*.[98] But, even as late as *Capital*, his critique of political economy operated as if a limited form of methodological nationalism—centered on Britain—

---

96 Unfortunately, much of this work has been produced by specialists working in parallel. We need a broader assessment of the ways these lines of inquiry relate to the natural-historical standpoint proposed in *Capital*.

97 Kevin Anderson, *Marx at the Margins: On Nationalism, Ethnicity, and Non-Western Societies* (Chicago: University of Chicago Press, 2010); see also Marcello Musto, *The Last Years of Karl Marx: An Intellectual Biography* (Stanford: Stanford University Press, 2020); Anderson, *The Late Marx's Revolutionary Roads*.

98 In *German Ideology*, for example, Marx writes that, with the development of global capitalist exchange based upon an international division of labor encompassing many nations, "history becomes world history":

> Thus, for instance, if in England a machine is invented which deprives countless workers of bread in India and China, and overturns the whole form of existence of these empires, this invention becomes a world-historical fact. Or again, take the case of sugar and coffee, which have proved their world-historical importance in the nineteenth century by the fact that the lack of these products, occasioned by the Napoleonic Continental System, caused the Germans to rise against Napoleon, and thus became the real basis of the glorious Wars of Liberation of 1813. From this it follows that this transformation of history into world history is by no means a mere abstract act on the part of "self-consciousness," the world spirit, or of any other metaphysical spectre, but a quite material, empirically verifiable act, an act the proof of which every individual furnishes as he comes and goes, eats, drinks and clothes himself: Marx and Engels (1845), MECW V: 50–1.

World history is therefore a product of the globalization of capital, but so, too, is communism, since "the mass of workers who are nothing but workers ... presupposes the world market. The proletariat can thus only exist world-historically, just as communism, its activity, can only have a 'world-historical' existence": Marx and Engels, *German Ideology* (1845), MECW V: 49.

was analytically legitimate. But, after finishing *Capital* I, Marx began sketching notes for what appears to have been a massive world-history, encompassing the entire world and thousands of years, aimed at telling the back story on the emergence of capital. (During 1881–1882 alone, he wrote over 1,500 pages of notes on world history.[99]) Third: Saitō has demonstrated that Marx's writings displayed a marked "ecological turn" from the 1860s, and that nature and history preoccupied him for the remainder of his days. Saitō notes that by "enriching the concept of metabolism,"

> Marx aimed at comprehending the physical and social transformation of the relationship between humans and nature from historical, economic, and ecological perspectives. He also came to study different ways of organizing human metabolism with nature in precapitalist and non-Western societies and to recognize the source of their vitality for building a more egalitarian and sustainable society beyond capitalism.[100]

Saitō's remark brings us to the fourth element that changed: Marx's political strategy. Marx's late writings—particularly his 1875 "Critique of the Gotha Programme" and 1881 letter to Vera Zasulich—reveal that he had abandoned all linearity and teleology in his conception of the potential emergence of communist society.[101]

If we consider these four elements independently, they appear merely incidental to one another. But we should grasp them as a differentiated unity, reflections of a fundamental change in Marx's thinking: The fruit of his antiteleological, natural-historical standpoint, the realization of his earlier conception of communism coinciding with natural history.[102]

---

99 Michael Krätke, "Marx and world history," *International Review of Social History* 63 no. 1, 91–125.

100 Kōhei Saitō, *Marx in the Anthropocene: Towards the Idea of Degrowth Communism* (Cambridge, UK: Cambridge University Press, 2022), 67–8. In the original passage, Saitō writes that Marx's theoretical enrichment occurred "after the 1860s," but his research demonstrates that it occurred *during and after* the 1860s, so I have modified it accordingly here. I thank Saitō for confirming this point.

101 Anderson, *The Late Marx's Revolutionary Roads*.

102 Marx, Paris MS III, §IX; *Economic and Philosophic Manuscripts* (London: New Left Books/Penguin, 1970 [1844]), 355; MECW III: 304.

PART III
Elaborations of Marxian Natural History

Preface to Part III

In parts I and II, we found that reading *On the Origin of Species* influenced Marx's critique of capitalist political economy. Now, Marx read widely and had many influences. One could produce readings of *Capital* that show his indebtedness to Aristotle, Dante, Hegel, Balzac, and others. By proposing to add Darwin's name to this list, I do not deny any others. My concern is not to curate the names of great men whom we should praise because they helped our hero Marx on his journey. Given our planetary crisis, such academic credit-giving would be pointless. Rather, I want to strengthen our response to this crisis. To this end, I contend, we need Marx *with* Darwin and the analysis of the natural history of economic forms enabled by their work. Such a project was not completed by *Capital*. It was only introduced. We still need to elaborate a Marxian natural-historical standpoint.

I offer the three chapters of part III as contributions toward this end. Unlike parts I and II, which unfold in a relatively linear fashion, the chapters of part II are effectively independent—three distinct branches growing from the same trunk. While they may be read separately, they share the goal of contributing to a Marxian natural-historical perspective by addressing some of the fundamental questions facing us today. Let me briefly introduce them.

Capitalism is commonly presented to us as if it were a logical and timeless economic system. Chapter 6 takes up the questions: Is capitalism natural? Why and how did it emerge in the first place? By bringing a Marxian natural-historical perspective to these questions, we can generate stronger answers to them.

Darwin's *Origin of Species* showed that each form of life is historical and the unfolding of evolutionary processes has no telos. His discoveries raised new issues for the philosophy of history, a field of philosophy that was largely defined by the search for a meaning of the movement of human history and the discernment of humanity's ultimate endpoint. In chapter 7, I consider the implications of adopting a Marxian natural-historical perspective for the philosophy of history, the concept of human nature, the dialectic of nature, and—since these are commonly defined in terms of the peculiarities of human consciousness—Marx's treatment of consciousness and labor in *Capital*.

Part III concludes with political speculations on the prospect of a radically other world. Of course, I cannot predict the future. Nonetheless, I owe it to my reader to sketch my views, based on the natural-historical standpoint developed here, regarding the prospect for revolutionary change. I present these thoughts in chapter 8. However bleak our prospects at present, we must remember one basic lesson of natural history: Change is a constant. The story of humanity on Earth is unpredictable and has not yet run its course. Conditions may worsen. Indeed, that is likely. Yet we can still search for a way around the dead end that humanity faces today. Every apparent end is also a potential beginning.

The chapters that follow are offered, therefore, as contributions to the critique of capitalist society from a Marxian natural-historical standpoint. The project is by no means completed here. I present them with the hope that others will follow these paths of thought and take them in new directions.

6

A Natural History of Capitalism

Is Capitalism Natural?

The title of this chapter implies that capitalism has a natural history, which presupposes that capitalism is in some sense natural. If we accept that capitalism is a way of organizing human life on Earth, then this is only logical, for humans and Earth are natural. However, in another sense, capitalism is not natural: It is an inherently expansionary, accumulation-driven economic form, destructive of nature. We have, therefore, an antinomy: Capitalism is natural; capitalism is not natural. This chapter will provide a fresh way to resolve this antinomy by proposing a Marxian natural-historical approach.

The first step is to recognize that capitalism has been *naturalized*. We have been taught to think of capitalism *as* something natural. We are led to believe that capitalism—understood here as a "free-market society," which is how non-Marxists tend to speak of capitalism—has no real history: It simply exists, or was at least always there, in potential. People have always exchanged things, so a free-market society was inevitable; it was only a question of removing barriers to it—freeing ourselves and our markets.

Since 1859, this sort of thinking has been justified by two sorts of appeals to Darwin's theory. One concerns the evolution of the human species. The argument is that humans evolved to thrive as a species under just the sort of conditions that capitalism provides. We evolved to produce and consume, to truck and barter, to compete and accumulate, and so on (the specific qualities vary). Regardless of the specifics, the idea is that humans evolved in such a way that it is only logical that we should live in capitalist societies today. Capitalism provides the true native habitat of the human species; this is proven by the huge increase in the human population after adopting the capitalist form of society.

From a Darwinian perspective, such thinking is flawed. Humans evolved under social and environmental conditions that differ significantly from what we live in today. Our species spread around Earth long before adopting a capitalist form; humans have only lived in capitalist societies for around 0.1 percent of our species's existence. We cannot presume that this particular moment—the one we happen to live in—reflects the telos of our evolution and judge all our varied adaptations retrospectively from that vantage. Even granting such a perspective, there are grounds for skepticism about the idea that prior adaptations prepared humans to thrive under capital. During the short period that humans have lived in capitalist societies, the world has seen the sixth great extinction event, rapid warming, and many regions have departed from the environmental niche within which the human species evolved.[1] Even if capitalism seems like a good fit for humanity today, it is likely to prove problematic for our survival.

Some invert the argument to make a different claim. Rather than suggest that humans are perfectly adapted to capitalist society, they claim that humanity tried out various social forms and that capitalism proved to be the one for which we were most fit. By this reasoning, what evolves are not individual humans but social formations. This line of reasoning typically presumes a teleology, putting it at odds with Darwin's theory. We might also wonder whether we may judge capitalism more successful than previous social formations: Capitalism will not last as long as hunter-gatherer societies did. Moreover, this style of thinking departs from Darwin's model, since there is no equivalent to the mechanism of natural selection (where gains in individual fitness lead to changes in the proportion of offspring that exhibit traits) in the evolution of social forms.

---

1   Chi Xu, Timothy Kohler, Timothy Lenton, Jens-Christian Svenning, and Marten Scheffer, "Future of the human climate niche," *Proceedings of the National Academy of Sciences* 117, no. 21 (2020): 11350–5.

These objections notwithstanding, I do not think we can simply dismiss all talk of social evolution: Better to redirect it into more analytically coherent and politically emancipatory directions.[2] This was Marx's approach. If we accept the proposition that capitalism is but one particular socio-ecological form through which our species has come to organize itself and its relations with nature (HH×HN), then we can better understand how we came to live in capitalist societies while avoiding some common errors of reasoning.

Capitalism—the highest stage of society?

As European societies first began to be transformed by capitalist social relations, intellectuals generated theories about the changes occurring and their origin. Adam Smith's four-stage model provides a well-known example.[3] Smith claims that there is a natural, logical sequence to the development of human societies, which passes through four discrete stages: After hunting, shepherding, and agriculture, human society adopted the highest stage, "market society" (Smith's name for what we call capitalism). Smith's evidence and reasoning are weak enough to ignore. The important point is that his framework is teleological. Smith essentially built an argument from a specific presupposition—that all societies would, or should, end up looking like Scotland in the 1770s—and then narrated a story about why this would occur, projecting it backward through all human history.[4]

Historians of capitalism today often produce a more sophisticated version of this narrative. Typically, the variables that explain the rise of

---

2 Cf. Yanis Varoufakis's defense of studying economic history through evolutionary models of systemic change: "The problem with evolutionary social theory is neither that its origins are biological nor that its predictions follow trivially from its assumptions. It is rather that, in accordance with neoliberal thinking ([including] mainstream economics . . .), it tends to stay firmly within the confines of a *canonical* (that is, primitive) *evolutionary model* in which the individual is frozen in time and change occurs only at the level of the social. The study of history requires a deeper analysis": Varoufakis, "Capitalism according to evolutionary game theory: The impossibility of a sufficiently evolutionary model of historical change," *Science and Society* 72, no. 1 (2008): 63–94.

3 Adam Smith, *Lectures on Jurisprudence*, eds. R. Meek, D. Raphael, and P. Stein (Oxford: Clarendon Press, 1978); cf. Giovanni Arrighi, *Adam Smith in Beijing: Lineages of the Twenty-First Century* (New York: Verso Books, 2009).

4 Decades later, Hegel did something similar with his philosophy of history: See chapter 1.

capitalism are *technology* (the Industrial Revolution) plus *political ideas* (which lead to states getting out of the way of markets). The heroes of the story are engineers, philosophers, and entrepreneurs: Men (almost always white men) who create capitalist society through their guile, risk-taking, technology, and competitive spirit. This sort of history may help to plot a certain course of events and provide empirical material to fill in some gaps in our knowledge. But it suffers a fatal flaw: While it can demonstrate *that* these changes occurred, it cannot explain *why* (without, as Smith does, presupposing the very thing it seeks to explain). In the words of Ellen Wood, practically all accounts of the emergence of capitalist society have "treated it as the natural realization of ever-present tendencies":

> Almost without exception, accounts of the origin of capitalism have been fundamentally circular: They have assumed the prior existence of capitalism in order to explain its coming into being. In order to explain capitalism's distinctive drive to maximize profit, they have presupposed the existence of a universal profit-maximizing rationality. In order to explain capitalism's drive to improve labour-productivity by technical means, they have also presupposed a continuous, almost natural, progress of technological improvement in the productivity of labour. These question-begging explanations have their origins in classical political economy and Enlightenment conceptions of progress. Together, they give an account of historical development in which the emergence and growth to maturity of capitalism are already prefigured in the earliest manifestations of human rationality, in the technological advances that began when *Homo sapiens* first wielded a tool, and in the acts of exchange human beings have practiced since time immemorial.[5]

For good reason, Wood's description of this narrative refers to the evolution of *Homo sapiens*: Since 1859, the older liberal narratives, like Smith's, have been recast in Darwinian terms. In these histories, capitalism's emergence is naturalized: "In most accounts of capitalism and its origin, there really *is* no origin. Capitalism seems always to *be* there, somewhere;

---

5   Ellen Meiksins Wood, *The Origin of Capitalism: A Longer View* (London: Verso Books, 2002), 4–5. Those "Enlightenment conceptions of progress" still run deep among liberals.

and it only needs to be released from its chains . . . to be allowed to grow and mature."[6] The basic problem with liberal historiography of capital, then, is akin to the problem of pre-Darwinian histories of species: They produce detailed *descriptions* of each species and how it lives in place of a substantive *explanation* of its becoming. In lieu of an analytical framework for explaining causality, they narrate backward from a telos (which seems to need no explanation).

Is a nonteleological history of capitalism possible? Certainly. But since we live within capitalist societies, it is difficult not to presuppose the very conclusion that we are trying to explain. (Partly for this reason, the history of capitalism is not generally taught in economics.)

Marxist historians have generated powerful criticisms of conservative and liberal histories of capitalism, but vigorously disagree among themselves about alternative explanations. One school of thought, inaugurated by *Capital*, emphasizes the story of "so-called original accumulation": Capitalism started whenever people were separated from the means of labor. Some in this first school emphasize so-called original accumulation within England; others expand the horizon of research to link capital's emergence in Britain with the plunder of the Americas, the colonization of Ireland, the transatlantic slave trade, the formation of early industrial relations in the sugar plantations of the Caribbean, and so on. Since the separation of human beings from their means of labor (paradigmatically: The enclosures and the plunder of indigenous lands) is always resisted by the victims, this process always entails state-coordinated violence. Hence these narratives implicitly support the view that capitalism emerged from non-capitalism, via force. Such thinking raises a difficult question: How could anything emerge from what it is not? Posing the question allows us to appreciate its proximity to the question Darwin faced—to explain the emergence of existing species from extinct ones.

One way to finesse this problem is to focus on explaining how capitalist social relations emerged where they took hold of a society *in toto* for the first time. Such reasoning leads one to examine changes in social relations in western Europe generally and rural England particularly. A school of thought that has followed this path of research de-emphasizes Marx's "original accumulation" narrative in lieu of studying how and

---

6 Meiksins Wood, *The Origin of Capitalism*, 5.

why producers in rural England came to systematically compete with one another to produce commodities that they could sell profitably. To follow a trend in this literature, I will call this second group "political Marxism."[7] A third approach emphasizes consumption and dependence on markets for the social reproduction of life and the reproduction of the labor power commodity. I will call this the "consumption approach."

Even with a simple presentation of these ideas, it should be intuitive that the differences stem from thinking about the emergence of capitalism at different scales and with distinct objects of analysis. Under such conditions, it will be impossible to find a consensus.[8] My premise is that, if we accept the proposition that capital is a socio-ecological formation—a form recently adopted by our species—it follows that the becoming of capitalism amounted to the reforming of social relations and human relations with nature (HH×HN). From that vantage, all three of the aforementioned approaches have some validity and we need to incorporate them coherently to understand capital's natural history. This chapter reflects my attempt to do so by applying three general principles:

[1] M-C-M'. It will be helpful to disaggregate capital into distinct processes that constitute it as a differentiated unity, permitting us to study the historical becoming of these processes at different scales of analysis. If we accept Marx's definition of capital as value in motion, and his general formula for capital, then we can appreciate that a social formation becomes capitalist when it combines commodity production (employing the commodity labor power) with consumption (by proletarians dependent upon the market to acquire commodities) and reinvestment of surplus value realized as productive capital (by capitalists).

[2] HH×HN. Too often, histories of capitalism leave nature out entirely. A generation of environmental history and ecological Marxism

---

7 The name "political Marxism" is odd because none of these schools of thought are unpolitical. Like many misnomers, this one started from an insult, when Guy Bois accused Brenner and Wood of giving Marxist historicism an unjustified political inflection by focusing on (ostensibly *political*) class relations rather than (ostensibly *economic*) forces of production: See Guy Bois, "Against the Neo-Malthusian Orthodoxy," *Past and Present* 79, no. 1 (1978): 60–9. I thank Maïa Pal for this insight.

8 Complete consensus in historical scholarship is neither possible nor desirable. However, for the sake of effective social and political communication, Marxist historiography could do better in countering conservative and liberal histories of capitalism.

has shown the inadequacy of such exclusions.[9] Capitalism has not only changed how humans live on this planet; it also came into being because humans changed how we live on this planet. Since these processes have played out differently in space and time, a multi-scalar approach is useful. And, since the processes that comprise capital have distinct effects on HH×HN, it helps to consider them separately (though in reality they intertwine).

[3] Being and becoming. Historicizing capital is complex because it is not a material thing that we can define objectively, but a socio-economic formation, comprising a set of form-determined relationships, with no final endpoint. Even societies that are obviously capitalist (such as the USA today) continue to become more thoroughly capitalist through time. By implication, it will be useful to track both the *being* of capital (a complete circuit of capital manifest in a predominantly capitalist society) and the *becoming* of capital (the immanent emergence and deepening of specifically capitalist social relations). This will help to address the geographical puzzle that while capital's becoming requires a *transcontinental* perspective, we can also assert that capital achieved hegemony first in *England*.

Before proceeding, a qualification. This chapter contains no new empirical evidence, and I do not expect to resolve the debates among Marxists on the history of capitalism. My aim is more modest: To present a conceptual, Marxist, natural-historical vantage upon the emergence of capital. By decomposing capital into three processes and examining each as an element in capital's natural history, I hope to contribute to a more coherent narrative of capital's becoming. Doing so should also help us to understand *Capital* better. *Capital* does not provide the final word on capital's emergence (Marx certainly did not see it that way). By my reading, in *Capital* there are two different styles of arguments about the emergence of capitalism, which differ not only in terms of temporality (when the transition begins, how long the process takes, and so on) but also—going back to Hegel—in terms of the category of history on offer (reflective-critical versus philosophical). On one hand,

---

9   As Jason Moore has noted repeatedly: "The Modern World-System as environmental history? Ecology and the rise of capitalism," *Theory and Society* 32 (2003): 307–77; "Transcending the metabolic rift: A theory of crises in the capitalist world-ecology," *The Journal of Peasant Studies* 38, no. 1 (2003): 1–46; *Capitalism in the Web of Life: Ecology and the Accumulation of Capital* (New York: Verso Books, 2015).

Marx presents the emergence of capitalism as a logical process that developed over the course of centuries and which may be explained rationally. On the other hand, Marx shows that the becoming of capital is essentially irrational and contradictory: For, by the time the world becomes capitalist, it will have undermined its conditions of possibility. (Marx's claim that the true barrier to capitalism is capital itself, and his prediction that capital would beget communism, express the second style of historical reasoning.) The distinctive politics of these styles of historical interpretation are complex and cannot be resolved here. I simply want to try to present both views coherently. Bringing nature into the picture—which the young Marx scolded Hegel for failing to do—will help us.

One Social Formation, Three Economic Processes

The emergence of capitalism as a social formation required the effective conjugation of three distinct social processes:

1. The formation of a social group dependent upon sale of the labor power commodity (aka the proletariat) of sufficient scope to provide for expanded commodity production;
2. generalized market competition between producers employing C_{LP}, generating advances in commodity production processes as well as a compulsion for producers to adopt these changes;
3. expanded reproduction of capital facilitated by the realization of surplus value through sale of commodities and reinvestment in expanded production.

These processes did not emerge simultaneously or in the same way everywhere. Let us consider them separately.[10]

---

10 A fuller analysis of these processes would require discussion of the role of the state, which is beyond the scope of this book. The reason: The major "ingredients" of capitalist society—laboring people, commodities (surplus grain, dried fish, status objects, and so on), markets, money (whether cacao or coins), regional trade—have all existed for thousands of years. They all emerged during periods when human societies were organized by the state form (empire). It is impossible to imagine that human societies could have become capitalist without this prior reorganization. In

Table 6.1. Three processes of capital

	Process name	Description of process	Object of analysis	Setting and scale
I	So-called original accumulation.	Separation of laborers from land and means of labor.	Relation of human labor to land and means of labor.	Separation occurs in different ways in different places and times. The classic case is the separation of people from the land in England through the enclosures; *Capital* also mentions the transatlantic slave trade and the plunder of the lands of indigenous people.
II	Market competition.	Market competition induces changes in commodity production processes (mechanization, reorganization of production, etc.) to generate surplus value.	Production relations vis-à-vis competitive markets.	Rural England, sixteenth and seventeenth centuries.
III	Mass consumption.	Generalization of consumption by laborers (spending wages to purchase commodities).	Consumption relations of proletarians qua consumers.	Molecular process operating globally (but particularly since ca 1945) whereby proletarians become entirely dependent upon purchasing commodities to meet their needs (with demand induced by relentless marketing).

[1] *So called original accumulation*

Commodity production of a capitalist type requires the existence of a significant social group of proletarians, workers who produce a livelihood by selling their labor power as a commodity. Whenever these workers have been effectively separated from the means of production (and are therefore dependent upon such a sale to survive), we can be

---

natural-historical terms, a capitalist society is a state society with a specifically capitalist form. Hence the desire for a stateless capitalist society is an illusion: States enable the capitalist form.

confident that so-called original accumulation has occurred within that social formation.

Marx adapted this concept from Adam Smith. For Smith, the original accumulation that sets market society going is that stock of initial capital an investor uses to initiate production. In Smith's account of the origin of market society, the owners of capital in Britain had originally accumulated this capital through thrift and savings, yet Smith provides no historical evidence for this claim. For Marx, original accumulation is a *class process*: the separation of the laborer from the means of production. He writes: "So-called original accumulation is thus nothing other than the historical process of separating the producers from the means of production. It *appears* as 'original' because it constitutes the prehistory of both capital and the mode of production that goes with capital."[11]

What Marx calls "so-called original accumulation," therefore, means the separation of people from the means of production to create a dependent pool of proletarians, facilitating purchase of the labor power commodity. As Marx emphasizes, this process can play out in varied ways. Nevertheless, the result is to make people worldwide ever more dependent upon selling labor power as a commodity. A capitalist form of society requires the separation of people from collectively owned means of production (as found in many peasant communities before their incorporation into capitalist economy, for instance). As pre-existing economic forms have been reshaped, re-formed, and homogenized by specifically capitalist social relations, the diversity of the economic relationships found around the world prior to the nineteenth century has declined dramatically.

What were the conditions facilitating the emergence of this era? Serfdom had "long since been abolished" and "the medieval city-state" was in decline. But the "epoch-making" processes defining "the history of this separation process occurred when large numbers of people were violently torn from their means of production and subsistence, and thrust . . . into the labor market as proletarians. The whole process was based on land being expropriated from the workers."[12] Here, the consolidation of capitalism as a social form is equated with the separation of workers from the land, the original means of labor. Marx then provides

---

11 Marx, *Capital*, 651, my italics.
12 Marx, *Capital*, 652.

a geographical qualification to this explanation: "The coloration of this history has varied from country to country, and it has passed through its different stages in various sequences. Only in England has it assumed its classic form."[13]

Two questions arise. If capitalism emerged because "large numbers of people were violently torn from their means of production and subsistence, and thrust . . . into the labor market as proletarians," then could it not be said that whatever caused *that* to happen was the real cause of capitalism's emergence? In other words, could we not claim that capitalism was created by some precapitalist or non-capitalist force that acted upon European society? And—question two—why Europe? What makes England the classic form?

Although Western European societies were the first where discernably capitalist social relations became generalized, the process driving this change (the "becoming of capital") was world-encompassing.[14] The colonization of the Americas played a fundamental role in the economic transformation of Holland, England, and France. The dominant classes in these colonial-capitalist societies benefitted from the plundering of indigenous lands and enslavement of human laborers, generating surpluses and dependent markets which stimulated the consolidation of a capitalist bourgeoisie in England. Marx writes apropos "so-called original accumulation" in chapter 24:

> What characterizes the dawn of the era of capitalist production? Gold and silver were discovered in America; the native population there was wiped out, enslaved, and entombed in mines; India was conquered and plundered; and Africa was turned into a commercial hunting preserve with dark-skinned people as the prey. These idyllic processes largely constitute original accumulation. Following right behind them was the European nations' commercial war, whose battlefield encompassed the entire globe. This war began with the revolt of the Netherlands against Spain, swelled to gigantic dimensions in England's

---

13 Marx, *Capital*, 652.
14 Marx clarifies that while "capitalist production sporadically took shape in Mediterranean countries as early as the fourteenth and fifteenth centuries, the capitalist era dates only to the sixteenth century" (652). This paragraph and the following draw from Joel Wainwright, "Capitalism qua development in an era of planetary crisis," *Area Development and Policy* 9, no. 3 (2024): 407–27. I thank the journal editors.

Anti-Jacobin War, and continues into the present day as the Opium Wars against China.

The different aspects of original accumulation were put in motion by, in particular, Spain, Portugal, Holland, France, and England, more or less in chronological order. In late seventeenth-century England, they were methodically combined in a number of systems: The colonial system, the national debt system, the modern tax system, and the system of protectionism. These methods . . . all employed the power of the state, the concentrated, organized violence of society, to quicken the transformation of the feudal mode of production into the capitalist mode . . . and shorten the transitions. Violence is the midwife for every society pregnant with a new one. It is in fact a kind of economic power.[15]

"Violence is the midwife": The competition between European powers to dominate passage points, trade routes, and productive lands generated positive feedback between the cultivation of *means of making war* (armies, weaponry, strategic points) and accumulation of new *means of capitalist production* (labor, material, land). Adam Smith rued the formation of this vicious cycle, calling it the "unnatural path of development."[16] Capitalist socio-natural relations were consolidated along Smith's unnatural path in England. Once established there, capital's inherent tendency toward accumulation on an expanding scale—capital's globalization—has brought these relations to every corner of the earth.

How long did this take? Marx answers vaguely: "The capital relation grows out of economic soil that is the product of a long process of development."[17] All right, asks the historian, exactly how long a process are we talking about? Marx replies, "Centuries":

> When a normal working day was finally established, this was the result of *a centuries-long struggle* between capitalists and workers . . . When capital is still in its embryonic form and relies to some extent on state power—not merely the force of economic relations—to secure the right to absorb a sufficient quantity of surplus labor, what it demands appears quite modest compared with what it will later, as an adult, grudgingly

---

15   Marx, *Capital*, 682.
16   Arrighi, *Adam Smith in Beijing*.
17   Marx, *Capital*, 469.

give up. *It took centuries for workers set "free" by an advanced capitalist mode of production to get to the point where they would sell—in other words, would be forced by society to sell—the entire active period of their lives, even their very capacity to work itself, for the price of their normal means of subsistence.*[18]

This expansive time horizon, in which the emergence of capital is defined by processes developing through "centuries," calls to mind the geological time-horizon upon which Darwin narrates the emergence of new species from old ones. As Marx writes in *Capital*, when examining the history of socio-economic formations, we should only concern ourselves "with striking and general characteristics; for epochs in the history of society are no more separated from each other by hard and fast lines of demarcation, than are geological epochs."[19]

A brief assessment: Marx's account of "so-called original accumulation" makes the crucial point that the proletariat emerged historically through the violent separation of laborers from their land and means of labor. While this fact is essential to recognize, it cannot in itself explain the emergence of capitalism. For various forms of violent dispossession have occurred frequently in the past (often coinciding with war or the emergence of a state) without generating capitalist social relations. What distinguishes so-called original accumulation of a capitalist type is that it produces a new social class: The proletariat. Since the formation of the proletariat requires separation of labor from the land or means of labor, which implies separation of humans from pre-existing relations with nature, then no account of the natural history of capitalism could be complete without considering this process.

---

18 Marx, *Capital*, 241. Elsewhere: "It took capital centuries to extend the working day to its normal maximum limit—and then to the point where its limit was the natural twelve-hour day" (247); "Where did the original capitalists come from? The expropriation of the rural population directly called forth just large landowners, after all. As for the genesis of capitalist farmers . . . it proceeded slowly, developing over many centuries" (675).

19 In a letter to Vera Zasulich (1881), Engels repeats this geological metaphor apropos the "history of the decline of primitive communities": "As in geological formations, these historical forms contain a whole series of primary, secondary, tertiary types, etc.)."

[2] *Market competition between producers*

Marx argues that capitalism was consolidated in the sixteenth century: "Although capitalist *production* sporadically took shape in Mediterranean countries as early as the fourteenth and fifteenth centuries, the capitalist *era* dates only to the sixteenth century."[20] From this, one might look for the origin of capitalism in, say, fourteenth-century Mediterranean trade patterns. But it would be an analytical error to presuppose the conclusion of the story—the consolidation of a capitalist society—provides proof of the logical necessity for capital's emergence from prior conditions. To do so would be to fall into tautology, teleology, or both. The approach called political Marxism addresses this challenge by historicizing the social conditions at the time of capital's emergence without presupposing that they met capitalism's requirements, that is, without making functionalist arguments. The first step is to specify the essence of capitalist production relations and then identify the specific conditions that generated them in England.

Earlier, we noted that human societies were not fully capitalist until the compulsion to produce and reinvest surplus value had been generalized. This is the meaning of the compulsion toward expanded accumulation. Robert Brenner's analysis of the emergence of capital in rural England emphasizes this process. In Brenner's formulation (refined in the work of Ellen Wood and others), capitalism exists only when multiple capitalists (or firms) compete with one another through markets, ramifying changes in relations of production, and leading to the accrual of larger profits to the most efficient or effective producers.[21] To grasp the breakthrough to capitalist production,

> it is necessary to lay bare *both* the conditions that prevented the sustained growth of agricultural labor productivity *and* the historical processes that brought about the transformation of those conditions

---

20 Marx, *Capital*, 652, my italics.
21 Robert Brenner, "The origins of capitalist development: A critique of 'Neo-Smithian Marxism,'" *New Left Review* 1 (1977): 104, 25–92; see also Robert Denemark and Kenneth Thomas, "The Brenner-Wallerstein debate," *International Studies Quarterly* 32, no. 1 (1988): 47–65; Robert Brenner, "The Low Countries in the transition to capitalism," Journal of Agrarian Change 1, no. 2 (2001): 169–241 Ellen Meiksins Wood, *The Origin of Capitalism: A Longer View* (London: Verso Books, 2002); Spencer Dimmock, *The Origin of Capitalism in England, 1400–1600* (Leiden: Brill, 2014).

so as to make such sustained growth of agricultural labor productivity possible. For the achievement of regularly increasing agricultural labor productivity has seemed to be the critical condition making possible not only the break beyond the Malthusian cyclical pattern, but also both the provision of a sufficient food supply and the creation of a sufficient domestic market to support industrialization, with the latter defined simply as the movement of an ever increasing proportion of the labor force out of agriculture and into industry.[22]

What facilitated these changes—increasing productivity of agricultural labor specifically and the growing efficiency of production generally—is not one particular historical break:

> Neither a revolution in technology (like "the agricultural revolution" or even the "industrial revolution"), nor an "original accumulation of capital" for investment (as was derived, for example, from the gold and silver mines of the Americas or the African slave trade), nor the rise of an elaborate interregional/international division of labor ... has in itself sufficed to catalyze self-sustaining development. Such things could—and often did—contribute to *already ongoing* processes of increasing agricultural productivity, or modern economic growth ... but they could in no way constitute it or bring it into being.[23]

This is an approach that rejects Marx's claim—postulated most clearly in his 1859 "Preface to the critique of political economy" (written shortly before he read Darwin's *Origin*)—that socio-economic changes are propelled by contradictions between the forces of production and the social relations of production.[24] What then explains the historical emergence of capitalism and its growth imperative? Brenner answers:

> What makes for modern economic growth, particularly in

---

22 Brenner, "The Low Countries in the transition to capitalism," 169–241, 172.
23 Brenner, "The Low Countries in the transition to capitalism," 169–241, 172.
24 Ellen Wood describes as "vacuous" the "proposition that history is propelled forward by the inevitable contradictions between forces and relations of production, contradictions that emerge as developing productive forces come against the 'fetters' imposed by production relations": Ellen Meiksins Wood, "Marxism and the course of history," *New Left Review* 1 (1984): 147, 102. Wood's argument is attacked by Alex Callinicos, "The limits of 'political Marxism,'" *New Left Review* 1 (1990): 184, 110–15.

agriculture, is, in my view, something more general and abstract: It is the presence throughout the economy of a systematic, continuous, and *quasi-universal drive on the part of the individual direct producers to cut costs in aid of maximizing profitability via increasing efficiency and the movement of means of production from line to line in response to price signals*. This phenomenon comes into existence, I submit, only when the individual direct producers are not only free and have the opportunity, but also are *compelled* in their own interest, to maximize the gains from trade through specialization, accumulation, and innovation, as well as the reallocation of the means of production among industries in response to changing demand.[25]

Capital is defined by this drive: The compulsion spurred by market competition to maximize gains through innovative, specialized commodity production. From this vantage, evidence of capitalism's historical existence is measurable in definite patterns of human behavior. One virtue of this approach is to train the historical eye to identify these socio-economic patterns.

To further clarify the distinction between this process and the "so-called original accumulation," it will be useful to return to Marx's general formula for capital, M-C-M'. So-called original accumulation is the process that divides a historical totality—a precapitalist socio-economic form—by separating labor from land, placing both into the commodity form, thereby generating labor power (C_{LP}) and means of production (C_{MP}):

$$\begin{array}{c} C_{LP} \\ \nearrow \\ \text{precapitalist} \\ \text{economic form} \\ \searrow \\ C_{MP} \end{array}$$

Such separations had occurred frequently in history without generating capitalist societies. If, in ancient times, one group of humans deprived another of access to a resource (abundant hunting grounds, for instance), the vanquished group did not automatically become

---

25 Brenner, "The Low Countries in the transition to capitalism," 169–241, my italics. At the conclusion of the passage cited, Brenner cites his earlier work along with Ellen Wood's.

proletarians any more than the plunderers became bourgeois. The "so-called original accumulation" process that generates the capitalist form of society requires a more or less complete separation of people from the means of labor; moreover, it generates a sum of money for reinvestment: The first M of M-C-M'. Still, as Brenner justly observes, it is difficult to understand how any such "so-called original accumulation of capital" for investment that was "derived, e.g., from the gold and silver mines of the Americas or the African slave trade" could "have in itself sufficed to catalyze self-sustaining development" of market-induced competitive production.

The process he emphasizes is different. Here, distinct capitalist producers or firms ($M_{a, b, \ldots x}$) compete by using money to purchase labor power ($M—C_{LP}$) and means of production ($M—C_{MP}$), combining them in distinct ways to produce commodities ($C_{a, b, \ldots x}$). Such competition reinforces the distinctly capitalist conditions whereby individual firms or capitalists compete against one another to expand surplus value (profit) and dominance of commodity markets. Formulaically:

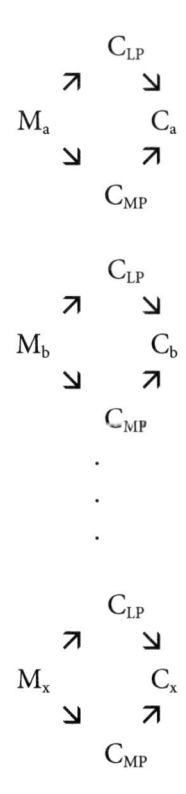

$$
\begin{array}{ccc}
 & C_{LP} & \\
 \nearrow & & \searrow \\
M_a & & C_a \\
 \searrow & & \nearrow \\
 & C_{MP} &
\end{array}
$$

$$
\begin{array}{ccc}
 & C_{LP} & \\
 \nearrow & & \searrow \\
M_b & & C_b \\
 \searrow & & \nearrow \\
 & C_{MP} &
\end{array}
$$

$$\vdots$$

$$
\begin{array}{ccc}
 & C_{LP} & \\
 \nearrow & & \searrow \\
M_x & & C_x \\
 \searrow & & \nearrow \\
 & C_{MP} &
\end{array}
$$

That is to say, capitalism is a socio-economic formation in which all individual producers are driven by market competition to constantly adapt, stimulating new production techniques that increase the efficiency of their production processes and capacity for expanded reproduction. The reader will recognize resemblances between this formulation and Darwin's description of the struggle for life causing perpetual adaptations of species.

A brief assessment: The political Marxist framework provides a strong starting point for explaining the historical emergence of capitalist social relations in rural England. Its analytical clarity, empirical support, and internal consistency make it a formidable theoretical framework. Nonetheless, Political Marxism often falls into the territorial trap of fixing early modern England as the spatial container within which capital emerged, when in reality, that society was connected to the wider world through empire. Surely the plunder of the Americas and the transatlantic slave trade contributed something to the emergence of capitalism in England. Moreover, Political Marxism has shown difficulty in incorporating perspectives from environmental history.

[3] *Consumption of commodities by proletarians*

Commodities are not produced for the sake of production but *for profit*. For the value congealed in the commodity to be realized—for its surplus value to be accumulated as profit—the commodity must be sold.

In *Capital*, Marx observes that in a capitalist economic formation, "intrinsically both C—M and M—C are mere conversions of given values from one form into another. But C'—M' is at the same time a realization of the surplus-value contained in C'. M—C however is not. Hence *selling is more important than buying*."[26] To be sure, production (M—C) is fundamental, but it is only the start of the process. The capitalist does not produce commodities in order to generate a pile of them to admire. The point is to sell them, to generate profit, to accumulate money—whence to begin the process again on an expanded scale.

Who buys these commodities? That depends on the type of commodity we are dealing with. In a capitalist society, commodities

---

26 Marx, *Capital*, volume II, MECW 36: 131, my italics. In Marx's notation, a modified commodity is C'.

may be divided into three types: Capital goods, consumption goods, and elite consumption goods. Capital goods are commodities that are purchased to facilitate production of other commodities. A bulldozer produced to clear the earth, sold to a firm that produces buildings for sale; computer software sold to a firm to manage their laborers: These are capital goods. By contrast, consumption goods are commodities purchased by people and consumed directly for sake of their use-value, which means for social reproduction (reproducing C_{LP}). Typically, they are bought with wages earned through the sale of the labor power commodity and, once consumed, cease to exist in their previous form.

Elite (or "luxury") goods are a subcategory of consumption goods, distinguished by the fact that their social function or use-value is principally to demonstrate high status (not to reproduce labor power). Luxury goods may well be consumed—the bourgeoisie pursues use-values too, including the pleasure of showing off their refined sense of taste—but they are just as often hoarded. In either event, the purchase of luxury goods constitutes *conspicuous consumption*.[27] Their purpose is principally to signal a power differential. However important luxury goods are for defining status hierarchies, I agree with Marx that, of all types of commodities, "luxury items proper are the least significant when it comes to comparing the technological capabilities of different epochs of production."[28] Nevertheless, as we see with yachts and private jets, the consumption of luxury goods has major consequences for HH×HN, and therefore cannot be ignored.[29]

---

27 On conspicuous consumption, see Thorstein Veblen, *Theory of the Leisure Class: An Economic Study of Institutions* (New York: Routledge, 2017 [1899]).

28 Marx, *Capital*, 155, note 5.

29 See Beatriz Barros and Rick Wilk, "The outsized carbon footprints of the super-rich," *Sustainability: Science, Practice and Policy* 17, no. 1 (2021): 316–22. The growing demand for such environmentally destructive elite goods demonstrates that Veblen was correct to claim (*pace* most Marxists of his time) that luxury goods are among the most significant lenses through which to study the operation of power. Veblen studied Darwin and Marx and encouraged an evolutionary approach to political economy. See Geoffrey Hodgson, "Veblen's *Theory of the Leisure Class* and the genesis of evolutionary economics," in *The Founding of Institutional Economics* (New York: Routledge, 2002), 170–200; William Dugger and Howard Sherman, *Reclaiming Evolution: A Dialogue Between Marxism and Institutionalism on Social Change* (New York: Routledge, 2000). A natural-historical standpoint clears ground for a rapprochement between ecological Marxism and Veblen-inspired approaches.

In a capitalist society, most commodities are purchased by ordinary people, proletarians qua consumers. Formulaically:

$C_{LP}—M_{wage}—C$ {consumer goods: rent, food, transport, etc.}

Whereas elite consumption:

$M'_{\text{profit (surplus value)}}—C$ {conspicuous consumption: private jet, yacht, expensive artwork, etc.}

The circuit of capital is incomplete unless the commodities which are produced can be sold profitably. *Expanded accumulation*, capital's *sine qua non*, therefore requires expanded consumption. Historically, this has meant that capitalism has *induced demand* by increasing the number of proletarian-consumers and establishing their dependence upon commodity consumption. The becoming-consumers of ordinary people is a process with a definite history that is inseparable from the generalization of capitalist class relations. We are accustomed to thinking of proletarianization ("having to work") as a sort of dehumanization and the consumption side ("getting to spend") as liberatory, but both dimensions of this historically entwined relation bear upon proletarian-consumers. To live by buying our required goods ($C_{LP}—M_{wage}—C_X$) is to be subjected to relentless marketing, surveillance, and other forms of social pressure. These practices—compelled by competition between producers for expanded sales—shape our very being and our desires. While this dynamic is not often emphasized in accounts of the emergence of capitalism, no capitalist society can exist without mass consumption.[30]

---

30 A pedagogical remark: Those who teach undergraduates about environmental issues, as I do, find that our students do not generally enter our classes with a coherent understanding of capitalism and its relationship with nature. What they *do* have is a subtle sense of themselves as *consumers*, capable of expressing manifold views about whether consumption practices are ethical or not, high or low status, and so on. Marxists often expend time and energy seeking to convince others that such views fail to provide a substantial basis for understanding capitalism; to understand capitalism, the thinking goes, we must focus on *production* (and labor). I regard these efforts as misdirected—a legacy of the emphasis on the production of surplus value in the only volume of *Capital* that Marx finished, volume I. Consumption is no less fundamental than production: Taken together, they form capital as a circuit through which value is generated and

Capitalism is often equated with the imperative to trade. This is a view associated with Ricardo, who advocated for free trade policies and specialization in the production of commodities based upon comparative advantage. Because volume I of *Capital* centers on the production of surplus value, Marx largely bracketed out the circulation of commodities and the movement of value between communities (these issues are central to volumes II and III). Yet Marx makes several important remarks in volume I about the role of commodity exchange in the becoming of capital, such as:

> Commodity exchange begins . . . at points of contact with foreign communities or their members. But *as soon as things have become commodities for the outside world, they turn into commodities inside a community as well*, due to a rebound effect. At first, chance alone determines the quantitative ratios at which commodities are exchanged. Commodities can be exchanged because of an act of will: Their owners agree to dispose of them reciprocally. In the meantime, *people gradually come to rely on use-objects produced by others*. Constant repetition makes exchange into a normal social process. And so, in time, at least some labor products must be created with exchange as their purpose, and there is a hardening of the distinction between the usefulness things have for meeting immediate wants and needs and their usefulness for exchange.[31]

Marx describes this process—an element of capital's becoming—in abstract terms. Historically, this process, whereby proletarians come to be more dependent upon markets for the goods that meet their essential needs, and whereby production becomes oriented toward exchange-value, involves violence. One well-known strategy of early capitalist empires (for instance, the British empire) was to create dependent markets, in which worker consumers had no choice but to purchase heavily taxed (British) goods. This aspect of colonialism was emphasized by Hegel—as well as Rosa Luxemburg, who recognized that the absence of a sufficiently large

---

realized. We should therefore meet our students as they are—consumer-proletarians inundated by marketing, surveillance, and debt.

31 Marx, *Capital*, 63–4, my italics.

domestic consumer class had generated problems in the emergence of capitalism, problems that led capital to try to find a solution *outside* the domestic market, through colonialism.[32] Such processes are sometimes conflated with so-called original accumulation, since it is through the separation of laborers from the means of labor (land) that they become dependent upon purchasing commodities (means to reproduce their labor power). Strictly speaking, the two processes are distinct and may play out in different rhythms. But the two processes are causally linked in another fashion. One way that peasants and small producers in precapitalist societies may be driven out of business is through shifts—often abrupt—in consumer demand: Away from locally produced foods and craft products, toward consumption of cheap commodities. In Luxemburg's words:

> Capitalism must . . . always and everywhere fight a battle of annihilation against every historical form of natural economy that it encounters, whether this is slave economy, feudalism, primitive communism, or patriarchal peasant economy. The principal methods in this struggle are political force (revolution, war), oppressive taxation by the state, and *cheap goods*.[33]

Our planetary crisis is widely experienced today by subaltern social groups through some combination of these same "principal methods in this struggle." Mainstream economics treats cheap goods as a boon for consumers. To confront the climate crisis we will need a more critical standpoint for evaluating the costs and consequences of cheap goods.

A brief assessment: The third approach has received the least emphasis in the Marxist historical literature. Research from environmental history and other fields, abutted by a natural-historical reading of *Capital*, provides strong support for the claim that the production of this new, world-encompassing social class, the proletariat qua worker-consumers, required the reorganization of HH×HN. The emergence of the proletariat was a major event in Earth's natural history.

---

32 Cf. G. W. F. Hegel, *Elements of the Philosophy of Right* (Cambridge, UK: Cambridge University Press, 1991); Rosa Luxemburg, *The Accumulation of Capital* (New York: Routledge, 2003 [1913]).

33 Luxemburg, *The Accumulation of Capital*, 369, my italics.

Bringing Capitalism Back Down to Earth

This chapter opened with the question, Is there any sense in which we can say that the emergence of capitalism was natural? We can now clarify that the answer must be answered affirmatively—but that its naturalness is *historical*. The specific ways that capital shapes the human relationship with Earth are inseparable from the historical processes of its becoming. This implies that the geographical variations in human relations with nature have played a role in the emergence of capitalism. Such thinking is often rejected by Marxists, but Marx makes several statements in *Capital* to emphasize the importance of geographical and environmental factors in the emergence of capitalism. For instance: A capitalist form of society, Marx argues,

> *presupposes that human beings have come to rule over nature*, and where nature lavishes its gifts upon a person too freely, "it guides him by the hand, like a child on leading strings." There, in other words, human development isn't a natural necessity. Thus *a temperate climate*—not the tropics, with their lush vegetation—*is capital's motherland*. The natural foundation for the social division of labor is the variety of the soil, or the diversity of its natural products, not its absolute fertility. For when people live in varying natural conditions, this motivates them to multiply their wants and needs, their skills, and the means and methods of their labor, while the need to collectively control a natural force and thereby work more efficiently, the need to appropriate or tame such a force by applying human labor on a large scale, has played a decisive role in the history of industry.[34]

I read this as a restatement of Hegel's claim, except Marx substitutes capital for spirit. Though both explanations hinge upon geography, Marx's claim has the virtue of historicizing human-environment

---

34 Marx, *Capital*, 470, my italics. Citing the Greeks, Hegel made a similar Eurasianist argument apropos human history: "The true theatre of History is therefore the temperate zone; or, rather, its northern half, because the earth there presents itself in a continental form, and has a broad breast, as the Greeks say": G. W. F. Hegel, *The Philosophy of History* (New York: Dover, 1956), 80.

relations. In the same chapter, Marx argues that "labor's productivity depends on natural conditions, which can all be traced back to the nature of human beings, such as their race, and to the natural world around them."[35] This sentence makes little sense unless read in light of *Origin*.

Marx's arguments about the evolution of human productivity and the specific role of geography in the formation of capitalist social relations remain speculative and incomplete. Nonetheless, they provide us with some clues about his standpoint. He emphasizes the importance of natural variation in geographical conditions, while acknowledging that these are not sufficient in themselves to explain capital's emergence:

> Favorable natural conditions bring about only the possibility—never the reality—of surplus labor and thus also surplus-value or surplus product. Labor's different natural conditions are responsible for the fact that the same amount of labor goes farther toward satisfying wants and needs in some countries than in others—they cause the necessary labor-time to vary from place to place when all other conditions are the same. But natural conditions affect surplus labor only as a natural limit that determines the point where labor for others can begin, and this limit recedes in proportion to the advance of industry.[36]

While I recognize that this claim must remain conjectural, I believe that we have strong grounds to claim that Marx's line of reasoning here was directly inspired by *Origin*. After all, Darwin not only destroyed teleological reasoning about the direction of the development of species—with the logical corollary that "stadial" (stage-of-history) theories of historical change are highly suspect—but he accomplished this in part by emphasizing geographically induced natural variations. Darwin's barnacles, finches, tortoises, and worms vary in part because of environmental variation; reproductive isolation (allopatric speciation), Darwin recognized, is one process through which new species emerge. In *Capital*, Marx transposes this lesson to human "labor's

---

35 Marx, *Capital*, 469.
36 Marx, *Capital*, 471.

different natural conditions" as a factor critical for explaining the emergence of capitalism as a social formation.

Although it may be uncomfortable for some Marxists to admit, the evidence of Darwinian inspiration is across the pages of *Capital*. Recall (from chapter 2) that, since Darwin could not directly *demonstrate* the descent of species via natural selection—the process takes too long—he used three sorts of empirical evidence to *illustrate* his theory: Artificial selection, paleontology, and biogeography. We have now seen that Marx also employs all three in *Capital*. I leave it to the reader to decide whether to call this inspiration or imitation. Ultimately, what matters to me is not the provenance of Marx's ideas, but how we can use these ideas to interpret our world and change it for the better.

Once we open the door to acknowledging that "labor's different natural conditions" was a factor critical for explaining the emergence of capitalism as a social formation, a lot of other animals can run out of the barn. Does the presence of certain crops explain the rise of capitalism? Is the continental breadth of Eurasia the key factor, since it permitted movement of seeds, animals, and cultural practices across relatively homogenous (latitudinally bound) climate zones, unlike Africa, the Americas, and Australia?[37] Or perhaps climate itself explains the rise of capitalism?

For some time, Marxists have dismissed these questions by noting, rightly, that this path of inquiry may lead to geographical or environmental determinism ("things are the way they are because of geography and nature"). History shows that such determinism usually results in political conclusions which are bad for subaltern social groups. Whatever the merits of the previous generations' ways of answering these questions, the climate crisis is prompting old questions anew: How do people respond to environmental change? What are the likely political-economic consequences of rapid climate change? Looking to the past for insights, many are asking: What were the consequences of earlier periods of climate change for social, political and economic life? What might these histories teach us about the present planetary crisis?

---

37   The answer is no: Angela Chira, Russell Gray, and Carlos Botero, "Geography is not destiny: A quantitative test of Diamond's axis of orientation hypothesis," *Evolutionary Human Sciences* 6 (2024): E5.

The urgency of this line of questioning goes some way to explain the renaissance since the 1990s in the field of environmental history.[38] For Marxian natural history, the most valuable subfield in this genre concerns the social, climatic, and environmental "catastrophe" of the long seventeenth century.[39] This literature has demonstrated that global climate change was a critical factor in the world-historical transformation we retrospectively call the birth of capitalist modernity. The promise of a research program that bridges environmental history with Marxist analysis of capitalism could not be greater—but, alas, this promise has been largely unfulfilled.[40]

Earlier we saw that the birth of capitalism is usually described as simple matter of liberation: As soon as people were freed from the shackles of feudalism, capitalism assumed its natural course as the logical form of human existence. But what if, instead, the capitalist form of society

---

38 Since William Cronon published *Nature's Metropolis: Chicago and the Great West* (New York: W. W. Norton & Co., 1991), the field has grown so rapidly (and extended beyond provincial US and European borders) that even specialists struggle to stay on top of the literature. New programs and research centers have emerged, along with new classes, teaching books like Hubert Lamb, *Climate, History, and the Modern World* (New York: Routledge, 2002); John McNeill, *Something New Under the Sun: An Environmental History of the Twentieth-Century World* (New York: W. W. Norton & Co., 2001); Steven Mithen, *After the Ice: A Global Human History, 20,000–5000 BC* (Cambridge, MA: Harvard University Press, 2006); and Neil Roberts, *The Holocene: An Environmental History* (Oxford: John Wiley and Sons, 2014); also, programmatic statements by Edmund Russell, *Evolutionary History: Uniting History and Biology to Understand Life on Earth* (Cambridge, UK: Cambridge University Press, 2011). None are remotely Marxist. To generalize, environmental historians provide useful empirical accounts of specific places, firms, commodities, species, or problems, but on the basis of a vague and liberal conception of capitalism.

39 I take "catastrophe" from the subtitle of Geoffrey Parker's *Global Crisis: War, Climate Change, and Catastrophe in the Seventeenth Century* (New Haven: Yale University Press, 2013), the pinnacle of this subgenre.

40 Indications of such promise are shown by the work of Jason Moore and Andreas Malm: For Moore, see "The Modern World-System as environmental history?" in *Theory and Society*; "The Capitalocene, part I: On the nature and origins of our ecological crisis," *Journal of Peasant Studies* 44, no. 3 (2017): 594–630; *Capitalism in the Web of Life: Ecology and the Accumulation of Capital* (New York: Verso Books, 2015); for Malm, *Fossil Capital: The Rise of Steam Power and the Roots of Global Warming* (London: Verso Books, 2016); *The Progress of this Storm: Nature and Society in a Warming World* (London: Verso Books, 2018). There are important philosophical differences between these two (for example, Moore's ontological commitment to monism distinguishes his project from most ecological Marxists, including my own: For an efficient critique, see Malm's *Progress*). Nevertheless, I admire and appreciate both projects.

was basically the result of chance? What if it emerged when some humans, amidst serious socio-natural crisis, trying to survive under difficult conditions, produced a new form of existence unconsciously and unintentionally? I recognize that my questions are leading and speculative. That is intentional. For, while I cannot develop the argument fully here, the hypothesis we must examine is that capitalism's emergence as a social formation was essentially an accident in Earth's natural history, one in which climate change was a contributing factor—an accident with potentially fatal consequences for our species, and which we therefore need to correct as soon as possible.[41]

Previous Marxists rejected such thinking by appeal to *Capital*, but as I have argued, the text solicits an aleatory natural-historical perspective. Let us pick up again where Marx is discussing nature and that central category of his critique, surplus value. Not only on the production (supply) side but also with respect to consumption (demand), Marx insists that "*surplus-value has a natural foundation*," insofar as humans are natural beings and the development of our capacity to labor on Earth is a process of natural history; but, qualifies Marx, this process "is natural only in the very general sense that no absolute natural barrier prevents a person from saddling someone else with the labor his own existence requires. It would be quite wrong to see a mystical something in this spontaneously arising productivity."[42] No, there is nothing mystical about human society adopting a capitalist form; it was an event in our natural history, spurred by the development of labor's productive power, the evolution of people's wants and needs, and (this part is still not entirely clear) variable natural and geographical conditions, producing a shift in the specific form of organization human social relations:

> Only once human beings have lifted themselves out of their earliest animal state, and their labor has become social to some extent, do relations emerge whereby the surplus labor one person performs becomes the condition of another's existence. Labor's acquired

---

41 To put it this way removes much of the agency that is normally attributed to the great bourgeois leaders and thinkers who supposedly led humanity into the capitalist era. Perfect.

42 Marx, *Capital*, 468, my italics. On this theme, see Nancy Fraser, *Cannibal Capitalism*, chapter 4.

productive power is still meager at the dawn of civilization, but so are the wants and needs that develop along with—and out of—the very means through which they are satisfied.[43]

There is a commonsense view that all people basically need the same things. While perhaps true at a high level of abstraction, once we bring it down to Earth, complications arise.

As a human society comes to be organized by capital (M-C-M'), *demand is induced* among worker-consumers. For value to be realized, capital needs workers to buy the commodities they produce, and individual capitalists will stop at little to stimulate expanded consumption of the commodity they have to sell. In simple commodity production (CMC), the producers are not indifferent to the particularities of the commodities that are produced and consumed. By contrast, under capitalist production, M-C-M', production of *any* commodity will do, so long as it may be sold at a decent profit. Similarly, a capitalist trader—a merchant—will buy and sell *any* commodity for which it can buy low and sell a little higher, typically by moving commodities between geographical regions where market conditions and the value compositions of commodities differ. Suppose that the value of x tons of maize = y AR-15s = z kg of fentanyl. If I can acquire x tons of maize and y AR-15s in one market (say, the US) and exchange them in another (say, Mexico) for 8z kg of fentanyl, which I then manage to bring into the US and sell, then I will have accumulated value without producing a commodity.[44]

I do not mean to suggest that *merchant* capital is the problem;

---

43 Marx, *Capital*, 468–9.

44 To change caliber: "Value is the tacit but astoundingly powerful relation that enables us to make an AK-47, an SUV and a field of grain exchangeable as 'equivalents': E.g., 100 AK-47s = 10 SUVs = 1 field of wheat ... My example is in no way absurd; indeed, the political economy underlying much of the current land-grab in Africa consists in exactly this exchange of equivalents: Arms, elite luxury goods, and agricultural land": Geoff Mann, *Disassembly Required: A Field Guide to Actually Existing Capitalism* (Oakland: AK Press, 2013), 30 (cf. Geoff Mann, "Give your mom a gun," *London Review of Books* 46, no. 5, March 7, 2024). The becoming of capitalism in western Africa entailed just such an "exchange of equivalents": Arms + elite luxury goods for enslaved human beings. See Walter Rodney, "African slavery and other forms of social oppression on the Upper Guinea coast in the context of the Atlantic slave-trade," *The Journal of African History* 7, no. 3 (1966): 431–43; and A. G. Hopkins, *An Economic History of West Africa* (London: Routledge, 2019). No coherent natural history of our world could be written without such equations.

A Natural History of Capitalism

productive capital also must prioritize profitability. Consider the most important commodity produced today: Electricity. Electricity is still largely produced from coal, gas, and oil—not because it is cheaper or better than solar and wind power (it is neither), but because doing so is more *profitable*.[45]

All this has major implications for the wants and needs of ordinary people. Whatever we may think we want out of life, the fact is that most people need to sell their labor power as a commodity in a market that is driven by accumulation of value. The great historical achievement of the capitalist form of society has been to create

> a systematic and seemingly natural set of social relations that uses labour not to produce useful or beautiful things for the good they provide—if it did, then the "value" of those three things [maize, AR-15s, and fentanyl] could never be rendered equivalent. Instead, capitalism condemns labour to produce "value" in this specifically capitalist sense.[46]

The capitalist form of society also condemns working people to take the wages they earn and make a life from what can be bought. The ability to purchase "anything that one desires" is limited by the quantum of the wage earned but also by the limitations of desire—even desire has a limit—and, crucially, the distortions of human existence inherent to a competitive social form. To put it in standard economic terms: On the supply side, capitalist relations of production and exchange induce general social indifference to the production of use-values; on the demand side, they induce general social indifference to the satisfaction of collective needs. Consider food. There is sufficient food in our world to feed approximately 11 billion human beings, but around 1 billion (out of a population of almost 8.5 billion) experience chronic hunger. It is difficult not to become numb to injustices which are logical and necessary consequences of the prevailing social formation.

The emergence of capitalist social relations therefore constitutes an epochal transformation in the human social relationship with Earth.

---

45 Christophers, *The Price Is Wrong*.
46 Mann, *Disassembly Required*, 30.

In all previous modes of production, the specificity of use-value was inseparable from the particularities of production on land. And when goods were traded in precapitalist social formations, the process came to an end with the consumption of the traded goods. The aim was to obtain use-value. By contrast, the circulation of commodities in capitalism is the outward expression of the relentless accumulation of value: The process begins and ends with the same thing, money (M-C-M'):

> A commodity is the starting point in the C-M-C circuit, whose endpoint is a second commodity. The second commodity falls out of circulation and is consumed. C-M-C's ultimate purpose is consumption or the satisfaction of wants and needs—in a word, use-value. In contrast, money is both the starting point and endpoint in the M-C-M' circuit. The motive that drives [M-C-M'] and the goal that defines it is exchange-value.[47]

The alpha and omega of the capitalist form of society "have the same economic form: Both are money."[48] The expansion of capital, this relentless pursuit of surplus value as money, knows no *logical* limit.

Whether there are *social* and *ecological* limits is another question. Natural scientists have sounded the alarm that capitalist society is breaking through the planetary boundaries within which our species evolved.[49] To provide a sense of where we stand today in Earth's natural history, consider the following facts. In April 2025, the concentration of CO_2 in the atmosphere reached 430 ppm. The last time that Earth's atmosphere had this much CO_2 was around 3 million years ago (long before *Homo sapiens* existed), during the Pliocene, when the world was much warmer than at present, sea levels much higher.[50] The

---

47 Marx, *Capital*, 124.
48 Marx, *Capital*, 124.
49 Will Steffen et al., "Planetary boundaries: Guiding human development on a changing planet," *Science* 347, no. 6223, DOI 1259855; Katherine Richardson et al., "Earth beyond six of nine planetary boundaries," *Science Advances* 9, no. 37 (2023): Eadh2458.
50 The mass melting of ice of that period—dubbed the "Pliocene deglacial event"—provides Earth scientists with a means to estimate likely sea-level rise given present atmospheric concentrations of greenhouse gases: Rachel Bertram et al.,

A Natural History of Capitalism 225

burning of fossil fuels has increased average temperature by more than 1.5°C several years before the IPCC models had expected the breaking of this limit agreed to in Paris.[51] The warming already generated by the increase in CO_2 concentrations has caused enough ice to melt at the poles—shifting water mass toward the equator—that Earth's axial tilt (obliquity) has changed measurably.[52] The warming has also changed the dynamics of every ecosystem in ways that are pushing more and more humans outside of the natural-historical envelope within which we evolved.[53] The world's wildlife populations are crashing (a 73% decline between 1970 and 2020)[54] and under the most likely warming scenario, vegetation will be radically altered everywhere, increasing stress on all remaining wildlife.[55] In around 2020, total anthropogenic

---

"Pliocene deglacial event timelines and the biogeochemical response offshore Wilkes Subglacial Basin, East Antarctica," *Earth and Planetary Science Letters* 494 (2018): 109–16.

51 To be precise: In 2023, mean global temperature was 1.54°C above the 1850–1900 mean: Robert Rohde, "Global Temperature Report for 2023," Berkeley Earth, January 12, 2024, at berkeleyearth.org. This book was written during a year-long period, August 2023 to August 2024, when *every consecutive month* broke the previous record for global mean temperature for that month. For an analysis of this conjuncture, see Andreas Malm and Wim Carton, *Overshoot* (London: Verso Books, 2025).

52 Over a twenty-year period (2003–23) the Greenland ice sheet lost 4,352 gigatons of ice: Shfaqat Khan et al. "Smoothed monthly Greenland ice sheet elevation changes during 2003–2023," *Earth System Science Data Discussions* (2024): 1–34. Recently, Greenland's ice is melting and flowing away at a rate of *30m tons per hour*: C. Greene et al., "Ubiquitous acceleration in Greenland ice sheet calving from 1985 to 2022," *Nature* 625, no. 7995 (2024): 523–8. On obliquity and polar drift, see Shanshan Deng et al., "Polar drift in the 1990s explained by terrestrial water storage changes," *Geophysical Research Letters* 48, no. 7 (2021): E2020GL092114.

53 Chi Xu et al., "Future of the human climate niche," *Proceedings of the National Academy of Sciences* 117, no. 21 (2020): 11350–55. Not only humans: By transforming the surface of the earth—directly, for example, by converting forests to maizefields, and indirectly, for example, by burning fossil fuels and heating the planet—we have degraded the habitats of undomesticated mammals to such an extent that they are in sharp decline, if not an extinction spiral: Lior Greenspoon et al., "The global biomass of wild mammals," *Proceedings of the National Academy of Sciences* 120, no. 10 (2023): E2204892120; John Damuth, "Wild mammals through the lens of biomass rather than biodiversity," *Proceedings of the National Academy of Sciences* 120, no. 1 (2023): E2301652120.

54 The Zoological Society of London, "2024 Living Planet Index": Zsl.org.

55 T. Conradi et al., "Reassessment of the risks of climate change for terrestrial ecosystems," *Nature, Ecology, and Evolution* 8 (2024): 888–900.

mass—concrete, commodities, and so on—surpassed the sum of all living biomass.[56] Earth today: Like the Pliocene, but covered in plastic.

One could go on and on: The natural sciences supply us with an endless supply of dizzying studies documenting the transformation of Earth in our time. The vital question is: *Why* are these natural-historical changes occurring all at once? There are remarkably few theories available to answer this question. The neo-Malthusian approach answers: "Because there are too many people." A humanistic, techno-critical approach says that the problems are caused by the human drive to master nature.[57] While both approaches can be used to make plausible claims, neither has, nor could, produced a coherent and effective answer to our climate crisis. Only an approach centered on the capitalist form of society can explain the underlying cause and orient us toward a just future.

As a socio-ecological form of organizing humanity, capitalism has definite limits. These can be clarified by considering the three processes inherent to capitalist society and their implications for HH×HN (see Table 6.2). Each of these processes shapes people's responses to ecological problems created by capital and introduces possibilities of economic crisis. Consider so-called original accumulation: The historical becoming of capital was a process driven by the creation of *new* proletarian-consumers, mainly by separating billions of peasants (independent producers) from their lands. Through this process, so-called original accumulation converts humanity into proletarian-consumers and Earth into a pile of commodities.[58] Then, what happens when there are no more peasants to make into proletarian-consumers? Capital's expansion reaches a limit.

---

56 Emily Elhacham et al., "Global human-made mass exceeds all living biomass," *Nature* 588, no. 7838 (2020): 442–4.

57 Cf. Martin Heidegger, *The Question Concerning Technology* (New York: Harper, 1977 [1954]); Andrew Feenberg, *Questioning Technology* (New York: Routledge, 2012); Pope Francis, *Laudato Si'* (*encyclical*), 2015, chapter 3, passim. To its credit, and unlike Heidegger, *Laudato Si'* also provides a critique of capitalism.

58 For a lucid discussion of the ecological consequences of so-called original accumulation, see Kōhei Saitō "Primitive accumulation as the cause of economic and ecological disaster," in ed. M. Musto, *Rethinking Alternatives with Marx: Economy, Ecology and Migration* (London: Springer, 2021), 93–112.

A Natural History of Capitalism

Table 6.2. Three processes of capital: HH×HN and limits to capital

	Process	HH×HN	Limits to capital (i.e., crisis)
I	So-called original accumulation.	Separation of people from the land (Marx: "the separation of people from nature is a historical process"); conversion of Earth into means of production and a pile of commodities.	The "completion" of so-called original accumulation means that there are no more large pools of humans in precapitalist communities yet to be proletarianized, threatening expansion of capital relation.
II	Market competition.	Competition among producers ("market forces") drives expanded production and the conversion of Earth into commodities, causing harm to most species (including our own).	Environmental changes wrought by capitalism (e.g., global heating) threaten the interruption of capitalism as a socio-natural formation.
III	Mass consumption.	Dependence of proletarian lives upon commodity consumption (coupled with capital's dependence upon the same for valorization) leads to the generation of *waste*: For example, CO_2 in atmosphere and plastic in oceans.	If commodities are not sold, their value is not realized; lack of sales (due to low wages, unemployment, or boycott) may cause crisis of value realization.

The break between humans and nature did not begin with capital. Nonetheless, it takes on new qualities and greater intensity as so-called original accumulation transpires:

> It wasn't enough that the things workers needed in order to work, and the things they worked with, were gathered as capital on one side of the capital relation, while people who had only their own labor-power to sell appeared on the other side. It also wasn't enough to force those people to voluntarily sell themselves. What developed as capitalist production advanced was a working class whose members see the demands of that mode of production as self-evident natural laws, having been brought up to do so and also as a result of tradition and habituation. Once the capitalist production

process had become highly organized, it broke all the resistance it encountered. The continuous generation of a relative surplus population kept the law of labor's supply and demand, and thus wages, on a track that fit with capital's valorization requirements. And the silent force of economic relations sealed the capitalist's domination over his workers. Direct extra-economic violence is still used, of course, but only in exceptional cases. *For the most part, the capitalist can entrust the worker to the "natural laws of production," i.e., count on the fact that the worker is dependent on capital, something that arises from the conditions of production themselves and is guaranteed and perpetuated by them. Not so during the historical genesis of capitalist production.*[59]

The becoming of capital is therefore in part a process of creating dependent human beings who genuinely believe in capital's own logic as "self-evident natural laws." As our social form becomes capitalist, the human being is transformed into a bearer of capitalist ideas.[60]

Marx's discussion of so-called original accumulation is therefore not only a historical analysis and a critique of capitalism. It can also be

---

59   Marx, *Capital*, 670, my italics.
60   I emphasize *social form* because Marx implies that social formations became capitalist well after the capitalist *mode of production* came into existence. To quote Tony Burns:

> Marx argues that although it is true that capitalism as a mode of production first emerged in Europe in the late medieval period, as early as the fourteenth century, nevertheless this is not true of capitalism as a social formation. For the first social formation which might properly be described as capitalist only came into existence later, after the capitalist mode of production had "taken hold" in medieval society, had expanded, developed and (eventually) become dominant. As Marx puts it in the *Grundrisse*, "although we come across the first beginnings of capitalist production as early as the fourteenth or fifteenth century, sporadically, in certain towns of the Mediterranean," nevertheless, "the capitalistic era" proper, containing the first capitalist social formation, in the strict sense of the term, "dates from the sixteenth century." Marx suggests that this transition first took place in sixteenth-century England and that it was associated with the development of the capitalist mode of production in agriculture there, especially but not only having to do with the trade in wool. Paradoxical though it might seem, a logical implication of these remarks is that in Marx's view, the capitalist mode of production may be present in social formations which are not capitalist (Burns, "Marx and the concept of a social formation," *Historical Materialism* 32, no. 3 [2024]: 158–87, 174).

A Natural History of Capitalism

read as an allegory about how to overcome capital and its ecological contradictions: By re-pairing of human society with the earth and the liberation of our relations and habits from capitalism. As Kōhei Saitō explains:

> Marx consistently argued that it is necessary to re-establish the "original unity" in the future society beyond this alienation from nature: "The original unity can be reestablished only on the material foundation which capital creates and by means of the revolutions which, in the process of this creation, the working class and the whole society undergo."[61] Indeed, Marx's famous remark on the "negation of the negation" in volume 1 of *Capital* corresponds to this reestablishment of the "original unity" as the process of overcoming the separation in the metabolic exchange between humans and nature.[62]

In chapter 2, we saw that Darwin's breakthrough was to grasp the evolution of species in terms of an undirected transformation of already-existing species. There is some basis for discussing changes in human societies in an evolutionary sense. The traces of this style of thinking are exhibited by Marx in his late writings. Yet, such thinking was largely repressed in the early twentieth century. In lieu of a natural-historical approach to capitalism, Marxism-Leninism converted Marx's writings into a catechism of revolutionary statism. This required repression of the analysis of the *becoming* of capitalism. If there is no real account of capitalism's becoming—if you think it did not exist until "so-called original accumulation," and then it did, *tout court*—it follows logically that every society must pass through that stage to get to communism: Needs must, however painful. Such thinking led to the horrendous errors whereby ostensibly communist states carried out original accumulation to create ostensibly communist societies. In the long run, such violence only generated capitalism by circuitous routes. Can we imagine another end?

---

61 Saitō cites Marx's *Economic Manuscript of 1861–1863*, MECW 33: 340, a manuscript written after Marx read Darwin.
62 Kōhei Saitō, "Primitive accumulation as the cause of economic and ecological disaster," 96.

From Stages to Multilinear Change

In Part II, we saw that Marx's organization of *Capital* and discussion of some of its core themes changed as a consequence of studying Darwin's *Origin*. Something similar could be said for the way he approaches the question of the emergence of capitalism, which we can detect by again comparing his pre-1860 economic writings, such as *Grundrisse* (1857–8) and *Contribution to a Critique of Political Economy* (1859) with *Capital*. In a word, Marx abandons the stadial form of his narrative of capitalism's emergence. Marx's abandonment of any sort of "mechanical succession of stages" in human history follows from the encounter with Darwin, whose theory undermines any sense of evolutionary progression.[63] The Darwinian framework treats the "origin" of a species not as an act of creation (being) but a process of emergence from already existing species (becoming).

The emergence of capitalism in the Grundrisse

In the *Grundrisse*, Marx begins his discussion of the emergence of capitalist social relations in Europe with the following remark:[64]

> One of the prerequisites of wage labor and one of the historic conditions for capital is free labor and the exchange of free labor for money, in order to reproduce money and to valorise it [. . .] Another prerequisite is the separation of free labor from the objective conditions of its realization — from the means and material of labor. This means

---

63 Maurice Bloch, *Marxism and Anthropology* (London: Routledge, 2004 [1983]), 66. Bloch presents a nuanced view of Marx's approach to history, but treats the late work (after *Grundrisse*) sparingly.

64 This subsection reproduces, with modifications, material from an earlier work: Joel Wainwright, "Uneven developments: From the *Grundrisse* to *Capital*," in *In Marx's Laboratory*, eds. Riccardo Bellofiore, Guido Starosta, and Peter Thomas (Leiden: Brill, 2013), 371–91 (a modified version of an earlier research paper published in *Antipode*). I thank the editors for their support and criticism. On the *Grundrisse* and *Capital* and multilinear developments, see also Anderson, *Marx at the Margins*, chapter 5).

above all that the workers must be separated from the land, which functions as his natural workshop.[65]

Marx thus posits two necessary conditions for the emergence of capitalist social relations: The exchange of "free" labor (neither slave nor serf) and the separation of these laborers from the means of labor, particularly the land. Marx elaborates by identifying the conditions needed to free the worker as "objectless, purely subjective labour capacity confronting the objective conditions of production as his *not-property*," that is, a person with nothing to sell but his labor, his own life. In a passage replete with dizzying Hegelian terminology, Marx summarizes the emergence of capitalist social relations as a set of presuppositions and dissolutions:

> A process of history which dissolves the various forms in which the worker is a proprietor . . . *Dissolution* of the relation to the earth—land and soil—as natural condition of production—to which he relates as his own inorganic being . . . *Dissolution* of the relations in which he appears as *proprietor of the instrument*. Just as the above landed property presupposes a *real community*, so does this property of the worker in the instrument presuppose a particular form of the development of manufactures, namely *craft, artisan work* . . . *Dissolution* . . . at the same time of the relations in which the *workers themselves*, the *living labour capacities* themselves, still belong *directly among the objective conditions of production*, and are appropriated as such—i.e. are slaves or serfs.[66]

Capitalist social relations emerge not as a pre-formed, external totality but come into existence through—Marx italicizes the word thrice—the *dissolution* of older social relations. Capitalist social relations emerge therefore in a way that is both whole—since the essence of capitalism is the hiring of labor as a commodity, which happens at the "beginning" of

---

65   Marx, *Grundrisse*, Wangermann trans., MECW 28, 399.
66   Marx, 497–8. Elsewhere in the *Grundrisse*, Marx explains that the fundamental quality of capitalism as a social relation lies in the "exchange of living labour for objectified labour—i.e., the positing of social labor in the form of the contradiction of capital and wage labour—is the ultimate development of the *value relation* and of production resting on value," Marx, *Grundrisse*, Nicolaus trans. (1973), 704.

capitalism—but also profoundly incomplete, since capitalist social relations must reproduce themselves elsewhere and beyond an initial purchase of labor. It takes time for everything to dissolve, so to speak. Earlier in the *Grundrisse*, Marx insists that capitalist relations of production "do not develop out of *nothing*," nor do they emerge "from the womb of the Idea positing itself" as for Hegel. No, capitalist social relations emerge

> within and in antithesis to the existing development of production and the inherited, traditional relations of property. While in the completed bourgeois system each economic relation presupposes every other in a bourgeois economic form, and everything posited is thus also a presupposition, that is the case with every organic system. This organic system itself, as a totality, has its presuppositions, and its development to its totality consists precisely in subordinating all elements of society to itself, or in creating ... the organs it still lacks. This is historically how it becomes a totality. The process of becoming this totality [constitutes] a moment of its process, of its development.[67]

The emergence of capitalist social relations—not all at once, but by positing relations that are then taken as premises for advance—is the counterpart to capitalism *becoming* totality, a process that is never complete.[68] In the paragraph immediately following the passage I just cited, Marx turns his attention to the moment when capitalist social relations encounter non-capitalist relations through geographical diffusion:

> if within one society the modern relations of production, i.e. capital, are developed in their totality, and this society then takes possession of a new terrain, as e.g. the colonies, it finds, or rather its representative, the capitalist, finds, that his capital ceases to be capital without

---

67 Marx, *Grundrisse*, Nicolaus trans., 278. Apropos the final word, we should read this in light of Marx's warning that any use of the concept "development" would require the destruction of its association with the concept of *progress*.

68 In the *Grundrisse*, there is nothing that resembles what most contemporary scholars of development studies call "capitalist development." There is only the becoming of capital and the development of capitalist social relations.

wage labor, and that one of the presuppositions of wage labor is not only landed property in general but modern landed property; landed property which, as capitalized rent, is expensive and which, as such, excludes the direct use of the soil by individuals. [Therefore] Wakefield's theory of colonization ... is immensely important for a correct understanding of modern landed property.[69]

Capitalism emerges through dissolution of precapitalist social relations *in Europe*, from whence it flows outward—a dynamic solvent for transforming precapitalist relations elsewhere.

In a famous passage in the *Grundrisse*, Marx asserts that the relations of production under capitalism are the most complex of any society: "Bourgeois society is the most developed and many-faceted historical organization of production."[70] In this passage, Marx argues that capitalist society is in some sense more complex than previous forms of society (I agree) and that we must study the completed structure of capitalist society to understand its formation (ditto); but he further claims that capitalist society is the "most developed" form (a dubious proposition), one that provides a key to all previous social formations generally (a baseless claim). To make the last two points, Marx provides an evolutionary metaphor: "The anatomy of man is a key to the anatomy of the ape ... Bourgeois economy thus provides a key to that of antiquity."[71] Marx's metaphor presupposes that "man" is the higher—completed—form of the "ape." Such thinking is totally incompatible with Darwin's framework. Recall Darwin's sketch of the tree of speciation from chapter 2: There are no linear developments; no species is ever complete. The teleological style of reasoning reflected in Marx's metaphor contributed to the Eurocentric, stadial thinking about the history of capitalism later enshrined as dialectical materialism and which generated flawed political strategies in the twentieth century. Which explains why, after reading Darwin, Marx left it out of *Capital*.

---

69   Marx, *Grundrisse*, Nicolaus trans., translation lightly modified.
70   Marx, *Grundrisse*, Wangermann trans., 42.
71   Marx, *Grundrisse*, Wangermann trans., 42. Recall that Marx wrote *Capital* during the era of the London "gorilla wars."

The emergence of capitalism in Capital

In an insightful essay on Marx's discussion of "forms which precede capitalism," Ellen Wood notes that, notwithstanding scattered references to "original accumulation," in the *Grundrisse* Marx did not seek to explain the transition from feudalism to capitalism as such. "His objective is rather to highlight the specificity of capitalism in contrast to earlier forms of property and labour."[72] Wood elaborates on the shift we find between *Grundrisse* and *Capital*. At the time of writing *Grundrisse*, Wood observes, Marx "has not yet entirely broken with the most common question-begging accounts of how capitalism originated." Marx treats the origin of capitalism largely as "a matter of allowing its already existing elements to grow" (reflecting Marx's Hegelian roots). By contrast, after reading Darwin, in *Capital* we find Marx "hinting at a very different explanation" for the emergence of capital: By 1867, Marx sought "the source of the transition not in the 'interstices' of feudalism but rather in [capital's] own internal dynamics, in its own constitutive property relations, which gave rise to an authentic social transformation."[73] When he wrote the *Grundrisse*, Marx knew that these constitutive property relations required the separation of living labor from the means of production. The shift that Wood identifies therefore emerges when Marx changes the way he conceives the separation of labor from the means of production in relation to capital's becoming.

The question of precapitalist formations returns in *Capital* in two ways, neither as extensive nor speculative as in the *Grundrisse*. In part 5 of the third volume of *Capital*, best known for its analysis of the role of finance in the production process, we find a chapter on "precapitalist relations" in which Marx argues that the transition from precapitalist relations into capitalism was made possible by two "antediluvian forms of capital" that long predate the emergence of capitalism *in toto*: Usurer's capital and merchant's capital.[74] Yet credit cannot take credit for creating capitalism,

---

72 Ellen Meiksins Wood, "Historical materialism in 'Forms Which Precede Capitalist Production,'" in *Karl Marx's Grundrisse: Foundations of the Critique of Political Economy 150 Years Later* (New York: Routledge, 2008), 79–92, 84.

73 Wood, "Historical Materialism," 85. Although I agree with Wood, I note that she does not provide a compelling hypothesis concerning the reason for the shift in Marx's perspective.

74 Marx, *Capital*, volume III (1981), 728–49. In Marx's discussion, credit appears as both the earliest and the highest stages of capitalism. On credit as a reagent in the emergence of capitalism and the dominant expression of its "highest stages," see also Lenin's

since "usury, like trade, exploits a given mode of production but [can]not create it; both relate to the mode of production from outside."[75]

Marx returns to these problems more substantively in part VIII of *Capital*, volume I, on so-called original accumulation, where he sketches the end of capitalism *(Capital's* second conclusion):

> The number of capitalist magnates falls continuously, and the remaining ones monopolize and usurp for themselves all the advantages that this process of transformation holds. Meanwhile, misery increases, as does the amount of pressure, subjugation, degradation, and exploitation inflicted upon the constantly growing working class. But the outrage felt by the members of that class also increases, and they are brought together and are trained and organized by the mechanism of capitalist production itself. Capital's monopoly now shackles the very mode of production that had flourished because of and under it. The concentration of the means of production and the socialization of labor reaches the point where neither process is compatible with its capitalist shell. This bursts, and now the bell tolls for capitalist private property. The expropriators are expropriated.[76]

What was supposed to bring about the end of capitalism is capital itself, or at least conditions created by capital, such as social divisions that reach a point where they become intolerable. Now that capitalism has remade our world and the centralization of the means of production and the socialization of labor have reached unprecedented extremes, one might legitimately ask if Marx was wrong. For, despite the entanglement of all peoples, the mass of misery, and the increasing numbers of proletarians, the expropriators have not yet been expropriated. Without making excuses for Marx, the period of history since *Capital* has demonstrated that capitalism is a pliable social formation, capable of temporarily neutralizing its contradictions. But this does not contradict Marx's identification of capital as the ultimate barrier to capital, and capitalism's contradictions are particularly acute today: See Table 6.2.

Development of Capitalism in Russia (2000 [1899]), chapter 6, section 6: "merchant's and industrial capital in manufacture" and his study of *Imperialism* (1939 [1915]).

75   Marx, *Capital* III (1981), 45.
76   Marx, *Capital*, 691.

After making his big prediction, Marx does something strange. *Capital* does not end with this *political-speculative* conclusion (see Table 6.1), the destruction of capitalism, but with a critique of a theory of colonialism by Edward Gibbon Wakefield (1796–1862). Marx found Wakefield's theory useful to illustrate that capitalism did not arise naturally, even in colonies of capitalist states.[77] The existence of the capital-labor relation is "initially sporadic," but capitalism posits it increasingly: Not only in theory, but also concretely, as is demonstrated by colonialism. Marx returns to this "immensely important" line of thinking later in his notebooks. Here is the key passage on Wakefield in Marx's economic writings before the final chapter of *Capital*:

> The merit of *Wakefield*'s new system of colonization is not that he discovered or promoted the art of colonization, nor that he made any fresh discoveries whatsoever in the field of political economy, but that he naively laid bare the narrow-mindedness of political economy without being clear himself as to the importance of these discoveries ... The point is that in the colonies, particularly in the earliest stages of development, bourgeois relations are not yet fully formed; not yet presupposed, as they are in old established countries. They are in the process of becoming. The conditions of their origin therefore emerge more clearly. It appears that these *economic relations* are neither present by nature, nor are they *things*, which is the way the political economists are rather inclined to view capital.[78]

For Marx, Wakefield demonstrates two intertwined truths: Capitalism is not a system of markets or capital, but an ensemble of social relations; and the becoming (not to say "origin") of capitalist relations can be found in the colonies. Marx's clearest statement in *Capital* that "capital, rather than being a thing, is a social relation between persons that is mediated by things," appears in the final chapter on colonialism—and he attributes this discovery to Wakefield.[79]

---

77 "At earlier stages of production ... an earlier working class may be present sporadically, not however as a *universal* prerequisite of production. The case of *colonies* (see *Wakefield* [. . .]) shows how this relation is itself a product of capitalist production": Marx, *Economic Works 1861–1863*. MECW 30: 74–5.
78 Marx, *Economic Works 1861–1863*. MECW 30: 256–7.
79 Marx, *Capital*, 694.

This attribution is not entirely ironic. Marx and Wakefield understood more clearly than any political economists of their time that colonialism was intended to resolve contradictions engendered by Britain's early advance as an industrial capitalist society. Moreover, like Marx, Wakefield saw in the growing British proletariat a new class that could bring about a political transformation. Yet their interpretations of these facts are fundamentally distinct. Wakefield examines the colonial situation as a would-be statesman, one fearful of the rising proletariat ("with the continuance of discontent and the spread of education amongst the common people, Chartism and socialism will have many a struggle for the mastery over a restricted franchise and private property: And in these struggles I perceive immense danger for everybody").[80] For Wakefield, colonialism can help to stabilize British capitalism, so he advocates reforming the Colonial Office's land policies. In stark contrast, Marx sees the colonial scene as a laboratory within which to examine the emergence of capitalist social relations.[81] Wakefield promotes colonialism to overcome two contradictions—the overaccumulation of capital, and labor strife—by *extending* capitalist social relations to the colonies. Yet these contradictions were not actually explained until Marx wrote *Capital*. Thus, both Wakefield and Marx saw, in their way, that colonialism would help save British capitalism. The difference between them is that only one felt British capitalism worth saving.

Conclusion: Capital and the Historical Metamorphoses of Nature

The becoming-capitalist of human society was an event in Earth's natural history. Everything has changed, even Earth's axis.

In the face of such a world-historical and all-encompassing change—where many different things happened at different times and places—it is less important to establish the correct empirical chronology of events than to obtain a standpoint that allows us genuinely to appreciate the fundamental dimensions of the event. The becoming-capitalist of

---

80 Wakefield, "A View of the art of colonization, with present reference to the British empire," in *Letters Between a Statesman and a Colonist* (Kitchener: Batoche Books, 2001), letter XI (1849).

81 Though Marx praises Wakefield as "the most important political economist of the period" around 1830 (*Capital*, 617), elsewhere he excoriates Wakefield's simplistic method.

human society entailed the conjugation of multiple distinct processes. Each of the three processes that I discussed in this chapter are explicated in *Capital*, though Marx gives priority to the first. Some rebalancing is therefore due. We also need to give greater emphasis to geography and HH×HN. Late in life, Marx took important steps in both respects, but fell well short of completing his natural-historical analysis of capitalism.

What distinguishes capitalism is not that it fully separates humans from nature. Total separation is impossible; we are natural beings living on Earth. Neither is capitalism distinguished (as is often said) by the way it makes people lose concern for environmental issues. Many people in capitalist society feel passionately about the importance of environmental problems, and thanks to the development of the natural sciences, humanity has never had such a complete understanding of Earth's natural history. Rather, capital is distinguished by its inherent drive for expanded accumulation of value that leads to the conversion of everything (including labor and Earth) into commodities. So long as we live in capitalist society, value accumulation will be imperative, commodity relations must expand, and the planetary crisis will worsen.

To date, most ecological Marxists have emphasized two points about capitalism's ecological crisis. First, capitalism's emergence required so-called original accumulation, that is, the separation of people (living laborers) from the means of production (the earth). This "freed" people as dependent proletarians: People condemned to sell labor power as a commodity, earn money, and buy commodities to survive. This process fundamentally changes the human relationship to nature, which becomes a means of production ("natural resources") for commodity production and consumption. Second, following Marx's analysis of metabolism in *Capital*, inspired by the scientific insights of Liebig, ecological Marxists emphasize that capitalism has caused a fundamental break in the metabolic relations of humans and Earth. The so-called "metabolic rift school" has argued that capitalist society has opened "metabolic rifts" within the basic biochemical and ecological cycles of the earth. While it is undeniable that capital's drive modifies and damages our natural environment, these two events, "so-called original accumulation" and "the metabolic break," do not coincide historically. So-called original accumulation is an enabling condition of the emergence of formally capitalist social relations. It has occurred at separate times in different places, but the process only began a few hundred years ago (and continues today). By contrast, science provides evidence of metabolic breaks thousands of years ago. As a rule, they can be found everywhere wherever empires developed urban areas. Consider Marx's example of the spatial separation of human waste from the site of agricultural production. This problem—how to manage human waste and soil fertility—was dealt with by, for instance, the ancient Roman and Chinese empires using elaborate engineering works. In every ancient imperial society, we can find other large-scale illustrations of "metabolic breaks," such as deforestation : see Table 6.3.

In November 1877, a decade after publishing *Capital*, Marx wrote a letter in French to some Russian readers about his analysis of so-called original accumulation in *Capital*.[82] These readers had objected that, if Marx's theory of history in *Capital* was correct, then countries like Russia were condemned to follow a European path through so-called original accumulation into capitalism. Marx challenges this interpretation, arguing that his analysis was limited to Western Europe and that, under certain circumstances, other places could avoid the same fate:

> The chapter on [so-called original] accumulation does not pretend to do more than trace the path by which, *in Western Europe*, the capitalist order of economy emerged from the womb of the feudal order of economy. It therefore describes the historic movement which by divorcing the producers from their means of production converts them into wage earners (proletarians . . .) while it converts into capitalists those who hold the means of production in possession . . . The basis of this whole development is the expropriation of the cultivators. "This has not yet been radically accomplished except in England . . . but all the countries of Western Europe are going through the same movement," etc.[83]

Thus, Marx objects that his critic

> metamorphose[s] my historical sketch of the genesis of capitalism in Western Europe into an historico philosophic theory of the *marche générale* imposed by fate upon every people . . . But I beg his pardon. (He is both honouring and shaming me too much.)

---

82 Marx, Letter from Marx to editor of the *Otecestvenniye Zapisky*, November 1877, in *Marx and Engels Correspondence* (New York: International Publishers, 1968); MECW 24, 196–201. Accessed July 1, 2024, at marxists.org.

83 Here, Marx cites his 1879 French edition of *Capital*, volume I, 315.

Table 6.3. A broader view: Four transitions in humanity's natural history

	1. Emergence of *Homo sapiens*	2. Emergence of state-organized societies
Estimated date of this natural-historical transition	Circa 400,000 years ago.	Circa 5–10,000 years ago
Is this transition detectable in the atmospheric record?	No	Yes
Did this transition cause a "break" in metabolic relations?	No	Yes
Does so-called original accumulation occur in this transition?	No	No
Paradigmatic environmental problems.	Holocene extinction of large mammal species throughout the Americas (12,000 years ago): Hunted to extinction after arrival of humans.	Deforestation and soil degradation at a regional scale.
The character of "nature" as a concept of religion and ideology.	Before the emergence of the settled communities, there was no concept of "nature"—only an immediate relation to all that existed around small, nomadic bands of humans.	The concept of nature emerges via separation of the human from rest of nature (individual as idea, soul): A decisive step in the "self-estrangement of [humans] from nature" (hence "the antithesis of nature and history is created": Marx).

3. Transition to capitalism as a social formation	4. Reformation of capitalist society as ecological communism
Circa 300 years ago.	Has not occurred: Must occur soon for humanity to overcome planetary crisis.
Yes	No, because the transition has not occurred.
Yes	Metabolic breaks would need to be repaired to facilitate emergence of ecological communism
Yes	Ecological communism must "reverse" original accumulation (relink people with their means of labor).
Planetary warming, nuclear radiation, and the sixth great extinction of species.	Paradox: Ecological communism cannot come into existence unless we overcome the planetary ecological crisis, but we cannot overcome the planetary crisis via capitalist states
Capitalism reduces nature to a pile of commodities plus means of production. Capital thereby seems to free humans from nature as spirit, but money becomes the new fetish. In reaction, bourgeois society romanticizes nature and nation (fascism). This marks the second decisive step in the "self-estrangement of [humans] from nature."	Our inherited concept of "nature" would have to be transcended: "fully developed naturalism, equals humanism, and as fully developed humanism equals naturalism" (Marx).

Marx emphasizes the potential multiplicity of paths into and through the capitalist form of society.[84] Notably, in the same letter, where he summarizes the historical lesson of the conclusion of *Capital* in one sentence, Marx employs Hegelian language to draw a parallel to "the metamorphoses of nature":

> At the end of the chapter [on so-called original accumulation,] the historic tendency of production is summed up thus: That *[capitalist production] begets its own negation with the inexorability which governs the metamorphoses of nature*; that it has itself created the elements of a new economic order, by giving the greatest impulse at once to the productive forces of social labour and to the integral development of every individual producer; that capitalist property, resting as it actually does already on a form of collective production, cannot do other than transform itself into social property.[85]

Note that it is capitalist production that negates capital; capital transforms itself, as nature does. Marx's expression, "the inexorability which governs the metamorphoses of nature," suggests, again, the trace of his encounter with Darwin.

---

84 Michael Krätke observes that: "In the 1870s Marx wanted to restrict the validity of his analysis of capitalist development (in the section on [so-called original accumulation]) to Western Europe and he said so several times (although in private letters or in drafts never finished and never sent). But what did that mean? Marx knew at that time that both the United States and Russia followed a different path of economic development. But *the real question to ask is whether Marx's account of the origins of capitalism in Western Europe is correct or not. It is not, as most historians will agree today.* The English example is rather exceptional, in other parts of Europe the transformation of land into private property, of landowners into capitalists and of peasants into free labourers followed very different roads. Marx had the right theoretical clue: Capital as a social and economic relation presupposes a separation between the owners of the means of production and the owners of nothing but human labour power, and this separation must have been achieved by some historical process, as it cannot be assumed to be a natural state. But wielding this clue was another matter, and playing with phrases and words will never suffice, even if those phrases have been coined by Marx himself": Krätke, "An unfinished project: Marx's last words on *Capital*," in *Marx and Le Capital: Evaluation, History, Reception* (New York: Routledge, 2022), 144–71, at 163, my italics.

85 Marx, Letter from Marx to editor of the *Otecestvenniye Zapisky* (see note 82).

A Natural History of Capitalism

To recognize this continuity in the late Marx's writings—their natural-historical qualities—does not commit one to the simplistic or teleological view that *capital evolved*. Marx's arguments about the origins of capital are more complex. As Fredric Jameson writes,

> On origins ... [Marx] will in effect and in his practice ... offer a genealogy, distinguishing between origins and preconditions. The sample narratives offered here—such as the terrible story of the expropriation of the English peasants—are not exactly given to us as causes; but rather a setting in place of one of the preconditions required for the emergence of that new thing called capitalism ... Compared with the other historical modes of production we can document, capitalism is as strange a species as aliens in outer space, and it is not exactly to be accounted for by what the doxa normally identifies as evolutionary theory.[86]

In a footnote that supports the just-quoted sentence, Jameson comments:

> It is worth noting the appearance of Darwin in two long footnotes in *Capital* ... [Darwin's] authority, although it serves famously to insert human history into natural history, is here associated with the multiplicity of other species and with Hegel's idea of the *geistiges Tierreich*, the multiplicity of secular trades and callings, of productive talents, rather than with such evolutionary stories as "the survival of the fittest."[87]

---

86 Frederic Jameson, *Representing Capital: A Reading of Volume One* (New York: Verso Books, 2014), 75.

87 Jameson, *Representing Capital*, 75. In this passage, Jameson refers obliquely to Marx's (1862) letter to Engels, in which he writes,

> It is remarkable how Darwin finds among the beasts and plants his English society with its division of labor, competition, opening up of new markets, "inventions" and the Malthusian "struggle for existence." It is Hobbes' bellum omnium contra omnes [war of all against all] and reminiscent of Hegel in his *Phänomenologie*, where bourgeois civil society is described as a "spiritual animal kingdom" [die bürgerliche Gesellschaft als 'geistiges Tierreich'], while in Darwin the animal kingdom figures as civil society. Marx, letter to Engels, June 18, 1862, in MECW 41, 381.

Jameson tries too hard here to separate *Capital* from *On the Origin of Species*. The references to Darwin in *Capital*—both explicit and implied—are faithful to the general enthusiasm shown by Marx to that work.

I concur with Jameson that *Capital* distinguishes "between origins and preconditions," being and becoming. Regrettably, the debate surrounding the emergence of capitalism has often fixated upon determining the origin point of distinct processes which, strictly speaking, have no origin. I have argued that it would be better to cultivate a conception of capital's emergence which emphasizes processes of becoming. Capitalism came into being not all at once or through some sort of divine creation, but through evolving human activities. The emergence of capitalism was neither conscious nor planned; people acted under conditions where it was not at all clear that they were creating something new. Over time, we became bearers of novel roles specific to capitalist society; human social relations changed form: The capitalist social formation came into being. Recall Darwin's poetical words from *Origin* where he describes species in the fossil record as "forms of life, entombed in ... consecutive, but ... separated, formations." *Capital* provides us with a view of humanity as a form of life entombed in consecutive, separated, socio-ecological formations. In this view, the natural history of humans on Earth entails the becoming of new forms from already existing material and social relations. These processes can be traced to time before the origin of humans as a distinct species. From the moment of our emergence, all we have ever been are humans, becoming: And all we are becoming are humans organized in distinct forms.

When stated in these terms, it becomes clear why Marx was so attracted to Darwin's breakthrough. Darwin provided an explanation of the origin of species which did away with the idea that species existed (being) because of some divine act of creation; he replaced this with a conception of species as emerging (becoming) out of already existing forms of life. From being to becoming, things to forms: This is the shift that Marx then applied to consider the natural-historical forms of

---

While Marx is criticizing Darwin, he is *also* criticizing Hobbes, Malthus, and Hegel: So for Jameson to write that Darwin is "associated ... with Hegel's idea of the 'geistiges Tierreich' ... rather than such evolutionary stories as 'survival of the fittest'" is somewhat misleading.

human societies. Herein lies the answer to our antinomy (capitalism is natural; capitalism is not natural). The antinomy is resolved once we conceive of capitalism as a socio-economic formation inaugurating a remarkable yet dangerous epoch in the natural history of human beings on Earth.

7
Philosophical Implications of Marxian Natural History

Natural History as Philosophy of History

Marx's analysis of capitalism is not only a critique of political economy; it also provides a philosophy of history. But of what sort? Readers of Marx have debated this question since the late nineteenth century, and the question still produces serious disagreement. Complicating matters, some on the Left reject the philosophy of history out of hand. The very idea seems like a throwback to the Enlightenment, with its notions of the unfolding of reason toward some goal (whether God, liberty, or peace). Anyway, even if a nonteleological theory of history could be presented today, what practical value would it have for us in the midst of our planetary crisis?

But it is this crisis that has made everyone into philosophers of history anew. For, whether we are trying to explain our historical condition (how did we get here?) or speculate upon possible futures (where are we going?), some intuitive notions about temporality, causality, and historical process are in play. All people are philosophers, since everyone uses language, concepts, and reason to make sense of their world.[1] Everyone has some sense of history's rhythms and an intuition about the aim of human existence (even if that feeling

---

1 Antonio Gramsci, Q11§12, in *Selections from the Prison Notebooks* (New York: International Publishers, 1971), 323–43.

is that there is none). A good deal of our political disagreements with one another are expressions of these views, which mainly come from religion, family, and education.[2]

There is a further reason that old questions about the philosophy of history have renewed urgency today. In almost every respect, the politics of climate change are no different from all the other political struggles of modern history. To generalize: The struggle over climate concerns whether most of humanity must suffer negative consequences simply to preserve the privilege and wealth of a small minority—or whether the majority can displace that wealth and privilege to create a more just form of society. That is an old story. But there is one respect in which climate politics differs fundamentally. This concerns the way that the climate crisis inverts the Left's relationship to time.[3] We who struggle to change society are usually motivated to stop some *present injustice* and are motivated by a vision of a *better future*. We perceive the "now" as inadequate, indict the past for generating this situation, and take hope from the promise of tomorrow. Even if we admit that we are unlikely to achieve that future, it inspires us to continue struggling ("I may not get there with you . . ."). By contrast, amid planetary crisis, the future is a source of fear, even hopelessness.

This apparently small shift transforms the character of political experience on the Left. The dread of something worse to come poisons the present. Since the "now" is not as inadequate as what we expect to come, thoughts of the future fill us with shame. Shame is weak motivator in struggles for collective emancipation. This helps explain why so many of us struggle to find our bearings amidst the planetary crisis and are drawn to meditations upon history. Disoriented, we yearn for a better way to make sense of our time.

Marx's philosophy of history after Darwin

One defining attribute of a philosophy of history is its capacity to formulate hypotheses about causality. A philosophy of history should help to answer questions like: Why do societies change? When might we expect

---

2 One reason that religion and class remain powerful predictors of political views.

3 To recapitulate an argument I made in *Climate Leviathan* with Geoff Mann.

one social order to break down or evolve into something new? These are not technical questions: Everyone carries some sort of answer to them. This does not mean, however, that every implicit philosophy of history is equally strong or valid. Quite the contrary: Some answers are better than others because they are more coherent and better rooted in evidence. To confront our planetary crisis, we need to ask whether the conventional implicit philosophy of history people carry in their heads is up to the task.

Religious traditions have provided most implicit philosophies of history. Consider the teaching that history has a particular end involving the transcendence of this world, a "Day of Judgment," for example. That so many people find this sort of philosophy of history compelling today suggests that people crave teleology for the confident orientation it provides. The orientation provided is reassuring in a religious sense (despite all its twists and turns, human history has an underlying goal that only God knows), but also epistemological (I do not need to worry about where I stand in history; my knowledge is true regardless) and ontological (historical processes do not determine me; my true being/spirit transcends time).

During the Enlightenment, as the Christian tradition of such thinking became secularized through liberal philosophy, a belief emerged about the naturalness of a "market society" (or what I have called capitalism). Liberals claim that market society regulates itself naturally through rational social mechanisms: A free market that efficiently distributes labor and commodities via competition, supply, and demand; a democratic civil society that effectively distributes power and status via competition; and so on. Liberalism thus explains history from the vantage of this endpoint, which is in turn naturalized. The result is a secular teleology.[4] Consider Adam Smith's four-stage theory of types of human society. Smith posits market society as more than a recent period of economic history, but as the telos of a direction in history that is only recognizable retrospectively; it is as if we discovered that we were made to truck and barter, but only after we had been hunters, shepherds, and farmers.[5] After Darwin, this secular teleology

---

4 Ellen Meiksins Wood, *Liberty and Property: A Social History of Western Political Thought from Renaissance to Enlightenment* (New York: Verso Books, 2012).

5 Adam Smith, *Lectures on Jurisprudence* (Oxford: Oxford University Press, 1978).

would be recast in evolutionary terms: Humans live in capitalist societies today because we evolved to fit just such a society.⁶ To believe such a teleology, it is helpful to ignore the violent history of the formation of a capitalist world. Indeed, the liberal tradition has long repressed the facts of capitalism's historical emergence (we glimpse this when the political demand is raised to address the plunder of indigenous lands or the transatlantic slave trade).⁷ There is also a crucial environmental consequence of this secular teleology. By engendering a popular certitude about market society, it naturalizes capitalism's destruction of our planet.⁸

Our planetary crisis, if nothing else, should destroy such illusions. Can Marxian natural history provide a coherent alternative?

Practically all historiography before Marx adopted some sort of teleological narrative whereby the explanation of "what happened" is predicated on "what will come" or "what had to be." Whereas traditional historiography treats "nature" as passive backdrop and "history"means human agency, Marxian natural history interrupts that "history" with "nature," and vice versa. The object of this standpoint is the dynamic interactions between social relations and human exchanges with the natural environment: HH×HN. This has political implications, for as Sean Sayers observes,

> Marxism is not only a theory of history, it is also a political programme . . . It regards the development towards socialism in moral terms as a progress in "civilization" . . . [Yet] the progress involved here cannot be understood in teleological terms . . . [The] very notion of a final human end must be rejected. There is no absolute ideal of "full

---

In fairness to Smith, some readers find a more nuanced theory of history and capital in his works: See, for example, Giovanni Arrighi, *Adam Smith in Beijing: Lineages of the Twenty First Century* (New York: Verso Books, 2009).

6 Versions of this argument can be found in, for example, Friedrich Hayek, *The Fatal Conceit: The Errors of Socialism* (Chicago: University of Chicago Press, 1988) and Stephen Pinker, *The Blank Slate: The Modern Denial of Human Nature* (New York: Penguin, 2003).

7 Domenico Losurdo, *Liberalism: A Counter-History* (New York: Verso Books, 2014).

8 Whether such a theory of history and capitalism constitutes an *eschatology* remains an open question among Marxist philosophers. (So far as I am aware, it is not debated by any other kind of philosopher.)

human development" or self-realization . . . in terms of which historical progress can be assessed. In this respect, Marx's account of history can be compared to Darwin's theory of evolution.[9]

I concur—but we can go further and say that Marx's account of history can be explained in part by Darwin's theory. For, though Marx anticipated his standpoint in 1844–45, it was only realized after 1860 when *Origin* provided grounds for a nonteleological, natural-historical approach.

Darwinism is often accused by religious thinkers of evacuating life of meaning. In truth, it clears space for us to appreciate life's grandeur anew (as Darwin tried to convey in *Origin*'s conclusion: See chapter 2). Like Marx, by shifting perspective to ask new questions and provide original insights, Darwin exposed the hollowness of metaphysical answers to questions about life's meaning. The social Darwinists seized that opportunity to naturalistically derive bourgeois morality by claiming that natural selection proved the validity of relentless individual competition and Hobbes's *bellum omnium contra omnes*. (As noted in chapter 2, Darwin occasionally fell into this trap himself.) Marxian natural history reflects the genuine negation of social Darwinism by historicizing the ostensibly naturalness of capital. The resulting critique of capital—particularly its analysis of the "spirit" of commodity fetishism in capitalist society—provides us with sense of direction and a way out of these traps. We find a means for moral and political arguments about what is to be done that are not derived from generalizations about nature, religious metaphysics, nor capital's expectations, but from our collective desire for emancipation.

In sum: With its critique of capitalism as an economic formation that re-formed HH×HN and thereby generated our planetary crisis, *Capital* provides grounds for a post-Darwinian philosophy of history. But it only goes so far.

Four hypotheses on historical causality

It would be convenient for Marxists if we agreed on the basic parameters of Marx's philosophy of history and could present it in a simple

---

[9] Sean Sayers, *Marxism and Human Nature* (London: Routledge, 1998).

and convincing fashion. This is not the case. Marx never wrote a book called *Philosophy of History*. His remarks on the theme are concentrated in his early writings and his final notebooks—texts that were incomplete and only published long after Marx died (raising complex hermeneutical questions).[10] Let me provide a bare-bones sketch. Marx argued that previous philosophies failed because they did not take the correct starting point for understanding human societies, which is their socio-economic organization. If you want to understand, say, an ancient society of which you know practically nothing—you cannot read their written language (if there was one; usually there was not), you do not know the names of their god (if there was one; usually there were many)—then you start by digging around to learn the basics of the organization of social life. How did people produce food? How did they obtain water? How did they shelter themselves? Since life requires meeting basic needs, then the way that human social groups organize themselves to meet these needs—the social form of production, exchange, and consumption—will shape that society. As forms of production and modes of exchange change, so does the social form; such changes shape and are shaped by class conflict.

Marx's texts provide a richly textured version of this idea; nevertheless, even as an incomplete starting point, I hope that you will accept that this is so obviously merited that we can register it as a valid presupposition for a theory of history.

There is much more to say about *how* we might apply this basic sketch of an approach to human history. But the trickiest point for a philosophy of history is not, at any rate, establishing premises and providing guidelines for social analysis. It concerns teleology. If there is neither an end nor a end-defined purpose to history, then can we say what it is all about? If in previous histories, the twists and the turns in the story could be explained (with appropriate mediations and particularities) by an underlying logic—the need to fulfill the end, for things to get to a particular state—then how would a nonteleological, materialist philos-

---

10   Marx's best-known statements about philosophy of history are written in manuscripts called *The German Ideology*, which have a complex history: Terrell Carver and Daniel Blank, *A Political History of the Editions of Marx and Engels's* German Ideology *Manuscripts* (London: Springer, 2014).

ophy of history ever explain *why* things happen the way they did, or for that matter, predict what will come?[11]

Marx did not provide a consistent (nor entirely convincing) answer to these questions. Rather than give us *one* coherent statement on causes of historical change (or the historical determination of forms of society), Marx provided, by my count, *four* distinct arguments. Stated briefly, these claims are as follows, from simplest to most complex:

[1] The history and form of society is determined by the prevailing economic *forces of production*. For instance, Marx wrote in 1847:

> Social relations are closely bound up with productive forces. In acquiring new productive forces men change their mode of production; and in changing their mode of production, in changing their way of making of living, they change all their social relations. The hand-mill gives you society with the feudal lord; the steam-mill society with the industrial capitalist.
>
> The same men who establish their social relations in conformity with the material productivity, produce also principles, ideas, and categories, in conformity with their social relations.[12]

These lines are clear and clever—and therefore widely cited—but amount to a weak theory of history. Granted, the social conditions of production influence people's conception of the world. Slave owners encouraged ideas and categories that support slavery for the same reason that the fraction of the bourgeoisie that owns the fossil fuels in the earth's crust encourages climate change denialism.

Nevertheless, Marx's claim amounts to an oversimplification: productive forces → mode of production → social relations → ideas. History and society are more complex than this; the arrows do not only go in one direction. Statements like the one just quoted have earned Marx the unenviable reputation as a "determinist" who thinks that technological forces of production determine human history. By citing such

---

11 For this reason, many late nineteenth-century and twentieth-century Christian theologians opposed both Darwin and Marx, and linked them together: They are the two most powerful modern critics of teleology.

12 Marx, *The Poverty of Philosophy*, MECW VI (New York: International Publishers, 1992 [1847]), 80–1; 166.

passages, one can write Marx off. But this is not all he wrote on the matter.

[2] The driving force of history is *class struggle*. In the opening of the *Manifesto of the Communist Party*, Marx and Engels write:

> The history of all hitherto existing society is the history of class struggles.
>
> Freeman and slave, patrician and plebeian, lord and serf, guild-master and journeyman, in a word, oppressor and oppressed, stood in constant opposition to one another, carried on an uninterrupted, now hidden, now open fight, a fight that each time ended, either in a revolutionary reconstitution of society at large, or in the common ruin of the contending classes . . .
>
> The modern bourgeois society that has sprouted from the ruins of feudal society has not done away with class antagonisms. It has but established new classes, new conditions of oppression, new forms of struggle in place of the old ones.[13]

This argument is elaborated in Marx's journalistic writings about class struggles playing out in different capitalist societies, particularly France.[14]

The second claim is stronger than the first but also has limitations. While we may grant, at a high level of abstraction, that all human history has involved class conflict, what this means remains extremely vague. The existence of perpetual class conflict does not tell us anything about the specific causes of change—nor why class conflict may result in certain cases in the emancipation of subaltern social groups, in other cases greater levels of repression.

Even without a theory of causality, from a left wing political perspective, the second claim becomes appealing for two reasons. First, it justifies relentless class struggle, for this alone holds the key to social transformation. Second, it provides a basis for hegemony of communist states. As a rule, once communist parties take power in any society, the second claim becomes official ideology: "We won because we correctly combined political strategy and class struggle." Note, however, that,

---

13 Marx and Engels, *Manifesto of the Communist Party* (1848), MECW VI: 482–5.
14 Marx, *The 18th Brumaire of Louis Bonaparte* (1852), MECW XI: 99–197; Marx, *The Civil War in France* (1871), MECW 22: 307–59.

after a few years, these so-called communist (in reality, state-capitalist) societies must change the official ideology. Celebrating class struggle will lead people to generate their own radical ideas. So, a cult of personality emerges to explain the communist revolution.

There are two more claims but let us briefly assess the merits of the first two. Stepping back, these theories of histories form a logical pair. In the first, Marx places too much emphasis on *economic and technical conditions*; in the second, on *political struggle*. The result of the first approach is called *economism*; the result of the second may be called *politicism*. Both result in overly one-sided oversimplifications about history. The existence of such statements in Marx's oeuvre have frequently been used to justify such oversimplifications (often with serious negative consequences in so-called communist societies). A reader will ask: Why not set these two against each other? Can we mutually negate economism with politicism? Certainly, but this is not as easy as it may seem. Consider the next two claims.

[3] History is shaped by the unfolding of *contradictions between the development of the means of production* and the *social forms of organization of production and consumption* attendant to those means of production. This argument is particularly associated with Marx's preface to the *Critique of Political Economy*:

> In the social production of their existence, men inevitably enter into definite relations, which are independent of their will, namely relations of production appropriate to a given stage in the development of their material forces of production. The totality of these relations of production constitutes the economic structure of society, the real foundation, on which arises a legal and political superstructure and to which correspond definite forms of social consciousness. The mode of production of material life conditions the general process of social, political, and intellectual life. It is not the consciousness of men that determines their existence, but their social existence that determines their consciousness. *At a certain stage of development, the material productive forces of society come into conflict with the existing relations of production or*—this merely expresses the same thing in legal terms—*with the property relations within the framework of which they have operated hitherto.* From forms of development of the productive

forces these relations turn into their fetters. Then begins an era of social revolution. The changes in the economic foundation lead sooner or later to the transformation of the whole immense *superstructure*.[15]

As in the first claim, Marx argues that the "material forces of production" shape society. But unlike the first (simpler) claim, here he clarifies that the cause of change in history lies not merely with the presence or nature of the "material forces of production," but, rather, in their specific relation to the prevailing social relations of production. When the existing form and nature of technical development are coordinated with the prevailing social forms of ownership of those means of production, social and political life will find some sort of relative harmony and hence stasis. But, where they fall out of synch—because the development of the material productive forces of society runs ahead of the prevailing social form of the relations of production, social transformation or revolution will follow.

This is a more robust conception of historical change, with a more complex theory of causality, but still open to criticism.[16] Essentially, Marx provides here a more subtle form of economism which could create expectations that, under specific economic conditions, the "era of social revolution" *must* begin; but this could inhibit political analysis of alternative possibilities—such as the possibility that amid crisis, capitalism would *adapt its political form* into something new, like fascism. In his prison notebooks, Gramsci critically reexamines these relationships. Reflecting upon the failure of the proletariat of northern Italy to forge a political alliance with the southern peasantry and defeat fascism, Gramsci developed an analysis of politics centered upon the struggle for hegemony, that is, the moral and intellectual leadership of society that allows for certain class alliances and not others. In Gramsci's view, the view that the development of the forces of production was on the side of labor (reflected in the confident economism of the Second and Third International) hurt the Italian communists:

---

15 Marx, preface to *A Contribution to the Critique of Political Economy* (New York: International Publishers, 1970 [1859]), 20–1, my italics.

16 It is also important to ask whether Marx adopted this approach (described in 1859 as a "guiding principle" of his studies) in *Capital*.

> Determinism, fatalism, and mechanical thinking are an ideological aroma emanating directly from the philosophy of praxis [Marxism], a kind of religion or stimulant (functioning like mood-altering drugs). This is made necessary and justified historically by the "subaltern" character of certain social strata. For those who do not have the initiative in the struggle and for whom, therefore, the struggle ends up being synonymous with a series of defeats, mechanical determinism becomes a formidable force of moral resistance, of cohesion, of patient and obstinate perseverance. "I am defeated at the moment, but in the long run history is on my side, etc." Genuine will takes on the guise of an act of faith in a certain rationality of history, a primitive and empirical kind of impassioned teleology that functions as a substitute for predestination, providence, etc.[17]

Gramsci's conception of the political and "absolute historicism" may be appreciated as a corrective to the limitations of Marx's philosophy of history.[18]

[4] A fourth hypothesis may be read in the political denouement to *Capital*, in which Marx posits that contradictions between means of production and social organization of production and consumption, mediated by class struggle, will bring capital to its end:

> Once it reaches a certain level, [capitalist society] brings into being the material means of its own destruction. New powers and passions begin to stir deep within the belly of society, of a society in which they feel themselves to be fettered. It must be destroyed; it is destroyed. The individual and scattered means of production are concentrated, and thus the diminutive holdings of the many are transformed into the giant holdings of the few, while the land and means of subsistence and instruments of labor are thereby

---

17 Antonio Gramsci, Q11§12, in *Subaltern Social Groups: A Critical Edition of Prison Notebook 25*, eds. Joseph Buttigieg and Marcus Green (New York: Columbia University Press, 2021), 83. I thank Marcus Green for calling my attention to this passage and for his encouragement of this project.

18 On "absolute historicism," see Gramsci, Q11§27; *Selections from the Prison Notebooks*, 465; on Gramsci's conceptions of the political, see Peter Thomas, *The Gramscian Moment: Philosophy, Hegemony, and Marxism* (Chicago: Haymarket, 2009).

expropriated from the great majority of the people. This, the old society's destruction, a frightful and difficult process of expropriation, constitutes the prehistory of capital. . . . The moment this process of transformation has broken down the old society widely and deeply enough; the moment workers are turned into proletarians, and the things required for their labor have been turned into capital; and, finally, the moment the capitalist mode of production stands on its own two feet, the process whereby labor becomes social is altered from then on, as is the transformation of the land and the other means of production into socially exploited and, thus, shared means of production. Hence a change likewise occurs in the way private owners are expropriated. These processes all have a different form from this point on. No longer is the self-supporting worker the target of expropriation. Its target is now the capitalist who exploits many workers. This expropriation is brought about by none other than the operation of capitalist production's own immanent laws, which entails the concentration of individual masses of capital. One capitalist kills off many others. . . . The number of capitalist magnates falls continuously, and the remaining ones monopolize and usurp for themselves all the advantages that this process of transformation holds. Meanwhile, misery increases, as does the amount of pressure, subjugation, degradation, and exploitation inflicted upon the constantly growing working class. But the outrage felt by the members of that class also increases, and they are brought together and are trained and organized by the mechanism of capitalist production itself. Capital's monopoly now shackles the very mode of production that had flourished because of and under it. The concentration of the means of production and the socialization of labor reaches the point where neither process is compatible with its capitalist shell. This bursts, and now the bell tolls for capitalist private property. The expropriators are expropriated.[19]

This is the strongest statement with which to interpret Marx's view of historical causality.[20] Reduced to its essence, it is a tale of inherent

---

19   Marx, *Capital*, 690–1.
20   Given its place in Marx's oeuvre (the second conclusion of *Capital* I) there are grounds to consider it the approach preferred by the "mature" Marx. Still, I do not think

processes going too far. What processes? The centralization of the means of production and the socialization of labor: On one hand, the concentration of wealth and ownership, on the other, growing surplus population. This fundamental motif of *Capital* is employed here to argue that the asymptotic dissolution of precapitalist social relations into capitalism will, together with the concentration of value, burst the social formation open. This is a clear enough theory of social transformation. There are issues, however.

The first—formerly the more serious for Marxist theory—is that these events have not yet come to pass: The bell is still not tolling.[21] Nevertheless, the multiplication of millions of would-be-proletarians into the global labor market has been accompanied by deepening inequality at every scale and an extraordinary concentration of wealth. Although we see a growing centralization of the means of production and socialization of labor, it is difficult to foresee when these might "reach the point where neither process is compatible with its capitalist shell." My point here is not to quibble in empiricist fashion, to demand better evidence, but to pose a Kantian question. What are the conditions of possibility for knowing that a better world is coming? It may come, but we do not know.

The second problem brings us back to Marx's elegant maneuvering around the subject. It is that *Capital*'s conclusion leaves few guidelines for practical revolutionary activity. Capitalism ends when "the centralization of the means of production and the socialization of labour reach a point at which they become incompatible with their capitalist integument." If we reduce Marx's theory of revolution to this, it is difficult not to fall into economism. We can and should, therefore, consider other texts by Marx; yet we must also recognize that *Capital*'s conclusion is consistent with its mode of analysis, one that should be reaffirmed. *Capital* leaves us with political ambiguity.

Moreover, there is a clear mismatch between Marx's statement about his natural-historical standpoint in his preface to *Capital* and this ostensible philosophy of history. Marx's statements about the end of

---

we should judge the merits of Marx's text based on his intentions; what is essential is what these texts mean for us today.

21   This paragraph and the next draw from Joel Wainwright, "Reading *Capital* with *Being and Time*," *Rethinking Marxism* 27, no. 2 (2015): 160–76. I thank the *RM* collective.

capitalism appear to lack an ecological or evolutionary basis.[22] In chapter 8, I return to Marx's fourth claim to reconsider our planetary crisis. For the moment, let us clarify the grounds for such an interpretation by reconsidering the concept of human nature.

Human Nature Redux

By "human nature" I refer to the idea that all humans share some fundamental qualities by virtue of our being human. This means that at least some characteristics that define *Homo sapiens* as a species have general consequences for the nature of being human. To recapitulate (chapter one), Marx believed in human nature.[23] He proposed explaining human nature historically through study of the ensemble of the social relations that constitute human life. Since this ensemble of social relations (HH) always already includes and is mediated by our relations with nature (HN), properly understood, human nature is conditioned by and manifest in the ensemble of socio-ecological relations.

Let me anticipate three valid claims against this conception of human nature. First: It is too vague and fails to define a fixed set of qualities. For some, this means that human nature lacks the substance to serve us as a useful concept. But the opposite is true. A certain openness and ambiguity about human nature allows us to affirm the concept.[24] Second: The

---

22 The one that is most amenable to an ecological/natural interpretation is the weakest one—the first.

23 For a defense of this claim see Norman Geras, *Marx and Human Nature: Refutation of a Legend* (London: Verso Books, 1983); cf. Paul Thomas, "Nature and artifice in Marx," *History of Political Thought* 9, no. 3 (1988): 485–503; Sean Sayers, *Marxism and Human Nature* (London: Routledge, 2013); Vanessa Wills, *Marx's Ethical Vision* (Oxford. Oxford University Press, 2024), chapter 3. If we take "humanism" to be an affirmation of human nature and a concept necessary to grasp history (in which humans are treated as historical actors), then, like most post-Enlightenment thinkers, Marx was a humanist.

24 The concept of human nature has often been avoided or repressed on the Left—social Darwinism, Stalinism, and World War II contributing to skepticism on the topic—while the Right has generally embraced it. Marxian natural history needs to sail between the Scylla of essentialism and the Charybdis of relativism: "Formulating human nature as a stable essence . . . means reducing it to an instrument by which politics can pursue a political realism that effectively naturalises the existing social hierarchy that is

diversity of humanity implies that human nature is indefinable. This is only true if we share nothing fundamental in common—but, clearly, we do. (Few books make this point more effectively than Darwin's study of human expressions of emotions).[25] Third: Arguments for human nature have often been exploited by powerful people to produce unjust outcomes. By this reasoning, the problem with human nature is not that it is invalid, but that it can be argued in any way imaginable, and any attempt to advocate for it in an emancipatory direction will validate uses of the concept toward unjust ends. This third objection is, for the Leftist, the most powerful, and we need to confront it. It will be useful to have an illustration.

Consider Martin Wolf, liberal commentator at the *Financial Times*. In 2023 Wolf wrote a book called *The Crisis of Democratic Capitalism*. Describing three major challenges facing the world (his version of what I am calling our planetary crisis), Wolf makes a liberal argument on the basis of a negative conception of human nature:

> The first [crisis] is . . . the rise of demagogic, autocratic, and totalitarian capitalism . . . The second is the rise of China as a superpower. The last is the need to manage the challenges created by humanity's emergence as the cuckoo in the planetary nest. We should want to preserve

---

the expression of that supposed essence. And yet, turning human nature into an entirely social construction means dissolving it into interminable disputes incapable of implementing policies that can bring about real change in the existing conditions of exploitation": Andrea Bardin and Fabio Raimondi, "Shall we forget human nature?," *Contemporary Political Theory* 22, no. 1 (2023): 24–45.

25 Darwin, *The Expression of Emotions in Man and Animals* (New York: Penguin, 2009 [1872]). In his study of Marx's philosophy of human nature, Geras writes: "With respect to some emotions—namely, anger disgust, fear, happiness, sadness, and surprise—there is compelling evidence not just of their universality, but for a thesis of Darwin's that the facial *expressions* of them are similar regardless of culture, being of biological origin" (Geras, *Marx and Human Nature*, 99). I concur. Darwin's *Expressions* was initially conceived as one chapter of his *Descent of Man*, but this research subproject grew too large, and the resulting chapter too long, so he separated it from *Descent* and published it one year later as a stand-alone volume. *Expressions* was a major publishing event and could be considered the first mass-consumed coffee-table book, due to its reproduction of large photographs of faces of humans and animals. For a review of the first century of the post-Darwinian literature on emotional expression, see Paul Ekman, ed., *Darwin and Facial Expression: A Century of Research in Review* (Los Altos: Malor Books, 2006).

freedom, peace, and cooperation. It is going to be a difficult task to do so, given our remarkable capacity for destruction and the authoritarianism, tribalism, and shortsightedness characteristic of our species.[26]

These three dynamics do indeed present serious challenges for the liberal, capitalist order; I further agree with Wolf that we should strive for emancipation, peace, and cooperation.[27] But is it really true that our species is characterized by "authoritarianism, tribalism, and shortsightedness"? Wolf provides no evidence for these claims. We must ask: If humans are essentially tribal and authoritarian, how did liberal society ever emerge to produce a thinker like Wolf? All bourgeois societies have produced iterations of this paradoxical line of thought: Recall Hobbes's description of social life as "war of all against all."[28] Liberalism seems to be organized around the presumption of a human nature at war with liberal norms.

In the wake of Marx and Darwin, Freud's theory of the psyche provides the most significant presentation of an argument that human nature is inclined toward authoritarianism (if not exactly tribalism).[29] I noted in the preface that Freud found it "strange" that Marx claimed to study the development of forms of society as a process of natural history; if nothing else, I hope that my book has made this claim seem a bit less strange. This may be the place to revisit Freud's criticism of Marx in his undelivered lecture.

Freud makes two criticisms of Marx, both with some validity but marking a missed opportunity. First, Freud claims that, when we look at the emergence of social distinctions, these cannot be attributed solely to the economic class nature of society. According to Freud, these

---

26 Martin Wolf, *The Crisis of Democratic Capitalism* (New York: Penguin, 2023), 347.

27 *Pace* Wolf, I would not say we need to "preserve" these things, since they have not yet been achieved.

28 See John Gray, *The New Leviathans: Thoughts After Liberalism* (Boston: MIT, 2023), a defense of Hobbes's contemporary relevance: Published simultaneous to Wolf's *Crisis*, Gray is more stoical about the end of liberalism.

29 Like that of Kant, Darwin, and Marx, Freud's achievement—whatever the limitations of his theories—was to discern certain limitations of human reason: In this sense, his work exemplifies the Enlightenment tradition. Freud viewed Darwin's theory as one of the three great assaults on human narcissism.

distinctions "were originally distinctions between clans or races" which emerged from conflict and variations in

> psychological factors, such as the amount of constitutional aggressiveness, but also by the firmness of the organization within the horde, and by material factors, such as the possession of superior weapons. Living together in the same area, the victors became the masters and the vanquished the slaves. There is no sign in this of a natural law or of a conceptual evolution.[30]

Whatever the merits of Freud's claim, it is difficult to evaluate this as a critique of Marx, for, as we have seen, as we have seen, Marx was not solely concerned with class. Still, let me try to represent Freud's position and provide a partial response.

Freud implies that Marx failed to provide sufficient evidence for his contention that natural laws exist in history. Freud's point here is fair, though, as we have seen, the sort of "natural law" that people have been looking for in *Capital* may have been based on the wrong style of science: Darwin's theory of evolution, not Newton's physical laws, provides us with a model for Marx's presentation of "laws" in *Capital*. Freud moreover felt that Marx overstated the importance of relations of production as economic factors:

> the factor of man's control over Nature, from which he obtains his weapons for his struggle with his fellow-men, must of necessity also affect his economic arrangements ... But it cannot be assumed that economic motives are the only ones which determine the behaviour of men in society.[31]

---

30 Sigmund Freud, "The Question of a Weltanschauung," lecture XXXV in *New Introductory Lectures on Psycho-Analysis*, ed. James Strachey (New York: W.W. Norton & Co., 1990 [1932]), 219. Recall that Freud claims that "There are assertions contained in Marx's theory which have struck me as strange: Such as that the development of forms of society is a process of natural history, or that the changes in social stratification arise from one another in the manner of a dialectical process" (1932), 219. He denies, then, that the division of society into masters and slaves constitutes a "natural law" "in the manner of a dialectical process." Freud's real target here is not Marx, but dialectical materialism.

31 Freud, "The Question of a Weltanschauung," 220–1.

Philosophical Implications of Marxian Natural History 263

These are reasonable comments with which Marx would have agreed. This critique of Marx therefore misses the mark. However, Freud further claims that Marx attended too little to the ways that social control of nature's forces contribute to conflict and war

> the influence exercised upon the social relations of [humanity] by progressive control over the forces of nature is unmistakable. For [humans] always point their newly acquired instruments of power at the service of their aggressiveness and use them against one another.[32]

It is true that Marx did not elaborate on the ways that evolving relations between competing social groups and their relations with nature (HH×HN) generate technological changes and reinforce unequal power relations. Granted that more work is needed to consider how this tendency is playing out today amid the planetary crisis. But *Capital* shows that Marx developed a powerful standpoint for analyzing these dynamics in capitalist society.

It would be difficult to disprove Freud's generalization that humans "always point their newly acquired instruments of power at the service of their aggressiveness and use them against one another."[33] Shortly after he wrote those lines, World War II reached its dreadful climax at Hiroshima and Nagasaki—revealing the prescience of Freud's hypothesis that our epoch would be defined by humanity's "latest tremendous victory over nature, the conquest of the air."[34] The state that came to dominate this "conquest of the air"—with its "newly acquired instruments of power"—has ruled this epoch of natural history. Today we face the prospect of fresh advances in "newly acquired instruments of power." What could prevent them from being placed "at the service of their aggressiveness and use them against one another"? That is an urgent question facing world society, an aspect of our planetary crisis.

Still, the question remains: Do we know what humans will *always* do? If so, how? Claims like Freud's and Wolf's about what humans "always" do—with weapons, for instance—fail to recognize the contingency inherent in natural history. They are not grounds to reject claims about

---

32 Freud, "The Question of a Weltanschauungv," 219.
33 Freud, "The Question of a Weltanschauung," 220.
34 Freud, "The Question of a Weltanschauung," 220.

human nature. For the Leftist, they should be taken as provocations to think in terms of Marxist natural history. Let us return to the wisdom contained in another of Gramsci's notes written, coincidentally, at the same time as Freud's 1932 lecture:

> one cannot speak of "nature" [or human nature—JDW] as if it were something fixed, immutable, and objective. One notices that "natural" almost always means "legitimate and normal" according to our current historical consciousness, but most people are not aware that their position in the present is historically determined—they [wrongly] take their way of thinking to be timeless and immutable.[35]

It follows that:

> Human "nature" is the ensemble of social relations that determines a historically defined consciousness; only this consciousness can indicate what is "natural" and what is "unnatural." Furthermore: The ensemble of social relations is contradictory at any given time and is continually developing, so that human "nature" is not something homogeneous that applies to all humans at all times.[36]

Thus, Gramsci crystallizes the perspective of Marxian natural history apropos human nature.

Discussions about human nature often lead to claims which are, upon reflection, not really derived conclusions, but, rather, clarifications of a priori presuppositions concerning consciousness, ontology, or political ideology. To conclude this chapter, I therefore briefly address two complicated issues (admittedly in a preliminary fashion): First, Marx's definition of labor in *Capital* and its dependence upon consciousness; second, the vexed ontological question of whether nature as such is dialectical. I turn to political questions in chapter 8.

---

35 Gramsci, Q16§12, "Natural, unnatural, artificial, etc.," in *Subaltern Social Groups*, eds. Buttigieg and Green, 125.
36 Gramsci, Q16§12, "Natural, unnatural, artificial, etc.," in *Subaltern Social Groups*, eds. Buttigieg and Green, 126.

On Metabolism and Human Consciousness

One of the attributes commonly used to distinguish humans from other animals is our consciousness. We humans think that our consciousness is distinct from that of all other animals and in some profound sense superior. For many linguists and biologists, only humans have anything remotely close to the faculty of conceptual language, for instance.[37] From a moral perspective, this line of thinking is of dubious value for supporting claims of the superiority of humans.[38] After all, it would be tautological to claim that humans are superior in holding *any* human property relative to all other beings. No doubt the crow feels that its capacity to reason is more crow-ific than other beings, and therefore superior.[39] Nevertheless, we can avoid this tautology simply by affirming that human reason (and therefore consciousness) is distinct from all other species, insofar as we can tell, and that some of the distinctive qualities of this consciousness are definitive for living labor of a human type. Without entering here into a deep discussion of animal rights, I want to suggest one implication of the question of human consciousness for Marx's ontology in *Capital* and then suggest how reading *Capital* with Darwin might allow us to avoid an important interpretive problem concerning human consciousness.[40]

Unlike in his earlier writings on political economy, in *Capital*, Marx defines human labor using the concept of material exchange or metabolism (Stoffwechsel). Humans act upon Earth, mixing our labor with matter, metabolizing as we go: This is living labor as activity. So far, so good. But Marx adds a crucial qualification to this definition. He says

---

37 Robert Berwick and Noam Chomsky, *Why Only Us: Language and Evolution* (Cambridge, MA: MIT Press, 2016).

38 This is not to deny the plausibility and importance of those Darwin-inspired evolutionary approaches to the formation of human morality. See Michael Tomasello, *A Natural History of Human Morality* (Cambridge, MA: Harvard University Press, 2016).

39 See Andreas Nieder, Lysann Wagener, and Paul Rinnert, "A neural correlate of sensory consciousness in a corvid bird," *Science* 369, no. 6511 (2020): 1626–9. I am indebted to Llama Wainwright for insightful conversations about corvid consciousness.

40 I agree with Marco Maurizi that Marxism provides an essential intervention since it can explain why our desire to do right to animals must take—within capitalist society—an errant path through the bourgeois conception of rights: Maurizi, *Beyond Nature: Animal Liberation, Marxism, and Critical Theory* (Leiden: Brill, 2021).

that the nature of this living labor is distinguished by human *consciousness* of its activity. I take it that this claim is analogous to what I wrote a moment ago, that many people distinguish humans from other animals by the superior character of our consciousness. Marx writes:

> Here [in *Capital*] we are presupposing a form of labor that human beings alone are capable of. Of course, spiders carry out operations that resemble a weaver's work, and bees produce honeycombs that would put some human builders to shame. What separates the worst builder from the best bee is that before the builder creates a structure in wax, he creates it in his head. The end result of the labor process already exists when the process begins; it exists as an idea—as something a worker imagines. The worker doesn't simply shape natural materials into a new form; he also realizes a goal in doing so: A conscious goal that functions as a law determining both the work he performs and how he performs it, and to which, moreover, he must subordinate his will. When the worker subordinates his will to his goal, this is no isolated act. The whole time he is working, he must orient his will toward the purpose of his labor. He must stay focused, in other words, while he also exerts himself physically. The less the worker is drawn to the substance of his labor and the activities it involves, and, in turn, the less he enjoys his labor as the free play of his physical and mental powers, the more he has to train his attention on his work.[41]

---

41 Marx, *Capital*, 153–4 (cf. Marx, *Capital*, Fowkes translation, 284). Apropos this passage in *Capital*, Fromm writes that "labor is the self-expression of man, an expression of his individual physical and mental powers. In this process of genuine activity man develops himself, becomes himself; work is not only a means to an end—the product—but an end in itself, the meaningful expression of human energy; hence work is enjoyable": Erich Fromm, *Marx's Concept of Man* (New York: Ungar Publishing, 1961). Fromm does not dwell on Marx's remark that the pleasure of labor comes when we are engrossed to the point that we lose consciousness of our attentiveness to the labor. Conversely, we become aware of the need to attend to work when it is unpleasant: The less a laboring human is "attracted by the nature of the work," writes Marx, "the less, therefore, [the worker] enjoys it as something which gives play to his [or her] bodily and mental powers, the more close his attention is forced to be." Fromm continues:

> Marx's central criticism of capitalism is not the injustice in the distribution of wealth; it is the perversion of labor into forced, alienated, meaningless labor, hence the

Marx claims that what distinguishes *human* labor is not just material activity—since the powerful muscles of horses, Marx reminds us in *Capital*, can drive motors—but material activity plus *consciousness* of the purpose of that labor's activity.[42] And, Marx's reasoning goes, only humans have this sort of consciousness.

Now, even if this point is granted—and before doing so, we might want to ask how Marx knew what crows or bonobos thought about the purpose of their activities—his definition is inconsistent with other claims in *Capital* (elaborated below). My wager is that if we read *Capital* in a way that affirms complexities within Marx's arguments about labor, nature, and consciousness, and try to address these on a natural-historical basis, we arrive at stronger position.

Let us consider some of the words and phrases that Marx uses to characterize the labor process (as distinct from the act of buying and selling the commodity labor power: The latter is akin to *buying or*

---

transformation of man into a "crippled monstrosity.". . . Since the aim of human development is that of the development of the total, universal man, man must be emancipated from the crippling influence of specialization. In all previous societies, Marx writes, "man has been a hunter, a fisherman, a shepherd, or a critical critic, and must remain so if he does not want to lose his means of livelihood; while in communist society, where nobody has one exclusive sphere of activity but each can become accomplished in any branch he wishes, society regulates the general production and thus makes it possible for me to do one thing today and another tomorrow, to hunt in the morning, fish in the afternoon, rear cattle in the evening, criticize after dinner, just as I have a mind, without ever becoming hunter, fisherman, shepherd or critic" [Marx and Engels, *The German Ideology*, trans. C. J. Arthur, (New York: International Publishers, 1970 [1845–46]), 53].

42 The figure of the horse marks an ambivalence in *Capital*. On one hand, Marx trots it out to show how capital treats human and animal labor as equivalent, to imply a moral critique: "when a capitalist pays what a day of labor-power is worth, he owns the use of the labor-power for a day, just as he would the use of any other commodity he rented for a day, say, a horse" (*Capital*, 160); "during a natural day of twenty-four hours, a human being can expend only so much vital power, just as a horse can only work eight hours a day" (201), "a slave owner buys his workers the same way he buys his horses" (236). On the other hand, Marx acknowledges that horses are *conscious* ("have a mind of their own"): "Of all the great motive forces handed down from the manufacturing era, horsepower is the worst. For one thing, a horse has a mind of its own" (346). Recent research suggests that horses make strategic plans concerning their labor: Louise Evans, Heather Cameron-Whytock, and Carrie Ijichi, "Whoa, No-Go: Evidence consistent with model-based strategy use in horses during an inhibitory task," *Applied Animal Behaviour Science* 277 (2024): 106339.

selling a pitchfork, whereas the former is akin to the activity of *using* the pitchfork). Consider the terms Marx uses to describe the labor process:

> [through labor, we] bring the play of [nature's] forces under <u>conscious</u> control
> The worker . . . realizes , . . a <u>conscious</u> goal
> [The worker is oriented toward] the <u>purpose</u> of his [or her] labor
> [The worker] must stay <u>focused</u>
> [The worker] has to train his [or her] <u>attention</u> on [their] work
> . . . working from the start with a specific <u>purpose</u> in mind.[43]

Several concepts align here: Consciousness, purpose, focus, and attention. Putting these together, Marx says that living labor is distinguished by the application of human *consciousness*, exhibited in *purposefulness* (goal-directedness) and *attentiveness* (focus upon the goal), for regulating our material exchanges with nature and producing use-values. In other words, our conscious purposefulness and attentiveness is brought to bear upon [a] the metabolic relations between humans and nature and [b] the production of use-values. I take [a] and [b] to be distinct phenomena, yet Marx implies that these two phenomena are unified materially through the object to which labor is applied.[44]

An example may be useful here. Suppose, for instance, that a woman is using a blowtorch to weld together two pieces of metal. Through her labor, she is [a] using heat energy to combine two pieces of metal (modifying nature metabolically) while simultaneously [b] producing a new object (with some use-value). Her consciousness of her goal and attention to her task is what marks her activity as distinctly human labor power. And this is consistent with the passage I quoted distinguishing human living labor from bee living labor: This particular human welder may not be as capable of building things with the elegance of a bee, but

---

43   Marx, *Capital*, 154–6, my underlining. Marx recapitulates here an argument from 1844: After equating labor with human "life activity," Marx posits that whereas the non-human animal is "immediately one with its life activity," "man makes his life activity itself the object of his will and of his consciousness . . . Conscious life activity distinguishes man immediately from animal life activity": Marx, *Economic and Philosophic Manuscripts* I (1844), §XXIV; MECW III: 276.

44   As I read Marx, these are two different things which constitute a differentiated unity—another point that deserves questioning.

only the human is conscious of her goal, purposeful, and attentive to it in her labor. Consider this recapitulation of Marx's claims from Sasaki Ryuji's excellent introduction to *Capital*:

> Like any other organism, human beings must live through interaction with nature ... Marx described this cycle between human beings and nature as their "metabolism" ... Human beings alter nature through their actions so that their metabolism with nature can be smoothly carried out. We can say that such activities "mediate" the metabolism between human beings and nature in the sense of regulating and controlling that metabolism.[45]

Sasaki then notes that Marx distinguishes humans from non-human animals by the conscious character of our labor:

> There is a decisive difference between human beings' mediation of the metabolism with nature and the mediation by other organisms, insofar as the former is done *consciously*, whereas the latter is only done *instinctively*. (Of course, animals other than human beings do have a certain consciousness, but there is a significant difference in degree compared to human consciousness.) In laboring, human beings first have a concept of what needs to be done, based on which action is taken to realize the concept. Therefore, the human mediation of the metabolism with nature is very much a conscious, and thus an intellectual, act. For Marx, "labor" refers to this conscious mediation of the metabolism with nature, unique to human beings. In other words, *labor is precisely the human mediation, regulation, and control of the metabolism with nature through conscious acts.*[46]

I have no quarrel with Sasaki's summary of Marx's position. But I believe that Marx's claims here deserve to be challenged, for they rest upon

---

45   Ryuji Sasaki, *A New Introduction to Karl Marx: New Materialism, Critique of Political Economy, and the Concept of Metabolism* (New York: Palgrave Macmillan, 2021), 134–5. I recommend chapter 3 for its lucid discussion of metabolism generally. A quibble: As I read *Capital*, I think it might be more correct to say that, under capitalism, human beings alter nature through their actions to facilitate accumulation of surplus value.

46   Sasaki, *A New Introduction to Karl Marx*, 134–5, my italics.

analytical presuppositions that are problematic in non-trivial ways.⁴⁷ Let us consider these and then ask whether *Capital* is consistent with itself on questions of consciousness, labor, and nature.

Should consciousness define human labor?

Human living labor is not always conscious of its goal, attentive, and purposeful. Quite the contrary. One of the things we often say about someone who has mastered any form of labor is that they "do it almost unconsciously." But set aside mastery—even amateurs can work unconsciously and inattentively. I do not think I am revealing any weird secret about myself if I confess that, when I am at work, I often feel like I am not sure about what I am doing. Sometimes I even feel purposeless. And I am very often inattentive and distracted. And yet while these words (purposeless, inattentive) carry a negative connotation, as a human, I do not always experience these states negatively. On the contrary, I often feel most human when I am working in ways that feel inattentive or purposeless.

Cognitive psychology—a discipline that did not exist in the 1860s—has clarified some qualities about human reasoning. When people are at work, they are often inattentive. This does not necessarily necessarily translate into bad work. Moreover, we know now that the presupposition of attentiveness (made in *Capital*) is complicated by the considerable diversity of ways of being attentive and a wide range of cognitive experiences. Consider ADHD.⁴⁸ Do those with ADHD who

---

47 Note that Sasaki does not define the "significant difference" between human and non-human consciousness. This is typical in Marxist writing. I agree that there are differences, but in light of our limited knowledge of non-human consciousness, they are not easy to explain definitively.

48 Russell Barkley, "Behavioral inhibition, sustained attention, and executive functions: Constructing a unifying theory of ADHD," *Psychological Bulletin* 121, no. 1 (1997): 65–94. An evolutionary school has proposed that traits associated with ADHD proved advantageous in hunter-gatherer conditions: "Nomadic lifestyles favouring exploration have been associated with genetic mutations implicated in . . . ADHD, inviting the hypothesis that this condition may impact foraging decisions in the general population": D. Barack et at., "Attention deficits linked with proclivity to explore while foraging," *Proceedings of the Royal Society B* 291, no. 2017 (2024): 20222584; yet randomized, peer-reviewed tests have not found clear evidence in support of this hypothesis: T. Arildskov, A. Virring, P. Thomsen, and S. Østergaard, "Testing the evolutionary advantage theory of attention-deficit/hyperactivity disorder traits," *European Child and Adolescent Psychiatry* (2022), Volume 31, 337–48. Part of the challenge here lies in

struggle to attend to their immediate task lack the ability to labor as humans?[49]

Marx's definition of human labor seems to presuppose a model case—the neurotypical, attentive person, conscious of what she is doing and why—which is exceptional, not the rule. In fact, I think this is an instance where Marx may have fallen for a bit of bourgeois propaganda. For it is capital's desire that labor should always be purposeful and attentive. Marx erred, therefore, in defining labor with such restrictive terms. Fortunately, we do not need to maintain them; we can make his definition less restrictive.

Let me anticipate a critical response to my reasoning here. Some Marxists might defend *Capital* here by saying: "Ah, but you are quoting *Capital* too selectively. Marx simply meant that humans are conscious and attentive in a particular way, that is, when laboring, they are purposefully producing use-values and are conscious of this." This argument has its merits, but it cannot square with Marx's example of the architect, and I am not sure it fits with his claim that humans are conscious of the mediation of our metabolism with nature: In various passages, Marx states that laboring humans are conscious of their aims not only in terms of *production of use-values* but of their *modifications to metabolic relations* (Sasaki is not alone in emphasizing this point). When I observe people laboring, I can see little evidence for conscious mediation of the metabolism with nature. In fact, in capitalist society, human consciousness and attentiveness of the metabolic consequences of labor typically arrive at a late stage. It is usually only present *post facto*: And, even there, not so much as a conscious plan, but something practically worked out via iterations. Generally, humans figure out metabolic relations iteratively, through experimentation. We poke a stick and see what flies out of the hive. Then we try to get the honey. Later, we reflect on how the bees knew how to make those perfect hives.

---

modeling behaviors of a nomadic social formation amidst a capitalist one (symptomatically, Arildskov et al. had their subjects search for *coins*). Marxism has much to contribute here: See Robert Chapman, *Empire of Normality: Neurodiversity and Capitalism* (London: Pluto, 2023); Jodie Hare, *Autism Is Not a Disease: The Politics of Neurodiversity* (New York: Verso Books, 2024).

49 As Chapman notes, "many of the questions on standard ADHD screening tests relate directly to work skills," meaning that one's potential proletarianization is implicated in ADHD diagnosis. Chapman, *Empire of Normality*, 116–7.

If one accepts my line of reasoning, it follows that the practical character of *human* living labor is not really so distinct from that of, say, crows or horses. I take this as good news, for it helps us to get beyond the impasse in the (often unproductive) debate between Marxists and "new materialists" about whether non-human natural beings are involved in commodity production.[50] Certainly, they are. The essential difference between human and animal labor in capitalism appears later: Not during production, but during consumption, that is, through the value relation. Only humans use money to buy commodities, ergo to realize value.

Marx's critique of capital does not need labor to be defined by consciousness

I have argued that human labor should not be defined by consciousness and attentiveness.[51] Fortunately, it is not necessary to maintain this definition and hold on to the best lessons of *Capital*.[52] To shift registers from an external to an immanent critique, let us bracket the question of whether Marx's conception of labor and consciousness was too restrictive to ask instead whether it is consistent with his own presuppositions in *Capital*.

One of the most powerful qualities of *Capital*, which I would like to affirm and emphasize, is that, through the course of his investigation,

---

50 For new materialism, see Rick Dolphijn and Iris van der Tuin, eds., *New Materialism: Interviews and Cartographies* (Ann Arbor: University of Michigan Press, 2012). For a critique: Andreas Malm, *The Progress of This Storm: Nature and Society in a Warming World* (London: Verso Books, 2018).

51 To take this position is not to deny the specificity of human consciousness or its relation to the evolution of language (on which see Berwick and Chomsky, *Why Only Us*).

52 Marx writes, somewhat vaguely, that in his analysis of labor in capitalist society he "is not speaking of the earliest forms of labor, namely, instinctual and animal-like forms," because "when a worker arrives in the commodity market to sell his [or her] own labor-power [as a commodity], he [or she] is operating under conditions very far removed from those in which human labor hadn't yet advanced past the instinctual form it had initially (in primordial times)" (*Capital*, 153). This presupposition is questionable, since human beings remain "instinctual and animal-like" today in capitalist society. Much depends on how we interpret Marx's vague expression, "the instinctual form" of labor "in primordial times." I do not think the validity of Marx's analysis of value and capital collapses if we challenge this presupposition. To the contrary, we can strengthen the critique by reinforcing the underlying presuppositions.

Marx shows—in what we might call a deconstructive reading of labor and nature—how his restrictive presuppositions regarding the laboring human cannot be sustained under the rule of capital. Although there are important differences between Freud and Marx, since some of these ideas later came to be associated with Freud and psychoanalysis, I will call these instances of "proto-psychoanalytic" analysis. Let me briefly mention three instances in which proto-psychoanalytic analysis can be detected in *Capital*, volume I:

[1] In the preface, Marx writes that, although he must criticize the bourgeoisie in the book, he does not seek to blame individuals for the problems of capitalist society. His explanation is that he does not treat these figures as conscious people but as the bearers of distinct class positions: "Individual persons play a role ... only insofar as they are the personifications of economic categories, or the bearers of particular class relations and interests."[53] Class processes are not defined by consciousness (the opposite is closer to Marx's position). It would seem contradictory for Marx to define the labor process itself by the consciousness of living labor. Perhaps it could be argued, but there are other grounds for skepticism.

[2] *Capital* shows that capitalism is a vampire-like social formation that gives over living labor to capital on an ever-expanding scale. His argument could be summarized by a sentence Marx wrote in his economic notebooks while preparing *Capital*: "The rule of the capitalist over the worker is the rule of things over man, of dead labour over the living, of the product over the producer."[54] This rule is not only impersonal but also largely invisible and unconscious. Through his analysis of commodity fetishism, Marx presents a theoretical account of social relations as mediated by commodities in which proletarians are unaware of the fundamental dynamics structuring our lives.[55] We participate in value relations *unconsciously*. This unconsciousness is why commodity fetishism is fetishistic: If we were conscious of it, the fetish would lose its power, its fetishistic character.

---

53   Marx, *Capital*, 8, my italics.
54   Marx, "The process of production of capital," (1994 [ca. 1861–63]), MECW 34: 339–466, 398. Marx returns to this theme in *Capital*: "Capital is dead labor that acts like a vampire: It comes to life only when it drinks living labor, and the more living labor it drinks, the more it comes to life" (*Capital*, 205).
55   Marx, *Capital*, chapter 1, §IV: See chapter 5 for a discussion.

[3] In chapter 5 Marx describes the fact that "dead labor"—the congealed form of previous investments of living labor that keep the valorization process humming along—is not seen *as* dead labor, that is, is not seen as the qualitative result of someone's life and labor.[56] Rather, we *only become conscious of the labor of others when things break down*; this implies that we are *not* conscious of the labor of others when things are flowing smoothly:

> When a product is put into new labor processes as a means of production, it thus loses its character as a product. Now it functions only as an objective factor, a thing, that aids living labor. A spinner treats the spindle merely as the means for his spinning and flax merely as the object of his labor . . . What doesn't matter for this process is that flax and the spindle are the products of previous labor, just as for the purpose of eating, it doesn't matter that bread is the product of the combined previous labor of farmers, millers, bakers, and so on. However, when the means of production in the labor process fail, their character as previous labor is keenly felt. A knife that doesn't cut, yarn that constantly comes apart—these things make it hard not to think of cutler A and spinner B.[57]

We become conscious of cutler A and spinner B only when things break down. Before this point, if we remain blissfully unaware of cutler A's labor and spinner B's labor, then we are equally unaware of their metabolism. And it follows that A and B are not conscious of the other, still less all the matter and energy—the stuff—that their labor regulates and manipulates.

It is beyond the scope of this chapter to weave these three passages into a coherent theory of consciousness and unconsciousness in *Capital*. Suffice it to say that Marx problematized consciousness in capitalist society to a significant degree and these passages sit awkwardly with a consciousness conception of human labor. The value of these insights for analysis of capitalism is far greater to us than anything gained by

---

56 This paragraph draws from my article, "Reading *Capital* with *Being and Time*" from *Rethinking Marxism*. I thank the RM collective.

57 Marx, *Capital*, 158. Apropos this passage, Fredric Jameson comments that Marx "anticipates the phenomenological doctrine of the relationship between consciousness and failed acts": Jameson, *Representing Capital* (New York: Verso Books, 2011), 97.

retention of the rigid definition of labor by consciousness. And Marxists cannot really have it both ways. We cannot hold on to these subtle and important claims about unconsciousness and mis-consciousness yet go on defining labor as "conscious" "purposeful" "attentive" activity.

Fortunately, it was sufficient for Marx to define human labor in terms of metabolic relations:

> the labor process . . . is an activity whose purpose is to create new use-values, the appropriation of natural materials to satisfy human wants and needs, and what universally allows the human metabolizing of nature to take place—the eternal natural condition of human life, which is therefore independent of all the ways people live, or common to all social formations.[58]

That is all we need to say to define human labor. Marx was right to define living labor in terms of metabolism and use-value production; these, not attentiveness and consciousness, define labor. The most fundamental category concerning labor is not consciousness or purposefulness but metabolism. Living labor is the name for active human changes to metabolic relations for the sake of the production of use-values. Darwin was right: The historical becoming of human labor occurred unconsciously.

On Dialectic and Marxian Natural History

For some Marxists, my arguments concerning nature, history, and capitalism will be concerning, for they are missing familiar expressions like "dialectical materialism" and "dialectic of nature." If I have avoided these, it is only partly because, in my experience as a teacher and activist, I find that they usually introduce greater confusion than clarity. Moreover, my concern has been to elaborate Marx's standpoint as described in *Capital*—studying socio-economic formations as natural-historical process—where he does not use the words "materialism" nor "dialectical" (still less "dialectic of nature"). Then, when his anonymous reviewer repeated back Marx's statements, Marx said that the reviewer was simply describing the dialectical method.

---

58 Marx, *Capital*, 159.

That comment, and subsequent arguments advanced by Engels after Marx's passing, have generated considerable confusion about Marx's conception of nature and dialectics. A few words on this matter are therefore due.

I have never found it valid to define nature by any sort of abstract philosophical concepts. It is not helpful to claim that "nature is dialectical" any more than it would be to claim that "nature is good" or "nature strives toward complexity." These are all metaphysical propositions, and the metaphysics of dialectics of nature have proven to be of dubious value for communist politics (just as the metaphysics of nature's goodness has contributed to conservative romanticism). Now, it may be that a dialectical style of thinking or writing is helpful for analyzing and describing certain natural phenomenon.[59] But the same could be said for mathematics, and we do not say "nature is mathematical." Still, much depends on the conception of dialectic in play. The term has its roots in ancient Greek thought and means different things to different philosophers. If "dialectic" is used to say that two things that are commonly treated as distinct and opposed form a unity of difference (and that we should aspire to grasp these phenomena as such), then Marxian natural history is dialectical, for it treats nature and history in just this way. But this is not the only way to conceive of dialectic and nature.

The philosophical debate over Marx's ostensibly dialectical interpretation of nature takes its cue from two principal sources: Engels's 1878–1882 text on *Dialectics of Nature* and the brief critique of them by György Lukács in 1923.[60] In *Dialectics of Nature*, Engels proposes that nature is dialectical; indeed, "nature is the proof of dialectics."[61] The dialectics of

---

59 Engels was surely right about this, and on this point, Marx agreed with him. Hence, when discussing the metamorphosis of commodities in *Capital*, Marx uses a metaphor from planetary motion that invokes Hegelian language: "We express a contradiction, for example, when we say that one body is constantly both falling toward another body and falling away from it. The ellipse is one of the forms of motion through which this contradiction is just as much realized as resolved" (78).

60 Engels, *Dialectics of Nature* (1987 [1878–1882]), MECW 25: 311–587; György Lukács, *History and Class Consciousness: Studies in Marxist Dialectics* (Boston: MIT Press, 1972 [1923]), 24.

61 Engels, *Anti-Dühring* ([1877] 1987) MECW 25: 23. I concur with Foster that few questions in Marxist philosophy have "been more contentious than the dialectics of nature," but I cannot agree (and find no evidence for his claim) that the rejection of an

nature, Engels posits, are manifest in three laws: (1) the law of transformation of quantity into quality; (2) the law of identity or unity of opposites; and (3) the law of the negation of the negation.[62] By my reading, Engels's laws are metaphysical, Hegelian speculations upon nature. I agree with Kōhei Saitō that Karl Marx "never really adopted the project of materialist dialectics that Engels was pursuing."[63] But the confidence and capaciousness of Engels's laws of dialectics proved reassuring for the generation of Marxists who followed him. Lukács' famous footnote in *History and Class Consciousness* was intended to correct this: "Engels—following Hegel's mistaken lead—extended the [dialectical] method also to knowledge of nature."[64] But then Joseph Stalin (no great

---

Engelsian "dialectics of nature" caused "an almost total abandonment of any connection to natural science . . . within Western Marxism" across the twentieth century: Foster, *The Return of Nature: Socialism and Ecology* (New York: MR Press, 2020), 16–17.

62   On these supposed laws, see Engels, *Anti-Dühring*, MECW 25: 110–32, and *Dialectics of Nature* (1987 [1878–82]), MECW 25: 492–503. For more affirmative interpretations of *Dialectics of Nature*, see Sven-Eric Liedman, *The Game of Contradictions: The Philosophy of Friedrich Engels and Nineteenth-Century Science* (Chicago: Haymarket, 2022 [1977]); Kaan Kangal, *Friedrich Engels and the Dialectics of Nature* (London: Palgrave Macmillan, 2020); J. B. Foster, *The Return of Nature: Socialism and Ecology* (New York: Monthly Review Press, 2020); and Sean Sayers, "Engels and the *Dialectic of Nature*," in *Friedrich Engels for the 21st Century: Reflections and Revaluations* (London: Springer, 2022), 33–51. Liedman concludes that Engels' path of inquiry culminates in "his three dialectical laws, which were an attempt at locking the changeable into fixed forms . . . There is a real dialectical tension between these perspectives: The realism of knowledge and the principle of changeability . . . Engels did not succeed in bringing this contradiction to a plausible, concrete conclusion" (455). I concur—and I suspect that Marx saw this end coming. So I respectfully part ways from Liedman, who claims that it is nonetheless "desirable and possible to realise Engels's line of thought" (455).

63   Kōhei Saitō, *Marx in the Anthropocene*, 67. Many readers have observed that Marx's presentation of concepts in *Capital* bears a stamp of Hegelian thinking. Yet, while Marx's mode of *exposition* is Hegelian, the character of the argumentation is not: Kōjin Karatani, *Transcritique: On Kant and Marx* (Boston: MIT Press, 2003), 9.

64   Lukács reasons that the dialectical method should be limited to "historical-social reality". Lukács, *History and Class Consciousness*, 24. I recommend Saitō's analysis of these themes in *Marx in the Anthropocene*, 73–100. For a contrary reading that affirms Engels's dialectics of nature, see J. B. Foster, *The Return of Nature: Socialism and Ecology* (New York: Monthly Review Press, 2020), chapter 6; more recently, J. B. Foster, *The Dialectics of Ecology: Society and Nature* (New York: Monthly Review Press, 2024), in which Foster writes: "It became customary in Western Marxist thought to refer to Lukács's footnote as a 'critique.' But . . . it could hardly be said that a critique of Engels on the dialectics of nature could be carried out, even by Lukács, in . . . a mere 110 words": Foster, "The dialectics of ecology: An introduction," *Monthly Review*, January 1, 2024, no page. *Pace* Foster, it could be done. The best critiques are often concise.

philosopher) upheld the opposite view, establishing dialectics of nature as the ontological foundation for communism in the 1930s.[65] Questioning Stalin could get you killed. Henceforth, the philosophical debate about the validity of Engels's views was badly obscured by Stalinism.

Lukács was forced to repudiate many of the things he had written in the 1920s. Thus in 1967, he criticized his 1924 study for "the tendency to view Marxism exclusively as a theory of society, as social philosophy, and hence to ignore or repudiate it as a theory of nature." *History and Class Consciousness*, he writes, "takes up a very definite stand on this issue. I argue in a number of places that nature is a societal category and the whole drift of the book tends to show that only a knowledge of society and the [people] who live in it is of relevance to philosophy."[66] In truth, Lukács overstates the problem. First of all, it is a fact that nature is a social category. Moreover, a Marxist perspective on nature must study "a knowledge of society" and social life. On the next page, Lukács writes that he tried "to explain all ideological phenomena by reference to their basis in economics"; his error, he implies, was that he failed to discuss what could be seen as a key concept from *Capital*: "labour as the mediator of the metabolic interaction between society and nature, is missing." By implication, he reasons, "the most important real pillars of the Marxist view of the world disappear and the real attempt to deduce the ultimate revolutionary implications of Marxism in as radical a fashion as possible is deprived of a genuinely economic foundation."[67]

This is a self-critique taken too far, the sort of hyperbolic statement we find only after "the most important real pillars of the Marxist view of the world" have been smashed by the hammers of Stalinism. For it cannot be said that Lukács completely missed labor's mediation of the metabolic interaction between society and nature. While it was not given the emphasis it perhaps deserved, that was even more true of the Stalinist ideological milieu to which Lukács's self-critique was a response. Lukács' elaboration:

---

65 Stalin's execrable *Dialectical and Historical Materialism* (Moscow: Foreign Languages Publishing House, 1938) begins (see §I) with a recapitulation of Engels's Hegelian speculations on the dialectics of nature. On Stalin's thought, see Isaac Deutscher, *Stalin: A Political Biography* (Oxford: Oxford University Press, 1949).

66 Lukács, "Preface to the New Edition," in *History and Class Consciousness*, xvi.

67 Lukács, "Preface to the New Edition," in *History and Class Consciousness*, xvii.

It is self-evident that this [deprivation or lack] means the disappearance of the ontological objectivity of nature upon which this process of change [i.e., history] is based. But it also means the disappearance of the interaction between labour as seen from a genuinely materialist [i.e., Marxist] standpoint and the evolution of the men who labour.[68]

This is an extremely vague formulation that could mean many different things (contributing to my feeling that we are reading something like a forced confession). The generous reading of Lukács here is the most plausible: His de-emphasis of labor's mediation of the metabolic relation made it difficult to generate a Marxian natural history of capitalism and laboring humanity. Fair enough. But then, that critique would also apply to most of the Marxist tradition since the publication of *Capital*. In an interview in 1967, Lukács is clearer:

We are using the fine word "ontology" . . . although one should really say that one is discovering the forms of being that new movements of the complex produce . . . being is a historical process . . . Marx wrote in *German Ideology* that there was only a single science, the science of history, and you will remember how enthusiastically Marx greeted Darwin, despite many methodological reservations, for discovering the fundamentally historical character of being in organic nature.[69]

Hold on to that line—Marx enthusiastically celebrated Darwin "for discovering the fundamentally historical character of being in organic nature."

Years before Lukács provided this lucidity, in 1932, Theodor Adorno delivered the annual lecture to the Kant Society in Germany. Of all things to discuss at a moment of extreme political tension—the Nazi party's electoral breakthrough came in the July elections—Adorno spoke on "The Concept of Natural History."[70] Much like Gramsci at the

---

68   Lukács, "Preface to the New Edition," in *History and Class Consciousness*, xvii.
69   Lukács, *Conversations with Lukács*, ed. T. Pinkus Boston: MIT Press, 1975 [1967]), 21.
70   Theodor Adorno, "The idea of natural history," *Telos* 60 (1984 [1932]): 111–24. See Susan Buck-Morss, *The Origin of Negative Dialectics* (New York: Free Press, 1984 [1932]), chapter 3. According to Buck-Morss (311), Adorno's natural history lecture was

same moment, Adorno sought to think of nature as historical and history as natural to enable a form of analysis adequate to the world political-economic crisis of the time. Noting that Adorno's essay "remains something of an enigma," Jameson reads it as "a methodological proposal, rather than a set of theses";[71] yet Adorno warns his audience that his conception of natural history should not be taken as "a synthesis of natural and historical methods," but as "a change of perspective."[72] What sort of change of perspective?

> [My] intention here is to dialectically overcome the usual antithesis of nature and history. Therefore, wherever I operate with the concepts of nature and history, no ultimate definitions are meant; rather, I am pursuing the intention of pushing these concepts to a point where they are mediated in their apparent difference.[73]

Shortly thereafter, Adorno clarifies:

> If the question of the relation of nature and history is to be seriously posed, then it only offers any chance of solution if it is possible *to comprehend historical being in its most extreme historical determinacy, where it is most historical, as natural being, or if it were possible to comprehend nature as a historical being.*[74]

This is an elaboration of Marx's natural-historical standpoint. Adorno's maxim—"No being underlying or residing within historical being itself is to be understood as ontological, that is, as natural being"—could be taken as a generalization or abstraction from Darwin's complex treatment of species. I read Adorno's call for the "ontological reorientation of the philosophy of history" toward "the idea of natural-history" as a gloss on the shift of perspective made by Marx in *Capital*.[75]

At around the same time that Adorno gave that lecture, Gramsci

---

only published in 1973.
71 Fredric Jameson, *Late Marxism: Adorno, or, the Persistence of the Dialectic* (New York: Verso Books, 2007 [1990]), 94.
72 Adorno, "The idea of natural history," §II, 118.
73 Adorno, "The idea of natural history," §I, 117.
74 Adorno, "The idea of natural history," §I, 117.
75 Adorno, "The idea of natural history," §I, 117.

penned a few lines in one of his prison notebooks which deserve rereading. In the note entitled "What is man?," Gramsci argues against one-sided conceptions of nature—as either dialectical or not-dialectical—on the grounds that we cannot presuppose an unhistorical distinction between nature and society:

> One must study the position of Professor Lukács towards the philosophy of praxis. It would appear that Lukács maintains that one can speak of the dialectic only for the history of men and not for nature. He might be right and he might be wrong. If his assertion presupposes a dualism between nature and man he is wrong because he is falling into a conception of nature proper to religion and to Graeco-Christian philosophy and also to idealism which does not in reality succeed in unifying and relating man and nature to each other except verbally. But *if human history should be conceived also as the history of nature* . . . how can the dialectic be separated from nature?[76]

Gramsci's note reflects the signature move of a Marxist approach to philosophy, namely, to insist upon historicizing the question rather than trying to resolve it on the basis of an abstract principle. If we accept a natural-historical perspective—in the terms just used by Adorno—then we could speak of a dialectic of nature, since nature includes human history: But it would no longer *be* a dialectic of nature in Engels's sense. It would be a natural-historical analysis of the sort proposed by *Capital*, which affirms the "fundamentally historical character of being" and "the subsumption of human history . . . under natural history—something henceforth indissociable from Darwin's own work and theorizing."[77]

In sum: Lukács, Adorno, and Gramsci remind us that we have no basis for defining nature qua dialectical—or any *other* metaphysical category—outside of an historical analysis of those categories and their relation to human practice. Thus, the question that caused Lukács such grief, and has generated so much heat within Marxist philosophy, was badly posed.

~

[76] Gramsci, *Selections from the Prison Notebooks*, Q10§54; 448, my italics. For Foster (2020), Gramsci was raising "the alarm with respect to the neo-Kantian character of Lukács's criticism," 18.

[77] The first quotation is from Lukács, *Conversations with Lukács*, 21; the second is from Jameson, *Late Marxism*, 94–5.

The coincidence of Gramsci's and Adorno's 1932 writings on nature and history provides us with a clue to answer two questions that will have come to mind: Why did they arrive at these conclusions simultaneously—at the same time that Freud commented upon Marx's "strange" idea? And, if Marx was inspired by Darwin when developing his natural-historical approach to capital, why has this has not been widely recognized in Marxism?

Freud, Gramsci, and Adorno were explicitly responding to fascism, a reactionary ideology that, in the face of capitalist crisis, promises a romantic return to natural order via ethno-racial-national supremacy. Fascism thus constitutes an anti-Marxian natural history. In their own ways, Gramsci and Adorno recognized that, to win the masses over from fascism, communism must provide an alternative conception of the world encompassing natural history. Hence, in 1932, they sought to revive Marx's conception of natural history, imputed to the preface to *Capital*.[78]

Philological factors also help explain the coincidence. First, the preface to *Capital* and Marx's comments about Darwin could not be properly understood until after his Paris manuscripts of 1844 were published (beginning in the late 1920s). Second, as we saw in chapter 2 almost immediately after the publication of *Origin*, Darwin's ideas became associated with social Darwinism and its affirmation of social hierarchy and inequality. Since practically all Marxists rejected social Darwinism, they were ambivalent toward Darwin.[79] Third, Marx's

---

78 Adorno's natural history lecture explicitly responds to Heidegger's *Being and Time* (1927); Adorno saw Heidegger's philosophy as fascist. Adorno never mentions *Capital* in his lecture, but he concludes by stating: "I wanted to speak about the relationship of these matters to historical materialism, but I only have time to say the following: It is not a question of completing one theory by another, but of the immanent interpretation of a theory . . . It could be demonstrated that what has been said here is only an interpretation of certain fundamental elements of the materialist dialectic": Adorno, "The idea of natural history," 111–24, §III, 124.

79 I write "practically all" because—astonishingly—some avowed socialists sought to marry Marxism to Spencer's social Darwinism: E.g., Enrico Ferri, for example, in a dreadful work (yet translated into multiple languages), *Socialismo e Scienza Positiva: Darwin, Spencer, Marx*, (Rome: Casa Editrice Italiana, 1894). As Engels wrote to Karl Kautsky (September 23, 1894): "that blatherer Enrico Ferri sent me all his recent writings . . . His book on Darwin-Spencer-Marx is an atrocious hotchpotch of insipid rubbish," MECW 50: 349. Ferri helped establish an influential school of racist sociology in Italy and became a follower of Mussolini. In "Some Aspects of the Southern Question,"

conception of natural history was out of synch with the approach to science imposed by Joseph Stalin. Marx saw that Darwin's breakthrough destroyed historical teleology. Yet, to generalize, the first two generations of Marxists wanted to find a specific telos in Marx's texts: Proletarian revolution via scientific socialism.[80] Not surprisingly, they found what they were searching for—particularly after Stalin took power in the USSR. In the face of counterrevolutionary attacks, the very survival of the Bolshevik revolution "required" the constitutive use of reason, and, by the late 1920s, Stalinism had devolved into a cult, reflected philosophically in so-called dialectical materialism. To pronounce Marx a natural historian would have been a death sentence in Stalin's USSR—which dominated the intellectual spaces of international communism.[81]

A turning point came in 1932, when the prospect for a Marxian natural history was essentially rediscovered independently by Adorno and Gramsci. The novelty of Marx's Darwin-inspired natural-historical turn was largely forgotten until, around five decades later, a trio of Marxist literary critics—Kōjin Karatani, Gayatri Spivak, and Fredric Jameson—independently raised the promise of Marxian natural history.[82] A vibrant literature on ecological Marxism subsequently blossomed and we are now in a stronger position to realize the (still largely undeveloped) project of Marxian natural history, which promises fresh insights into

---

Gramsci criticizes Ferri for contributing to racist stereotypes of Southern Italians as biologically inferior: *Selections from Political Writings*, 441–62, at 444 (cf. Q1, §44). I thank Marcus Green for his help on this point.

80   "Insofar as Engels . . . sought to counterpose the 'correct' positions of a 'scientific socialism', he laid the foundations for the worldview of Marxism, which was appreciatively taken up in Social Democratic propaganda and further simplified. . . , What dominated the Social Democracy at the end of the nineteenth century under the name of Marxism consisted of a miscellany of rather schematic conceptions: A crudely knitted materialism, a bourgeois belief in progress, and a few strongly simplified elements of Hegelian philosophy and modular pieces of Marxian terminology": Michael Heinrich, *An Introduction to the Three Volumes of Karl Marx's Capital* (New York: Monthly Review Press, 2012), 24.

81   Marxists were not the only ones to turn away from Darwin. Limiting ourselves to Western philosophy, the two most significant fields to flourish in the early decades of the twentieth century—the analytical philosophy inspired by Wittgenstein (Carnap, Quine, et al.) and phenomenology (encompassing here Husserl and Heidegger)—rejected Darwinism in foundational moves.

82   I thank these cherished mentors for their inspiration.

various theoretical problems.[83] Yet what we need most urgently are not philosophical treatises but practical solutions to our planetary crisis. So let us turn, finally, to the political challenges that make Marxian natural history vital today.

---

83   I agree with Fredric Jameson that "We are better placed today, after the extraordinary reinvigoration of evolutionary thought and the powerful rereading of Darwin . . . by Stephen Jay Gould and others, to grasp what might be at stake in the strategic but unclearly motivated act of repositioning this problem [of natural history] at the heart of the Frankfort School project": Jameson, *Late Marxism*, 94. After recapitulating the *Capital* dedication myth (an odd lapse, since *Late Marxism* was published more than a decade after its debunking), Jameson writes that *Capital* affirms "the subsumption of human history . . . under natural history—something henceforth indissociable from Darwin's own work and theorizing" (Jameson, *Late Marxism*, 94–5).

8
Prospect of an End

Climate crisis, incipient fascism, and war: Our world is in serious trouble. For some of us, the fundamental ground of these problems is the capitalist form of society. Yet ours is a minority view. Outside of Latin America and a few scattered pockets of resistance, the political Left is in bad shape. It is difficult not to feel a dark mood, if not dread—even a foreboding sense of the end of civilization. Facing these conditions, I have proposed a turn toward Marxian natural history. To some this may seem like an abstract gesture or a distraction from the present horrors. But if we accept the proposition that our planetary crisis stems from capitalism, and that our fundamental challenge lies in expanding a collective sense of alternative possibilities, then nothing would be more urgent than the cultivation of imagination and a shared perspective on our crisis. Still, I grant that this book—like most written in the spirit of Marx—remains abstract. It may be helpful, therefore, to summarize my argument in relatively practicable terms.

Marxian natural history teaches some basic lessons. Humans evolved and expanded our range and influence until we came to dominate Earth. The subsequent transformation of nature by humans is inseparable from social domination, of hierarchies of power that place some humans over others. Human societies have driven Earth's natural history for thousands of years. This process implies no specific denouement; natural history has no fixed point, or telos, toward which the logic of social struggles or environmental change must bend. Looking retrospectively, however, we can see certain turning points. The emergence of the state produced a major shift in the form of human society; the emergence of capitalism started with modifications of one state-organized society. Once the capitalist form of society emerged, it generated dynamic, global transformations until it enveloped lives everywhere. The resulting change in the form of human society transformed the historical relations of nature and society. Previously existing relationships between humans and nature were reorganized by and made subordinate to the drive of capital to commodify Earth and labor power to derive profit. The burning of fossil fuels, driven by the accumulation imperative, has brought this process to a breaking point. The liberal response—to promise a more "sustainable" version of capitalism qua development—is an empty piety. Capital won its battles against labor during the twentieth century and capitalists in fossil-fuel-producing regions have shown that they will use the state to repress efforts to end fossil fuel production. Moreover, capitalism is not an object to be reengineered, but an expansionary and accumulation-directed social form. This clarifies the failures of twentieth-century attempts to use the state to overthrow capital. Doing so led to various re-formations of social life in state-centered societies, which was never Marx's aim.[1] Nevertheless, the emergence of powerful "communist" state-centered societies generated the most dangerous conflicts in the history of our planet: Previously USA vs. USSR, today USA vs. China. By stimulating the drive to create weapons with the potential to destroy humanity, capitalism has brought us to the brink of disaster—even extinction. We must exit this epoch as soon as possible.

The fundamental contradiction of our time could be stated as follows: An immediate transition from capitalism is necessary to end our planetary crisis, yet there is no clear way out. The capitalist form of human society creates grave problems yet remains powerful enough to repress world revolution and undermine most experiments in creating alternatives. This contradiction is leading us into another world war.

There are a limited number of ways to solve this contradiction. One common "critical realist" line of thought goes like this:

---

[1] Cf. Kōjin Karatani, *Transcritique: On Kant and Marx* (Boston: MIT Press, 2003); Soichiro Sumida, *Marx Against the State: On the Heteronomy of the Political* [in Japanese] (Tokyo: Horinouchi, 2023).

a) Our planetary crisis cannot be solved within the capitalist political order;

b) Nevertheless, our planetary crisis is so severe that human civilization will likely be destroyed if we cannot prevent destruction very soon. Specifically, we must rapidly reduce carbon emissions and prevent world war. An honest political assessment is that (outside of a few exceptional places in Latin America, e.g.) the Left is too small to create a world revolution in the next decade;

c) Ergo, the "ecological deadline" has passed for revolutionary alternatives to capitalism. We must try to prevent the destruction of civilization and fight to build the ecological conditions of survival for human society *within* the present order, knowing that this is contradictory and doomed to fail.

I find this line of thinking compelling, but the conclusion, c), is self-contradictory. So long as humanity remains within the framework of capitalist society, the outcome will be war and destruction. Simply stated, we must find an alternative to c). It is as if the contradictory character of our world situation is so profound that the only form of reason adequate to grasp it must also be aporetical, even self-negating. I offer Marxian natural history as a means to shift the position from which we regard human affairs. As Kant surmised: "If the course of human affairs seems so senseless to us, perhaps it lies in a poor choice of position from which we regard it."[2] To conclude, I will try to sound out passages through our aporias in three steps: First, by clarifying four things which we should *not* do; second, by sketching an alternative; finally, by assessing the contribution of Marx's writing, after Darwin, to the study of the direction of natural history.

What Not to Do

[1] *Reject Darwin for atavism*

Many on the Right hate Darwin and Marx for the same reason: They challenged the comfort of teleology. A softer iteration of the same

---

[2] I. Kant, *Der Streit der Fakultäten*, Akademie-Ausgabe (1917 [1789]), 7, 83, cited in Théogène Havugimana, "Kant on the history of humankind: The invisible hand of nature behind the progress of the realization of freedom," *Estudos Kantianos* 5, no. 2 (2017): 159.

prejudice can be found where certain post-Enlightenment thinkers—Darwin, Marx, Nietzsche, Freud—are lumped together and rejected for their criticisms of agency. The implicit claim is that by complicating the theory of human reason and history, they deny humans agency or the capacity to make well-reasoned decisions about our lives. Yet the critical insistence of their thought clears paths toward a more capacious understanding of who humans really are and could yet become. It would be unwise to reject such contributions.

To say that Marxists should not reject Darwin amounts to accepting and embracing the naturalism and humanism of the Marxist tradition. What is usually called "historical materialism" places most of the emphasis on the latter term, materialism, without explaining what this really means. Any historical materialism that denounces human nature lacks a coherent basis. Consider, for instance, this statement by the Invisible Committee:

> When one asks the left of the left what the revolution would consist in, it is quick to answer: "placing the human at the center." What that left doesn't realize is how tired of the human the world is, how tired of humanity we are—of that species that thought it was the jewel of creation, that believed it was entitled to ravage everything since everything belonged to it. *"Placing the human at the center" was the Western project. We know how that turned out. The time has come to . . . to betray the species. There's no great human family that would exist separately from each of its worlds, from each of its familiar universes, each of the forms of life that are strewn across the earth. There is no humanity, there are only earthlings and their enemies, the Occidentals, of whatever skin color they happen to be.* We other revolutionaries, with our atavistic humanism, would do well to inform ourselves about the uninterrupted uprisings by the indigenous peoples of Central and South America over the past twenty years. Their watchword could be "Place the earth at the center."[3]

Set aside the fact that these anonymous writers lazily lump together millions of people under the racial-geographical headers "Occidentals"

---

3   The Invisible Committee, *To Our Friends* [À Nos Amis], trans. Robert Hurley (Los Angeles: Semiotext(e), 2014), my italics. Accessed October 5, 2023, at theanarchistlibrary.org.

and "indigenous peoples of Central and South America."[4] We humans cannot "betray" our species any more than bacteria could betray their own. We find here a category error: Shifting a term (betrayal) from one realm where it makes sense (alliances) to another (biology) in ways that invalidate the thought.

The underlying question is how we conceive the human relationship to the rest of nature. Like Kant, Marx thought that we could only reason *as* humans. I agree. However, humans reason differently than we did before our lives adopted a capitalist form. Among other things, as humans adopt distinctive class positions, capitalist or proletarian, our conception of the world and behavior become more separated from Earth. Capitalism is a social form that makes us more anthropocentric.[5] Rather than embracing atavism, let us take steps toward ending this damaging social formation.

[2] *Darwinian liberalism: or, capitalism as adaptation*

Some on the Left contend that any appeal to Darwin will lead to counterrevolutionary politics. Marx disproves this: He did not give up on revolutionary politics after reading *Origin*. This is not to deny the possibility of Darwinian liberalism, which clearly exists, only to claim that we cannot expect to displace it with arguments that fail to address evolutionary science or the influence Darwin had on Marx's thought.

Consider this example. In 1999, the utilitarian philosopher Peter Singer wrote a book calling for *A Darwinian Left*.[6] In the preface, he writes:

> My focus here is not so much with the left as a politically organized force, as with the left as a broad body of thought, a spectrum of ideas about achieving a better society. The left, in that sense, is urgently in need of new ideas and new approaches. I want to suggest that one

---

4 Apparently, the indigenous peoples of North America were not indigenous enough to make their list. Send a letter of complaint to: Invisible Committee, France, Europe, Earth.

5 In this sense, the Anthropocene is a suitable name for the epoch of the materialization of the logic of capital.

6 Peter Singer, *A Darwinian Left: Politics, Evolution and Cooperation* (New Haven: Yale University Press, 1999).

source of new ideas that could revitalize the left is an approach to human social, political, and economic behavior based firmly on a modern understanding of human nature. It is time for the left to take seriously the fact that we are evolved animals, and that we bear the evidence of our inheritance, not only in our anatomy and our DNA, but in our behavior too. In other words it is time to develop a Darwinian left.

After developing this line of thinking, Singer asks, "Can the left swap Marx for Darwin, and still remain left?"[7] I trust it will now be clear that this question is misplaced. To develop a Darwinian Left does not mean swapping Marx for Darwin: It means carrying out the promising but as yet unfulfilled project of a Marxian natural history.

Liberalism is by no means defined by Darwinism. The political philosophy of liberalism has deep roots in the thought of the seventeenth and eighteenth centuries and was well developed before Darwin wrote *Origin*. And liberalism changed a good deal through the twentieth century, with the emergence of Keynesian, feminist, neoliberal, and other variations. Given the planetary crisis, where does liberalism stand today, and how does Darwinism figure into it?

In *Climate Leviathan*, Geoff Mann and I argue that liberalism is evolving toward capitalist planetary management, legitimated through the need to address the planetary crisis while relaxing its historical attachment to the nation ("the people") in lieu of saving life on Earth.[8] It follows that only a theory that examines capitalism *and* sovereignty could orient us toward the coming struggles. Working from a set of basic presuppositions about the organization of politics and economic life in capitalist societies, we offer a speculative analysis of four potential future paths for humanity. These paths are defined by two conditions: On one hand, whether economic life will remain defined by capital and value form or be transcended into something like socialism or

---

7 Singer (*A Darwinian Left*) acknowledges that he has no "answers to the weakening of the trade union movement," nor, indeed, to any of the political-economic challenges of our time. For a thoughtful critique of Singer's position, see David Stack, *The First Darwinian Left: Radical and Socialist Responses to Darwin, 1859–1914* (Cheltenham: New Clarion, 2000) 3–6.

8 Joel Wainwright and Geoff Mann, *Climate Leviathan* (New York: Verso Books, 2018).

communism; on the other, whether the existing form of the political will be transcended to generate planetary sovereignty, or not. The most likely scenario is that the world will remain capitalist—indeed it is becoming more fully capitalist every year—and that the form of political sovereignty will adapt to become planetary. We call this path "Climate Leviathan" in homage to Hobbes.

In the face of the planetary crisis, the concept of adaptation is increasingly central to liberalism.[9] The central talking point of the 2023 IPCC AR6 synthesis report, for instance, is that "the way forward" is "climate-resilient development," specifically by "integrating measures to adapt to climate change with actions to reduce emissions in ways that provide wide benefits."[10] The keywords here are "development" and "to adapt." Loans that previously would be demarcated for economic development are today granted to facilitate adaptation to climate change; programs that previously would have been organized for social development are redirected to help people adapt to global heating. The accent shifts from economic growth to adaptation to environmental conditions. This change is motivated and directed by capital's drive for expanded accumulation.

In one sense, to claim that humans must adapt to a changing environment is a truism. That is what all species do all the time. And it is understandable that liberals who want to reduce the suffering of the poor would call for strategies that provide certain pathways to more successful adaptation by the poor on a warmer and weirder planet. Moreover, we need to retain the concept if we are to write statements like this (plucked at random from an IPCC report): "Current global financial flows for adaptation are insufficient for, and constrain implementation of, adaptation options, especially in developing countries."[11] Translated into common language, this says that rich people are not giving enough money to poor people to allow the poor to adapt to the environmental damage caused by the rich.

---

9   The next two paragraphs draw in part from Joel Wainwright, "Capitalism qua development in an era of planetary crisis," *Area Development and Policy* 9, no. 3 (2024): 407–27. I thank the editors of ADP.

10   The IPCC 2023 synthesis report, for example, says that "the way forward" is "climate resilient development ... to adapt": IPCC, "Synthesis report AR6, Headline statements." Retrieved September 20, 2023, ipcc.ch.

11   IPCC, "Synthesis report AR6, Headline statements," 8.

Nevertheless, we must be wary of adaptation-talk. The metaphor of adaptation—with its roots in Darwin's evolutionary theory—is both inadequate to explain our planetary crisis and a poor guide for what we must do about it: Re-form human society. Worse, as often occurs, the shift of a concept from a biological to a social register introduces dangers.

The concept of adaptation has an honorable history. As we saw in chapter 2, Darwin changed how we conceive of species; by his thinking, species are the capricious and temporary result of ongoing evolutionary processes and that adaptation was inherent to and ceaseless for all life. As Ernst Mayr explains,

> Darwin taught us that seemingly teleological evolutionary changes and the production of adapted features are simply the result of variational evolution, consisting of the production of large amounts of variation in every generation, and the probabilistic survival of those individuals remaining after the elimination of the least-fit phenotypes. Adaptedness thus is an a posteriori result rather than an a priori goal setting.[12]

What has changed is that adaptation has become the goal for liberal-capitalist societies.

Further note that, for Darwin, what adapt are *populations* of a *species*. Today, however, we often hear calls for adaptation applied to nations, cities, or specific social groups (for example, indigenous people and racialized minorities). In a world divided by myriad forms of discrimination, calls for capitalism qua adaptation are not only inadequate (since they will not generate mass action and just solutions); they are, in fact, likely to produce new injustices. Talk of climate adaptation also contributes to an illusion that the coming changes are readily manageable within the prevailing liberal capitalist order—they are not. Hence, if we are going to bring about a just response to climate change, we need a stronger call to arms than "We must develop; we must adapt!" From the

---

12   Ernst Mayr, *What Makes Biology Unique? Considerations on the Autonomy of a Scientific Discipline* (Cambridge, UK: Cambridge University Press, 2004), 58, cited in Mauricio Vieira Martins, *Marx, Spinoza, and Darwin: Materialism, Subjectivity and Critique of Religion* (London: Palgrave Macmillan, 2022), 200.

vantage of the two main political blocs in the capitalist world, most conservatives reject the IPCC's call for development as adaptation (with its attendant green Keynesianism); by contrast, liberals seem certain to embrace it as the only possible "solution" to the present crisis. At a moment when our world requires a new direction, we should not tell the poor to adapt to this capitalist world on fire. We should link arms with them to end it.

[3] *Accelerationism and catastrophism*

For a certain segment of the Left, the doomsday scenario just described is already immanent if not actually present, and the only logical response is to confront it. Bring on the global civil war! On the other side of the battle, we can reorganize life on Earth. Any action that hastens this course of events is rational; anything that mitigates against it is irrational. This is accelerationism. While coherent, I find it seriously misguided, for it can easily contribute to reactionary political positions.

Accelerationism is often indistinguishable from a distinct but overlapping style of thinking, catastrophism. This desperate line of thought says: Capital will destroy life on Earth *unless* we are lucky enough that some grave cataclysm, a horrible disaster, leads capital to self-destruct. There is a fatal flaw to this line of thinking. Capital is not a conscious thing; it is a socio-economic formation. We might as well ask a triangle to think benevolent thoughts. As Marx writes:

> Capital has "good reasons" to ignore how generations of workers all around it have suffered, and in its actual movement it is affected by the prospect of humanity's coming ruin and unstoppable depopulation just as much or as little as by the possibility that the earth will fall into the sun. Every time some swindle causes a stock to soar, everyone knows that the stock will eventually crash, and every person hopes that before this happens he will manage to collect the rain of gold and store it safely while someone else is caught outside in the lightning and thunder. "*Après moi le déluge!*" is the watchword of every capitalist and every capitalist country. Capital takes into account the well-being and mortality rates of its workers only when society forces it to ... This behavior doesn't come down to the individual capitalist's

will, to whether his will is good or bad. Free competition makes the immanent laws of capitalist production operate for individual capitalists as external laws that they are forced to obey.[13]

Capital teaches that, if we want something other than a deluge, we must force capital to confront the well-being of the proletariat and all those populations made surplus to capital. This requires transforming our social formation.

The psychic space from which accelerationism and catastrophism emerge was examined a century ago by Freud in his late works.[14] Freud argues that civilization—an effect of the collective renunciation of instinct—existed, in the first instance, to defend us against the forces of nature. However painful it may be to repress an instinctual drive, compensatory pleasures are gained when, for example, we humans unify to rebuild after a flood or storm:

> With these forces nature rises up against us, majestic, cruel and inexorable; she brings to our mind once more our weakness and helplessness, which we thought to escape through the work of civilization. One of the few gratifying and exalting impressions which mankind can offer is when, in the face of an elemental catastrophe, it forgets the discordancies of its civilization and all its internal difficulties and animosities, and recalls the great common task of preserving itself against the superior power of nature.[15]

What Freud describes is neither planning nor adaptation per se, but the spontaneous unification of discrete social groups—a unification compelled through the recognition, in the face of nature's threat, of our common humanity.[16] The prospect of such responses to the threats presented by the climate crisis provides hope for many who believe that

---

13 Marx, *Capital*, 239–40.
14 Sigmund Freud, *The Future of an Illusion* (1999 [1927]) and *Civilization and Its Discontents* (1999 [1930]), in *Standard Edition of the Complete Psychological Works of Sigmund Freud*, XXI, trans. James Strachey, (New York: Vintage, 2001).
15 Freud, *The Future of an Illusion*, 19.
16 Freud's insight helps explain a phenomenon elegantly documented by Rebecca Solnit in *A Paradise Built in Hell* (New York: Penguin, 2010): under certain circumstances, to quote the book's subtitle, "extraordinary communities [may] arise in disaster." Cf. Seymour, *Disaster Nationalism*.

if only the crisis arrives in the right form (storm, flood, drought) or with the right degree of fearsomeness (so many dead, so much turmoil), humanity will be subjected as beings with a common planetary interest. This prospect, so widely anticipated if not yet realized, holds our future in suspense. It is withheld not for lack of catastrophe, for there are plenty, but by a politics of reaction that promises the conquering of nature for some or another particular social group defined by nation, race, creed, or faith.

[4] *The dream of command and rational planning of nature*

Part of the current debate among ecological socialists concerns our goal. For those who interpret Marx and Engels as communists who sought the command and domination of nature, and who saw the state as the mechanism to bring this about, it is understandable that ecological socialism requires some sort of world state, "modernization" of the human relationship with Earth, and planetary sovereignty. In desperate times, a powerful state can be attractive. Logically, to rationally plan a planetary transition—whether to Climate Leviathan or ecological socialism—would require a planetary agent. I find no basis for such a vision in Marx's later thought; however, the late writings of Fredrich Engels proved to be more significant than those of Marx for the development of Marxism. This remains a touchy point for some Marxists, but it must be addressed.

There has been a tendency recently to reply to criticism of Engels by arguing that his work cannot be completely separated from Marx and, at any rate, it is unfair to pin the problems of twentieth-century communist states on a nineteenth-century thinker. I concede these points. I also recognize that Engels studied the natural sciences to analyze human-environment relations with sophistication. Nevertheless, Engels's influence complicates ecological socialist thought and politics because his writings contributed to a powerful tendency to imagine that we can overcome the capitalist planetary crisis through rational, socialist domination. This theme runs through his late writings like a red thread, beginning from his explanation of the distinction of humans from nature: "The animal merely *uses* its environment, and brings about changes in it simply by its presence; man by his changes makes it serve his ends, *masters* it. This is the final, essential distinction between man and other animals."¹⁷ In fairness, Engels qualifies this statement by describing, in prescient terms, the ecological consequences of human "mastery": "Let us not, however, flatter ourselves overmuch on account of our human victories over nature. For each such victory nature takes its revenge on us." "We by no means rule over nature like a conqueror over a foreign people, like someone standing outside nature—but that we, with flesh, blood and brain, belong to nature."¹⁸ Nevertheless, betraying the influence of Enlightenment thinking, Engels elaborates:

> our mastery of [nature] consists in the fact that we have the advantage over all other creatures of being able to learn its laws and apply them correctly. And, in fact, with every day that passes we are acquiring a better understanding of these laws … After the mighty advances made by the natural sciences in the present century, we are more than ever in a position to realise, and hence to control, also the more remote natural consequences of at least our day-to-day production activities.¹⁹

This passage invites a conception of socialism as a science to control and master nature. Of course, such a conception requires an agent to coordinate the rational mastery of nature: Here comes the party-state. (Today, the Communist Party of China comes closest to the realization of this ideology. Yet China is today a thoroughly *capitalist* society and the CPC arguably the single largest capitalist actor in the world.)

Consider a passage from what was arguably the most influential text of late nineteenth and early twentieth century Marxism, Engels's *Anti-Dühring*:

> With the seizing of the means of production by society, production of commodities is done away with, and, simultaneously, the mastery of the product over the producer. Anarchy in social production is replaced by systematic, definite organization. The struggle for

---

17   Friedrich Engels, "The Part Played by Labour in the Transition from Ape to Man," MECW 25 (1987 [1876]): 460. Cf. Maurizi, *Beyond Nature*.
18   Engels, "The Part Played by Labour in the Transition from Ape to Man," 460–1.
19   Engels, "The Part Played by Labour in the Transition from Ape to Man," 460–1.

individual existence disappears. Then, for the first time, man, in a certain sense, is finally marked off from the rest of the animal kingdom, and emerges from mere animal conditions of existence into really human ones. The whole sphere of the conditions of life ... which have hitherto ruled man, now comes under the dominion and control of man, who for the first time becomes *the real, conscious lord of nature*, because he has now become *master of his own social organization*.[20]

This is a tragic vision that the ecological socialist tradition must confront and transcend.[21]

Note Engels's repetition of "products" and "production": Engels treats emancipation as a matter of overcoming scarcity. On those grounds, capital always wins. Our great problem, to recall *Capital*'s first conclusion, is not lack of commodity production, but the generation of a large population that is surplus to value.[22] By giving too much emphasis to production (relative to circulation, consumption, and reproduction, for example), Engels exhibits a simplistic style of thinking, sometimes called "productivism" or "Prometheanism." Marx overcame Prometheanism in the 1860s.[23] Engels did not.

The second problem concerns the political strategy implied here. Engels unifies two thoughts and promulgates them systematically. The first concerns the *nature* of a revolution against capital; Engels equates the *seizing*, or taking, of the means of production, with the *simultaneous* overcoming of capital's domination over life: "With the seizing of the means of production by society, production of commodities is done

---

20 Friedrich Engels, *Anti-Dühring*, MECW 25 [1877]: 270, my italics.

21 Cf. Sven-Eric Liedman, *The Game of Contradictions: The Philosophy of Friedrich Engels and Nineteenth-Century Science* (Chicago: Haymarket 2022), 323–6, 437, and 450; Paul Heyer, *Nature, Human Nature, and Society: Marx, Darwin, Biology, and the Human Sciences* (London: Greenwood, 1982), 85–7; Kōhei Saitō, *Marx and the Anthropocene*, 60–1.

22 For this reason, *pace* Leigh Phillips and Michal Rozworski, Walmart is definitely not "laying the foundation for socialism": Phillips and Rozworski, *The People's Republic of Walmart: How the World's Biggest Corporations Are Laying the Foundation for Socialism* (New York: Verso Books, 2019).

23 Paul Burkett, "Was Marx a Promethean?," *Nature, Society, and Thought* 12, no. 1 (1999): 7–42; Kōhei Saitō, "Marx's ecological notebooks," *Monthly Review* 67, no. 9 (2016): 25–42. The timing again suggests Darwin's influence on Marx.

away with, and, *simultaneously*, the mastery of the product over the producer."[24] The twentieth century flatly disproved this hypothesis. When communists seized the state and means of production in China, North Korea, Cuba and Ethiopia, to name four illustrations, the production and circulation of commodities were not at all "done away with."[25] The form of value that organizes our lives cannot be swept away by an armed group, no matter how well-disciplined. That is because, as Marx emphasizes in *Capital*'s third conclusion, capital is not a *thing* that a party or state can dispose of, but a *form* of socio-economic relations organizing people's lives.

Engels missed this lesson of *Capital*. His desire for humanity to become "marked off from the rest of the animal kingdom," and emerge "from mere animal conditions of existence into really human ones [so that nature] now comes under the dominion and control of man, who for the first time becomes the real, conscious lord of nature," is also foreign to *Capital*. Engels derived this from early political economists and their interpretation by Hegel, for whom the anarchy of market society would be overcome by reason, custom, and the state (exhibiting spirit's progress in history). Whereas Marx realized his critique of Hegel after reading Darwin, by contrast, Engels remained a Left Hegelian: "Engels' conviction that we should control the anarchic drive of capitalist production and transform it into a planned economy was little more than an extension of classical economists' thought. And Engels' stance was, of course, the source of centralist communism."[26]

Pace Engels, the opposite of central planning is not necessarily chaos. As Darwin showed, nonteleological evolutionary processes produce complexity, change, *and* order. In chapter 3, we noted Marx and Engels did not read Darwin the same way and that some of the misunderstandings about Marx and Darwin were introduced by Engels, beginning with his speech at Marx's graveside.[27] As Paul Thomas observes, Engels's

---

24 Engels, *Anti-Dühring*, MECW 25: 270, my italics.
25 I limit myself to examples which I have had the privilege to study firsthand, but I suspect that my claim applies to other cases. This is not the place for a critical account of state communism and the prospects of another path to communism. Suffice to say that I am persuaded by the diagnosis of Kōjin Karatani, *Structure of World History: From Modes of Production to Modes of Exchange* (Durham: Duke University Press, 2014).
26 Karatani, *Transcritique*, 8.
27 See also Maurice Bloch's critique of Engels's anthropological theories persuasive: *Marxism and Anthropology* (London: Routledge, 2004 [1983]).

views run "against the fact that Marx read Darwin's *Origin of Species*," a book that has nothing to do with

> human efforts to transform nature, efforts which pale into relative insignificance in comparison with the sheer scope and scale of what Darwin was concerned to characterize. The idea that nature begins to exist for man only with the advent of active, human intervention in natural processes is an idea that owes nothing to Darwin and nothing to Marx, though it does play a part in Engels's speculations.[28]

The ideology of early twentieth-century Marxism that emerged from the study of works by Engels taught that the proletariat should seize the means of production, bring about the end of commodity production, and make the proletariat the conscious lord of nature. In theory, the agent that would act *as* the proletariat in this process was the *party*, which would coordinate the proletariat through a communist project. In reality, wherever revolutions occurred, it has been the *state*, unified around a subgroup of *party leaders*. The failures of twentieth-century strategies for communism are thus linked to Engels's vision of human domination of nature. The former is expressed through the latter, and vice versa, in his influential late works. Pace Engels, I do not believe we will ever mark off humanity from the rest of the animal kingdom, become conscious lords of nature nor masters of a planned society.[29] That is a hopeless delusion, a refusal to embrace the contingencies of natural history. On this point, we should confirm Darwin, Marx, and Freud—against Engels.

Engels certainly does not bear all responsibility for the failures of Marxism qua state socialism. Nevertheless, we cannot avoid recognizing that his conception of the human relationship with nature, and the role

---

28 Paul Thomas, *Marxism and Scientific Socialism: From Engels to Althusser* (New York: Routledge, 2008). Carver comes to a similar assessment ("it is questionable whether . . . the common methodology attributed to Marx and Darwin really fits" either thinker): Terrell Carver, *A Dictionary of Marxist Thought* (Oxford: Blackwell, 1983), 113.

29 In an epoch of climate crisis and species extinctions, "No sensible argument is advanced by easy talk about rolling back nature's barriers, or the Faustian or Promethean imposition of human purposes on natural processes": Thomas, *Marxism and Scientific Socialism*, 8.

of planning for overcoming the problematic relationship between humans and nature in capitalist society, played a central role in shaping the interpretation of Marx during the twentieth century. It was only after the 1990s—with the dissolution of the USSR and the capitulation of the Communist Party of China to a capitalist economic order—that ecological Marxist thought has flourished. We have been able to revisit Marx's texts and articulate an alternative view.

I wrote earlier that it would be absurd to claim that Darwin influenced Marx's political views. That is true, to a point. Marx and Darwin had no explicit influence upon each other as political thinkers. However, in one subterranean fashion, Darwin's scientific writings contributed to a shift in Marx's political thought.

In the 1840s—most famously in the *Communist Manifesto*—Marx felt confident enough to predict the inevitable fall of capitalism and its replacement by communism. After the Bolshevik revolution of 1917, those statements become foundational to the ideologies of actually existing state-socialist societies, the proverbs of a secular eschatology. By contrast, Marx's conclusions of *Capital* present an open and ambiguous vision of capital's futures—one reason the book was downplayed during the twentieth century in those ostensibly socialist societies. Some interpret this marked change in Marx's tone about the future as a decline in his confidence after the failure of the 1848 revolution; others attribute the conclusion of *Capital* (particularly its turn to colonialism) to Marx's desire to write an homage to Hegel.[30] Summarizing his assessment of Marx's late research notebooks, Kōhei Saitō memorably describes "Marx sweating blood as he moves away from his convictions about history as progress, arriving finally at a completely new conception of history."[31] Saitō attributes the change to Marx's "research into ecology and the communal societies of the non-Western and precapitalist world," and it is true that this research contributed to Marx's ideas. But then the question arises: What motivated Marx to conduct that research in the first place? Why sweat all that blood? No thinker as serious as Marx would devote decades of research into obscure topics unless they were already driven by a desire to learn something in particular. What was Marx after?

---

30 For the former, see Eric Hobsbawm; for the latter, David Harvey.
31 Saito, *Slow Down*, 110.

To recapitulate, after reading *Origin* in 1860, Marx recognized that the movement of natural history has no telos. I contend that this *scientific-philosophical* clarification, coupled with his growing *political-geographical* appreciation of the myriad struggles against imperialism throughout the 1860s, led him to positively reevaluate the prospect of multiple conflicts against capitalism and imperialism—albeit without the prospect of any particular end. To be sure, Marx remained committed to overcoming capitalism, but he became more uncertain about where and how this process would occur. Hence Marx's capacious studies into world history and the anthropological literature of his day—as well as his revised proposals about revolutionary paths in his letters to Vera Zasulich.

This ambiguity is crystallized in *Capital*'s first conclusion. Marx correctly predicted the emergence of a worldwide, relative surplus population (surplus to capital's valorization), but he could not anticipate its precise coordinates—how and where this population would live and struggle and with what political results. We still lack a fulsome account. Tracking the demographic and socio-natural conditions of the mass of humanity made surplus to capital will be fundamental to any Left program to come.

While the conclusion of *Capital* is therefore more *politically* ambiguous than Marx's earlier writings, it is stronger *analytically*. With inspiration from Darwin, Marx's critique of capitalist political economy became the first study of the natural history of a socio-economic form of society. *Capital* stands as the first analysis of human society that explains its development within the economic form of capital. The place of *Capital* in the canon of human sciences is thus assured. Yet what matters today is not the determination of that canon, but how we read it, and to what ends: What we can learn about ourselves and how we might change our form of society.

Toward a New Social Formation

In the face of our planetary crisis, we need to cultivate a practical set of responses to build genuine alternatives. In *Climate Leviathan*, Mann and I sketched conditions for a pathway we call "Climate X," a world that transcends capitalism without planetary sovereignty. We called for mass boycotts and general strikes, steep taxes on wealth, the renewal of mass movements for democratic socialism, and so on. This was the part of our book that received the most criticism: Many readers felt that we were too vague. I certainly respect that our readers expect clearer ideas from those who propose a new social formation. Still, I also agree with Marx that the role of the intellectual is not to draw up blueprints for the future, nor to "confront the world in a doctrinaire way":

> We do not say to the world: Cease your struggles, they are foolish; we will give you the true slogan of struggle. We merely show the world what it is really fighting for, and consciousness is something that it *has to* acquire, even if it does not want to.[32]

In that non-doctrinaire spirit, I should offer a statement of what I think "the world . . . is really fighting for."

There are many people in the world today—and many more to come—whose lives are so precarious that the immediate task is to secure survival. Our world's fate is tied to their struggles to survive. Yet our ultimate goal is not bare survival, but collective flourishing; not only confronting capital's domination but achieving genuine emancipation; neither helplessness nor Earth-mastery but common wealth and a virtuous, collective drive to reimagine and repair social and ecological relations. Achieving these goals will require that we move beyond a social form defined by relentless valorization of the commodity labor power and the implicit doctrine—which mistakenly seeks legitimacy from Darwin—*omnis vir enim sui*: Every man for himself. But what is the alternative? Surely it lies with the principle *omnia sunt communia* (all wealth shall be common). This does not mean poverty, nor denial of individual property. As Marx writes in the second conclusion of *Capital* (chapter 23):

> In a process that has the necessity of any natural process, capitalist production . . . produces its own negation—the negation of the negation. This restores individual property, which, however, is now based on the achievement of the capitalist era: Namely, the cooperation of

---

32 Marx, Letter to Ruge, 1843, MECW 3: 133–45, 144.

free workers and their collective ownership of the land and the means of production that are produced by labor itself.[33]

How do we realize this principle amidst radical uncertainty about the world's future, mass alienation, and social isolation? We might as well admit that we find ourselves in a struggle for sanity amidst these challenges. Within this admission lies an important source of inspiration (which should take pride in its Darwinian provenance): Humans evolved as highly social primates. We need friends, kin, and comrades with whom to build collectivities of solidarity and joy. But it must also be acknowledged that, within these sorts of goals, lie individual wills and desires, and it is often difficult to know *why* we want what we want. The religious character of many of the Left's slogans and ideas shows that a metaphysics of belief lies behind every political goal and program. This goes some way to explain the crisis the Left has faced globally since the 1990s: After the fall of actually existing socialism, the Left has passed through a difficult period of reconstituting a metaphysics of communism. I do not think this period has ended; indeed, this book could be taken as a contribution to the reconstitution of ecological communism, which could only be realized through struggles along a path like Climate X.[34]

For this political project, what is the virtue of a Marxian natural

---

33 Marx, *Capital*, 691. Reitter translates "die Erde" as "the land"; this could also be translated as "the earth". Elsewhere in *Capital*, Marx notes in passing that "economically speaking ... the land ... includes water" (*Capital*, 154). For that matter, "the land" would also encompass wind, our atmosphere, incoming solar radiation, geothermal energy, and so on. The point is that, by Marx's reasoning, under communism, what will be held as common wealth will include Earth, not just land. In personal correspondence, Reitter indicated that translating this passage as "collective ownership of the earth" would have gone "too much in the direction of later discourses of environmentalism," and I take his point. Nevertheless, for my Marxian natural-historical reading of *Capital*, it is important to keep the capaciousness of the original German term (Erde) in mind. (See also chapter 4, note 60.)

Apropos this passage, Saitō notes that Marx "aims at rehabilitating the commons at a higher scale for the sake of the full development of the richness of wealth for most working individuals. Against the logic of commodification by capital, communism seeks the commonification of wealth": Saitō, "Primitive accumulation as the cause of economic and ecological disaster," in ed. Musto, *Rethinking Alternatives with Marx* (2021), 93–112, 105.

34 For a fictional depiction of struggles to realize Climate X in the USA, see Stephen Markley, *The Deluge* (New York: Simon & Schuster, 2023).

history perspective? At a minimum, it reminds us of the need to repeatedly reveal that the torments of nature tearing through the world—heat waves, hurricanes, fires, floods—are not "natural disasters," for they are neither natural nor misfortune, but the logical consequences of capital's drive. Capital's drive exists *not* because of the private motivations of individuals nor as some abstract principle of our species ("humans are greedy"), but as a consequence of the socio-economic form that humans have adopted for a few centuries. So long as we remain in this form, humans will behave greedily.[35] The planetary crisis reveals the urgent need to transform the prevailing socio-economic form of society, establishing a new epoch in Earth's natural history. No social group has ever faced up to such a challenge. It is only reasonable to ask how in the world it could be enacted.

But how?

Nothing in this book changes the basic tools that the Left has at its disposal for confronting capital. We are still the ones who have the capacity to refuse to sell the labor power commodity (strike) and to refuse to buy (boycott) and thereby to interrupt the circuit of capital. We could call a general strike at any time on ecological grounds—calling for a halt to business as usual and for the restructuring of our social formation. Of course, just as unions do not go on strike without making strategic preparations, it would be foolish to try to shut everything down without thinking clearly about our aims. The risks and virtues of such thinking can be identified by contrasting two recent experiences from Europe. After the initial upsurge in enthusiasm for Extinction Rebellion—a network whose audacity and clever name exceeded the rigor of its political analysis—the movement faced grave difficulties after some of its adherents decided to interrupt the movement of mass transit (commuter trains and buses in London). Since we need more mass transit, not less, and the people who suffered delay were ordinary people, not elites or capitalists, this action was a dramatic self-defeat for the climate justice

---

35 "The problem is not that capitalism is a conspiracy of greedy people. The problem is that capitalism, as a way of organizing ... life, does its best to force us to be greedy": Geoff Mann, *Disassembly Required: A Field Guide to Actually Existing Capitalism* (AK Press: Oakland, 2013), 4–5.

movement. By contrast, when Greta Thunberg initiated her school strike, it was an event with potential world-historical implications. The reason was that her act presented an essential, existential question through a powerful image: If rapid climate change is going to destroy the planet, why should a child bother going to school? (The desire of liberal politicians to have their photograph taken with Thunberg signified that they had no answer to her question—only a desire to siphon some of her popular appeal as a moral prophet for their electoral gain.)

School strikes—even thousands of them, as we saw in the months before the Covid lockdown—cannot alone interrupt capital. But such an interruption *could* be done via mass boycotts and general strikes, if large and well-coordinated; add steep taxes on wealth, bans on the burning of fossil fuel, and steps toward meeting collective needs and ecological socialism become imaginable. Humanity could apply an emergency break to natural history.[36]

Executing such processes would require coordinated political action and the renewal of mass movements for democratic socialism.[37] And, if we are to take these steps while keeping ourselves fed and sheltered—in economic lingo, suspending harmful commodity production while sustaining social reproduction—then we need to lay the groundwork for revolutionary transformation while building some alternative forms of economic life.[38] Marx anticipated this and gave the name "associationism" to the post-capitalist mode of economic organizing.[39] Practically speaking, we can imagine associationism taking the form of networks of producers' and consumers' cooperatives, creating a grand social lattice for meeting peoples' needs: For producing and distributing public wealth, not accumulating value privately.

Saitō's proposal for "degrowth communism" provides us with a helpful clarification of how these classical ideas might play out in an

---

36 A reference to Walter Benjamin's famous remark: "Marx says that revolutions are the locomotive of world history. But perhaps it is quite otherwise. Perhaps revolutions are an attempt by the passengers on this train—namely, the human race—to activate the emergency brake," Benjamin, "Paralipomena to 'On the Concept of History,'" in *Selected Writings, Volume 4, 1938–1940*, eds. H. Eiland and M. W. Jennings (Cambridge, MA: Harvard University Press, (2003 [1940]), 402.
37 Nicos Poulantzas, *State, Power, Socialism* (London: Verso Books, 1978).
38 Alyssa Battistoni, "States of emergency," *Nation*, June 21, 2018.
39 Peter Hudis, *Marx's Concept of the Alternative to Capitalism* (Leiden: Brill, 2012); Paresh Chattopadhyay, *Socialism and Commodity Production* (Leiden: Brill, 2018).

ecological transition beyond capital. Saitō locates the kernel of this idea in the writings of the late Marx:

> Marx, having discarded a progressive view of history [what I have called his post-Darwinian destruction of teleology—JDW] incorporated the principles of sustainability and steady-state economics from communal societies into his revolutionary thought ... What Marx achieved at the end of his life was a vision of degrowth communism. This is the shape of society's future that no one had pointed out before, that constitutes the truly new analysis presented here. Even Engels, his comrade, was unable to grasp it. As a result, Marx's ideas were misunderstood after his death as founded on a unilinear view of history as progress—that is, as a leftist paradigm for productivism.[40]

Marx never prepared a detailed program for such a socio-ecological transformation; he wrote no manifesto for degrowth communism. However, in one of his final writings—his 1875 critique of the draft program of the United Workers' Party of Germany—he provided the following clarifications:

> In a higher phase of communist society, after the enslaving subordination of the individual to the division of labor, and therewith also the antithesis between mental and physical labor, has vanished; after labor has become not only a means of life but life's prime want; after the productive forces have also increased with the all-around development of the individual, and all the springs of cooperative wealth flow more abundantly—only then can the narrow horizon of bourgeois right be crossed in its entirety and society inscribe on its banners: From each according to his ability, to each according to his needs![41]

---

40   Saitō, *Slow Down*, 121–2.
41   Marx, "Critique of the Gotha Program," in Saitō, *Slow Down*, 124; "cooperative wealth" translates Marx's expression der genossenschaftliche Reichthum. Saitō explains: "The German word *genossenschaftlich* carries with it the connotation of both cooperatives understood as unions and cooperatives understood as free associations, and he tended to use it in phrases like 'cooperative production' and 'cooperative mode' of production." "'Cooperative wealth,' though, is a phrase that only appears in this line from the 'Critique of the Gotha Program.'"

There is no room in this vision for the fetishism of production and growth. Communism means replacing subordination with *associationism* and the capitalist accumulation of value with *cooperative wealth*.

As the concept has gained more attention, some Marxists have criticized the strategy of degrowth communism.[42] Essentially, the criticism is that degrowth is inherently a professional-middle-class goal that is set against the interests of the working class because, *ceteris paribus*, degrowth means austerity. As austerity is bad for workers—and workers hate austerity—socialists should hate degrowth. It will be unpopular and will not work anyway.

This line of criticism calls for some clarifications. First, "degrowth" is a neologism that means different things to people (this is a weak defense of the concept, but an important point to bear in mind). Unfortunately, the word "degrowth" is not only vague but also badly represents the goal of ecological communism. Properly understood, degrowth would mean overcoming capitalism's growth (profit) drive by reorganizing social life on a more just basis. But, since the term uses de-growth, not "value" or "capital," it displaces what is essential. I take it that this is why Saitō wisely made "degrowth" an adjective to qualify "communism."[43] Second, since "growth" is always an increase of something in particular, we should clarify what ecological communism aims to *increase* as well as *decrease*. It seeks persistent growth of good things: Trees, solidarity, love, freedom, fulfillingness—in a word, the growth of common wealth via collective liberation. What ecological socialism must end is growth of value and carbon in Earth's atmosphere. Since these points should be easily understood by ecological Marxists, what is the real cause of the debate? Partly it is personal, but if we put this aside, the confusion stems from certain underlying assumptions. For instance, that "economic growth" exists abstractly, outside of the capitalist social relations; second,

---

42   See, for example the—weak—polemic against Saitō by Matt Huber and Leigh Phillips, "Kōhei Saitō's 'start from scratch' degrowth communism," *Jacobin*, March 9, 2024, jacobin.com. Saitō has presented numerous responses to such criticisms; see also Michael Levien, "Climate change as class compromise? On the limitations of Huber's Marxism and climate politics," *Historical Materialism* blog, May 30, 2023.

43   If we accept the slogan "degrowth communism," then—as Gramsci wrote in his notebooks apropos the expression, "materialist conception of history"—the emphasis should be on the latter term (communism, history), not the former (degrowth, materialism).

that this "growth" is potentially politically progressive: For ordinary worker-consumers, an end to growth sounds terrible. In this sense, the anti-degrowth positions find a direct parallel with those who refuse to accept any criticism of the concept of development. But communism was never about "development" any more than "growth" or "progress": These are keywords of liberal, bourgeois ideology. By contrast, communism expresses the social drive for justice, dignity, and collective emancipation.

Marxian natural history further offers a standpoint from which to overcome an unfortunate opposition—reinforced by Marx's architectural metaphor, "base and superstructure"—between productive and reproductive activities.[44] Nancy Fraser astutely observes that we should treat this distinction as *internal* to capital.[45] The strategy that follows would entail not simply trying to bring greater attention to reproductive labor (a task that remains necessary, of course), but of practically tying forms of labor that are coded as productive to social reproduction, and vice versa. On the defensive side of the ledger, this means—to use a concept from economics—adopting strategies to internalize the externalities: Refusing to allow capital's accumulation drive to cast off costs while privatizing surplus labor. More positively, it means demanding that our lives may be lived for our communities—for building the commune: For transforming, inch by inch and hour by hour, the emancipation of space and time from capital and for the common good.

---

44 On the limits of Marx's architectural metaphor, see Kōjin Karatani, *Architecture as Metaphor: Language, Number, Money* (Boston: MIT Press, 1995); Kōjin Karatani, *The Structure of World History: From Modes of Production to Modes of Exchange* (Durham: Duke University Press, 2014). For assessments of Karatani's argument, see Joel Wainwright, "The spatial structure of world history," *Journal of Japanese Philosophy* 4, no. 4 (2016): 33–59; Nadir Lahiji, *Kōjin Karatani's Philosophy of Architecture* (New York: Taylor & Francis, 2024).

45 Nancy Fraser, *Cannibal Capitalism: How Our System Is Devouring Democracy, Care, and the Planet—and What We Can Do About It* (New York: Verso Books, 2023). It follows that we should refuse to grant the distinction ontological status, that is, treat it as merely methodological.

Simultaneous World Revolution

Practically all writing on degrowth, including Saitō's degrowth communism, remains vague on the political strategy for its achievement. Obviously, some sort of strategy is needed to coordinate strikes, boycotts, and the disruptions of the labor process which would necessarily accompany any degrowth policies. As Marx warns in *Capital*:

> Activated labor-power—that is, living labor—naturally has the capacity to preserve value as it adds new value. This natural gift, which doesn't cost a worker a thing, is a boon for the capitalist: It preserves his existing capital value. If the capitalist's business is running smoothly, he will be too focused on turning a profit to notice labor's generosity. But if the labor process is disrupted in some dramatic way, if a crisis occurs, he will be painfully aware of it.[46]

When the capitalist class discovers a dramatic disruption in the labor process, what does it do? Why, it uses the state to enforce labor discipline and conformity to power.

For generations, the response by the Left to this challenge has been to negate it directly—to try to seize the state, both for defensive reasons (to prevent the state apparatus from crushing the revolt) and offensive (to discipline capital—to stop the burning of fossil fuels, for example). Naturally, I recognize the attractiveness of this ostensible solution. But there are problems with this approach. The capitalist form of the state is ill-equipped to play the roles that the Left would like to assign it.[47] Most Marxists have underestimated Marx's contempt for the state and overestimated the likelihood that the state would wither away after a communist revolution; Marx recognized that the state cannot merely be "seized" one day and used for revolutionary ends the next. Second, the historical record of actually existing state-centered communism (Leninism and Maoism, for instance) has produced, in my estimation, poor results. However we draw up the balance sheet, they failed to generate communist societies. But where does this leave us? Are we doomed to remain stuck within the

---

46 Marx, *Capital*, 181.
47 Nicos Poulantzas, *State, Power, Socialism* (London: Verso Books, 1978).

capitalist form of society, falling deeper into planetary crisis and repressed by state apparatus? Is this the end?

Marx once made a remarkable claim: "communism is only possible as the act of the dominant peoples 'all at once' and simultaneously, which presupposes the universal development of productive forces and the world intercourse bound up with them."[48] The only coherent attempts to build a theory of revolution upon this conception comes from the vertiginous conclusion of Kōjin Karatani's study of *The Structure of World History* with his theory of "immanent and ex-scendent" struggles.[49] By immanent struggles, Karatani refers to the path of "associationism," the struggle to build a society based upon free and reciprocal exchange, organized via the networking of non-capitalist cooperative associations. Practically speaking, this means the creation of a broad, diverse social movement of worker-consumers:

> Counter-movements against the state and capitalism in each nation are cause *sine qua non*. Concretely, this requires the creation of a non-capitalist alternative economy based upon reciprocal exchange at the level of transnational networks—that is, without state dependence.[50]

Now, if we create non-capitalist economic networks that seriously threaten capital's expanded accumulation, we can be sure that capitalist states will seek to interfere or destroy our alternative networks. In Karatani's terms: "if these movements were to reach a certain level of development, they would certainly face disruption by the state and capital; transnational networks would be blocked and divided. Therefore, counter-movements from 'outside' or 'above' are just as necessary." I will return to these "counter-movements from 'outside' or 'above' " (which Karatani elsewhere refers to as "ex-scendent" struggles) momentarily. For the moment, I wish to stress that Karatani's reading of Kant and Marx aims to clarify a new path to conceptualize and realize alternatives

---

48 Marx and Engels, *The German Ideology*, MECW 5: 49. This sentence was written in Marx's hand: see Carver and Blank, *A Political History of the Editions of Marx and Engels's German Ideology Manuscripts*, 99.

49 See Karatani, *Transcritique*, §7; on simultaneous world revolution, Karatani's *The Structure of World History*; cf. the more pessimistic conclusion of Karatani's *Power and Modes of Exchange*.

50 Kōjin Karatani, "Beyond capital-nation-state," *Rethinking Marxism* 20, no. 4 (2008): 569–95, 591.

to capital. Drawing on a distinction from Kant's practical ethics, Karatani argues that we should not treat the task of transcending capital-nation-state as *constitutive*, but as a *regulative* idea. This implies living so as to reorganize society around a form of exchange that is at once free and reciprocal. Such a form of exchange would exist in parallel to and partly outside the circuit of M-C-M' and would not treat labor-power as a commodity. The model here is that of associations of combined consumer-producer cooperatives: A society organized as an ensemble of associations of free and equal producers (to recall Marx's expressions for communism in *Capital*). Building such relations would require transcending the value form and the organization of our lives around money.[51] All this is unlikely to occur soon on a global scale; hence the need to articulate a conception of practical ethics, so that we may live *as if* it were possible to transcend capital-nation-state (knowing that such transcendence is effectively impossible at present).

In chapter 12 of *Structure of World History*, Karatani examines this problem to argue that this antinomy lies at the heart of the challenge of transcending the capitalist state today:

> If by chance the revolution should occur in one country, it would immediately encounter interference and sanctions from other countries ... A socialist revolution that really aimed to abolish capital and state would inevitably face interference and sanctions. A successful revolution that wants to preserve itself has only one option: To transform itself into a powerful state. In other words, it is impossible to abolish the state from within a single country.
>
> The state can only be abolished from within, and yet at the same time, it cannot be abolished from within. Marx was not troubled by this antinomy, because it was self-evident to him that the socialist revolution was "only possible as the act of the dominant peoples 'all at once' and simultaneously."[52]

---

51 Thus, to facilitate the growth of an association of free and equal producers, we must "establish a financial system (or a system of payment or settlement) based on a currency that does not turn to capital, namely, that does not involve interest": Karatani, *Transcritique*, 297.

52 Karatani, *The Structure of World History*, 292. The inset quote is from Marx and Engels, *The German Ideology*, MECW 5: 49.

However, to say the least, "It is difficult ... to unite movements from various countries whose industrial capitalism and the modern state exist at different stages of development." Karatani concludes that Marxism must dislodge itself from the party-centered strategy that has dominated its political theory since at least the Second International. The fundamental problem arises from a conception of capitalist society derived from Marx's base-superstructure metaphor, in which the economy constitutes an autonomous base upon which various superstructural elements (state, law, religion, nation) stand. It follows from this conception that if one could turn the state against capital—after a communist party takeover, for example—and smash the economic base, a new superstructure would arise. Karatani's explanation for the failure of this strategy is that the nation, religion, and state arise from and are driven by distinct modes of exchange which predate the generalization of capitalist social relations; ergo, nation and state are endowed with their own reason for existing and they form autonomous powers that will support capital when capital is attacked.

The peculiar difficulty we face in overcoming global capitalism lies in this mutuality of nation, state, and capital, a triadic structure that Karatani likens to interlocking Borromean rings: An attack on any one fails to break the whole because the strength provided by the other rings. Think of the way that nationalism and the state resurged in China during the Maoist period, eventually (after the Great Cultural Revolution, a kind of civil war within the revolution) facilitating the hegemony of capital. We can also see this in the way that the influence of nation and state have resurged in the wake of the 2007 global economic crisis. Karatani explains:

> One frequently hears today that the nation-state will be gradually decomposed by the globalization of capitalism (neo-liberalism). That is impossible. When individual national economies are threatened by the global market, they demand the protection (redistribution) of the state and/or bloc economy, at the same time as appealing to national cultural identity [as indeed has occurred throughout the world since 2008—JDW]. So it is that any counteraction to capital must be one targeted against the state and nation (community).[53]

---

53 Karatani, *Transcritique*, 15.

Under these circumstances, it may well appear that we are stuck. However, Karatani finds a solution in simultaneous movement "from above and below," that is, immanent and ex-scendent struggle, for simultaneous world revolution.

Let us return again to the immanent movement from "below," or associationism. Here, the fundamental task is "to create a form of production and consumption that exists outside the circuit of M-C-M'," and the model is of the combined consumers/producers cooperative. The aim is to organize society around such cooperatives in an "association of free and equal producers," to repeat one of Marx's expressions for communism, without treating labor-power as a commodity. Building such relations requires uprooting capital's value form and money. Thus, to facilitate the association of free and equal producers, Karatani reasons, we must "establish a financial system (or a system of payment/settlement) based on a currency that does not turn to capital, namely, that does not involve interest."[54] In other words, we must be able to conduct exchange with a means other than money as we know it. Yet this is unlikely to occur soon. So where does this leave us?

As I understand Karatani, for world revolution to be possible, two conditions must be met. First, we would need to see the generalization of free and reciprocal exchanges, which, for Karatani, implies the return of nomadism.[55] Indeed, Karatani writes that "communism depends less on shared ownership of the means of production than on the return of nomadism."[56] Putting aside the challenges of reforming society along nomadic lines, again, if we were to successfully reorganize numerous communities for degrowth communism and Climate X—through movements from below for dignity and autonomy—the capitalist nation-state would attack.

This leads to the necessity of the second condition, namely the creation of an entirely reformed United Nations, a movement "from above" that would facilitate the consolidation and generalization of the movements "from below." Such a movement would mean the realization of Kant's argument for a World Republic of Nations. "Kant located the way

---

54 Karatani, *Transcritique*, 297.
55 In Karatani's terms, economic life organized by mode of exchange D.
56 Karatani, *Structure of World History*, xii.

to perpetual peace not in a world state but in a federation of nations."⁵⁷ Perhaps we could say that only movements from below could provide a basis for the World Republic; yet only a World Republic would protect the many radical experimental communities from being destroyed by state and capital. Hence the two elements of the Kingdom of Ends, the "above" of a radically reformed UN and the "below" of a multitude of free and reciprocal communities, must emerge "simultaneously." Ergo simultaneous world revolution.

However impossible this may seem, it is urgent and necessary. But what could impel people all around the world to unite toward such radical ends? One answer is that the coming cataclysms, the old threat of nature, could trigger a change in human consciousness. For instance, writing in the aftermath of the March 11, 2011, Tōhoku earthquake and tsunami—an event that raised fundamental questions about the socio-ecological form of Japan, generating major protests—Karatani wrote:

> Global capitalism will no doubt become unsustainable in twenty or thirty years. The end of capitalism, however, is not the end of human life. Even without capitalist economic development or competition, people are able to live. Or rather, it is only then that people will, for the first time, truly be able to live. Of course, the capitalist economy will not simply come to an end. Resisting such an outcome, the great powers will no doubt continue to fight over natural resources and markets. Yet I believe that the Japanese should never again choose such a path. Without the recent earthquake, Japan would no doubt have continued its hollow struggle for great power status, but such a dream is now unthinkable and should be abandoned. It is not Japan's demise that the earthquake has produced, but rather the possibility of its rebirth. It may be that only amid the ruins can people gain the courage to stride down a new path.⁵⁸

It may be. There are no guarantees of progress, no certainties apropos human nature, and no telos in natural history.⁵⁹

---

57 Karatani, *Structure of World History*, 302.
58 Kōjin Karatani, "How catastrophe heralds a new Japan," *Counterpunch*, March 24, 2011.
59 Marx thus provides a new answer to the old question of philosophical history. Kant proposed that humanity is directed (as it were by nature), independent of

Some outcomes are more likely than others. In my estimation, this strategy is a long shot. It will not be easy.[60] Predictions are fraught, but I believe that the most likely medium-term course of events is world war followed by the consolidation of a Climate Leviathan path.[61]

After observing that *Capital* examines the capitalist economy from the standpoint of natural history, Karatani notes that Marx's great work "only presents the necessity of crisis and not of revolution, and that revolution is a practical problem. And this 'practical,' I insist, must be interpreted in the Kantian sense: That the movement against capitalism is an ethical and moral one."[62] I agree. Our recognition of the necessity of crisis, coupled with the absence of telos, compels an ethical and moral movement against capitalism. Then the chief virtue of Marxian natural history lies not in epistemology, but in ethics, in our relationship to ourselves. To put it otherwise, Marxian natural history holds out the

---

individual will, toward the advance of collective interests. He called this the "cunning of nature," because it was as if nature directed humanity—without its rational understanding or even its awareness—toward an end. What drives the progress for Kant is what he called humanity's "social unsociality," the paradoxical character of human nature, whereby we express our social being through a drive to individuate ourselves (hence act "unsocially"), and this very drive solicits socialization. *Pace* Kant, Hegel proposed that history would unfold by the "cunning of reason," driven by human passion toward the development of spirit qua reason. The ostensibly apolitical, pessimistic assumption made by Kant (like Hobbes before him) concerning human nature implies potentially progressive political futures: Their "realist" presuppositions are "idealist" speculations, written in the future anterior tense. Hegel was more positive about the fate of human reason. By contrast, Marx—after Darwin—provides neither certainty about human nature nor assurances of historical progress. Humanity is "directed" by natural history via class struggle without telos. If we wished we could call Marx's position "cunning of natural history," but this would be, perhaps, too cunning.

60   Mann, *Disassembly Required*, chapter 7, passim.

61   One of the many curious qualities of Fukuyama's *The End of History and the Last Man* (1992) is its somber conclusion that "[t]he end of history will be a very sad time," with little for humans to do apart from "economic calculation, the endless solving of technical problems, . . . the perpetual caretaking of the museum of human history." Fukuyama's melancholic prediction is faithful to the right-Hegelian tradition he inherited. How fortunate that we will be spared this sadness. For what we face is not boring labor in a museum of human history, but the excitement of breaking out of the museum of natural history in which we find ourselves.

62   Karatani, *Transcritique*, 287. Karatani attributes the recognition that *Capital* promises only crisis to his teacher, Kōzō Uno: See Kōzō Uno, *Theory of Crisis*, trans. K. Kawashima and S. Brown (Leiden: Brill, 2022).

possibility of catharsis—or at least the specific sort of catharsis that allows us to see what we truly are and thereby grapple with the prospect of an end to this form of society that induces an endless repetition of violence and persistent separation from being on Earth. Although natural history makes no guarantee of collective emancipation (or any other end), it nevertheless provides a rational ground for humanity to transcend its present form so we may realize ourselves as free social beings. For that is how I understand communism, as fully developed naturalism = fully developed humanism. Then, the final word should go to Marx, jotting enthusiastically in his Paris manuscripts:

> *Communism* as the *positive* transcendence of *private property* as *human self-estrangement*, and therefore as the real *appropriation* of the *human* essence by and for man; communism therefore as the complete return of man to himself as a *social* (i.e., human) being—a return accomplished consciously and embracing the entire wealth of previous development. This communism, as fully developed naturalism, equals humanism, and as fully developed humanism equals naturalism; it is the *genuine* resolution of the conflict between man and nature and between man and man—the true resolution of the strife between existence and essence, between objectification and self-confirmation, between freedom and necessity, between the individual and the species.[63]

Here is the inspirational vantage that Marx arrived at early in life, as a Left Hegelian, one that could only be realized as his standpoint in *Capital* after reading Darwin, whose breakthrough allowed Marx to appreciate that there is no telos in natural history nor guarantees in life. Embracing these truths will help us to clear a path out of our planetary crisis toward Climate X.

---

63   Marx, Paris MS III, §*ad* XXXIX; MECW 3, 296; New Left Books/Penguin, 348.

A Marx-Darwin Timeline

A Timeline of Key Dates in the Lives of Darwin and Marx

July 1844: Darwin begins sketching his "Essay" on natural selection

1838: Darwin initially conceives and sketches his theory of natural selection in notebooks

1831–36: Darwin's voyage on the *Beagle*

February 12, 1809: Darwin born in Shrewsbury

1826: Darwin takes taxidermy lessons from John Edmonstone

May 5, 1818: Marx born in Trier

1836: Marx enrolls as law student in Berlin, studies Hegel

1841: Marx successfully defends his PhD in absentia

Late 1844: Marx begins writing his Paris manuscripts (unpublished until 1932)

Summer 1846: Marx and Eng[els] complete *The German Ideolo[gy]*; they find no publis[her] (unpublished until 19[32])

April 19, 1882: Charles Darwin dies in Down House

October 12, 1880: Darwin rejects offer of dedication of a book, thought to be *Capital Volume II*, but in fact a work by Edward Aveling

1864: Spencer coins the expression "survival of the fittest" in his book *Principles of Biology*

1863: Evidence supporting Darwin flows: Huxley's *Evidence as to Man's Place in Nature*, for instance

1868: Darwin publishes *The Variation of Animals and Plants under Domestication*

Darwin's **1862** *Fertilisation of Orchids*

1871: Darwin publishes *The Descent of Man*

November 24, 1859: publication of *On the Origin of Species* by Darwin

1872: Darwin publishes *The Expression of the Emotions in Man and Animals*

June 18, 1858: While writing on natural selection, Darwin receives paper from Alfred Russel Wallace describing natural selection

October 1873: Darwin writes to Marx: "I thank you for the honour which you have done me by sending me your great work on Capital"

|1850|1860|1870|1880|

1849: Marx and his family move to London

September 16, 1867: Marx publishes *Capital Volume I*

1848: Marx and Engels write *The Communist Manifesto*; revolutionary wave across Europe

November–December 1860: Marx reads *On the Origin of Species*

1872: Marx publishes 2nd German edition of *Capital Volume I* (version quoted in this book)

March 1851: Marx uses "metabolism" several times in his London notebooks after reading Roland Daniel's work

December 1860: Marx writes to Engels that *Origin* is "the book which contains the natural-historical foundation of our outlook"

1873: Engels begins preparatory notes for the study that came to be called *Dialectic of Nature*

1856–7: Marx begins writing his critique of capitalist political economy in notebooks that became known as *Grundrisse*

June 1862: Marx writes Engels attacking Darwin for "applying the social Victorian model to nature"

1870–82: Marx spends his final years writing extensive notes on world history, natural science, ethnology and related topics

March 14, 1883: Karl Marx dies in London

Index

A
absolute historicism, 7, 7n7, 256, 256n18
Absolute Spirit, 17
abstraction, 17, 18, 34, 35, 38, 118, 118n11
accelerationism, 293–5
accumulation
    capital, 95, 136, 106, 121, 123, 136, 143, 149–53, 162, 166, 166, 168, 174, 176–7, 182, 195, 206, 208, 210, 214, 223–4, 238, 286, 291, 307, 308, 310
    Darwin's metaphorical use of, 159
    see original (so-called), chapter 6 passim
adaptation
    Adorno and, 139
    capitalism and, 3, 160, 178, 196, 212, 255
    capitalism as, 140, 289–93
    Darwin on, 3, 53–5, 60–2, 69, 75, 88, 113, 196, 292

ADHD, 270–1
Adorno, Theodor
    and natural history, 279–83
    Minima Moralia, 139
Akimoto, Yusuke, 84n2, 97–8
alienation, critique of, 26, 114, 131–2, 229
Althusser, Louis, 24n19, 35, 107–8
Ancient Society (Morgan), 104–5
Anderson, James, *Enquiry into the Nature of the Corn-Laws*, 157
Anderson, Kevin, 105–6, 141n58, 188n97, 189n101, 230n64
animals (non human), 22, 28, 54n30, 113, 125, 126, 129, 140n57, 159, 166, 265–72, 295–9, *see also* bird, cat, crow, gorilla, horse, pigeon
Anthropocene, 289n5
"Après moi le déluge!," 293
Aristotle, 8n8, 31n37, 32–3, 68, 98n36, 193
artificial intelligence (AI), 138
artificial selection, 58, 69, 131

associationism, 305, 307, 310, 313
Autobiography (Darwin), 45, 46, 47, 74
Aveling, Edward, 100

B
Bailey, Samuel, 164–5, 167, 169
   Critical Dissertation on the Nature, Measures, and Causes of Value, 164–5
base and superstructure metaphor, 103n49, 254, 255, 308, 312
HMS *Beagle*, 43, 46n12, 47–9, 53n29, 63n56, 69, 79n95
Beck, Naomi, 13
bees, 166, 266–8, 271
Benjamin, Walter, 305n36
Der Beobachter (journal), 186
birds, 46, 62, 76, 80, *see also* crow, pigeon
Blanqui, Louis, 106
Bober, M. M., 102–3n48
Bottomore, Tom, 24n19
bourgeoisie, Marx on, 39, 133–4, 178n74, 253, 306
bourgeoisie, revolutionary character of, 133–7
Brenner, Robert, 200n7, 208–11
British Library, 89–90
British Museum, 89–90
Brown, Wendy, 98n36
Burkett, Paul, 4, 87n8, 112n4, 132n44, 169

C
capital
   accumulation, *see* accumulation, capital
   emergence of in rural England, 199–212
   formula for, 116n6, 119–22
   historicizing, chapter 6 *passim*
   mutuality of nation, state, and, 312–13
   valorization of, 114, 154, 160–2, 227, 274, 301–2
Capital: A Critique of Political Economy (Marx)
   on capital as a distinctive social form, 10, 88, 92–4, 98, 108, 113–14, 286, 289, 294
   fetishism in, 123n20, 163–82, 250, 273
   general law of capitalist accumulation in, 149–55
   horse in, 267n42
   human consciousness, 265–75
   material exchange, *see* material exchange
   natural-historical standpoint and, 4, 9, 23, 26, 28, 69, 88, 90–9, 182–4, 188n96, 189, 218, 249, 250, 258, 275, 280, 308, 315, 316
   object of analysis in, 5, 91, 92n20, 94, 95, 114, 151, 166, 173, 203
   political ambiguity of, 258, 301
   proto-psychoanalytic analysis, *Capital* as, 272–5
   socio-economic formations, 177–81
   theory of surplus population, 147–62
   value theory, 152, 163–74, 176–7, 181–2, 221–4, 238, 272, 307, 309
   writing of, 88–90

Index

capitalism
 as adaptation, 289–93
 criticisms of conservative and liberal histories of, 197–9
 emergence of in *Capital*, 233–7
 emergence of in *Grundrisse*, 230–3
 money in, *see* money
 natural history of, chapter 6 *passim*
 naturalization of, 195
 qua development, 286
 as a socio-economic formation, 10, 88, 92–4, 98, 108, 113–4, 286, 289, 294
capitalist development, *see* capitalism qua development
capital-labor relation, 119n12, 153, 12, 236
capital-nation state, 311–12
carbon emissions, 178, 213n29, 287, 307
Carnegie, Andrew, 75
cat, 67, 127, 129, 166–7
catastrophism, 293–5
cell theory, 118–n11
Chapman, Robert, 278n48, 271n49
China, 152n11, 188n98, 206, 260, 286, 296, 298, 300, 312
Christianity, 22, 36–40, 71, 71n74, 81, 248, 252n11, 281
class struggle, 21–2, 86, 93, 98, 102, 142, 152n11, 153n12, 177, 253–4, 256, 315
Climate Behemoth, 5n4
climate change/crisis, 3, 5–7, 62, 138–9, 154, 179, 183, 216, 219–21, 247, 252, 291, *see also* planetary crisis, global heating

Climate Leviathan, 291–2, 295
Climate Leviathan (Wainwright and Mann), 5n4, 18n3, 247n3, 290, 301
Climate X, 301, 303, 303n34, 313, 316
cognitive psychology, 270–1
colonial policy, 106, 215–16, 236–7
commodification, 115, 123n20, 303n33
commodities, consumption of by proletarians, 203, 212–17, 222
commodity fetishism, *see* fetishism
communism
 degrowth, 305–7, 309, 313
 described, 16, 21, 40n52, 106, 123, 181, 253, 254
 ecological, 241, 303, 307
 emergence of, 189, 202, 229, 241, 291, 298, 299, 309–16
Communist Party of China (CPC), 296, 300
conspicuous consumption, 213–14
consumers, *see* commodities, consumption of by proletarians
consumption, 199–200, 212–16
 of commodities by proletarians, 203, 214–6, 222
Contribution to the Critique of Political Economy (Marx), 38n46, 66, 99n39, 111n2, 255n15
cooperative wealth, 306–7
Creation, 37, 41, 63n54, 65, 70n72, 244, 288
The Crisis of Democratic Capitalism (Wolf), 260–1
Critical Dissertation on the Nature,

Measures, and Causes of Value (Bailey), 164–5
"Critique of the Gotha Programme" (Marx), 189
crow (*Corvus*), 265, 265n39, 267, 272

D

Darwin, Charles. *See also Origin of Species* (Darwin)
- achievements of, 3, 7, chapter 2 *passim*, 68
- on adaptation, 3, 53–5, 60–2, 69, 75, 88, 113, 196, 292
- *Autobiography*, 45n7, 46–7, 55n34, 63n56, 74
- HMS *Beagle* and, 43, 46n12, 47–9, 53n29, 63n56, 69, 79n95
- biogeography and, 69, 218–19
- *Capital* dedication myth, 100–1, 284n83
- *Capital*, quotations of in, 140–6
- critiques of from the Right, 10, 283n81, 287
- critiques of from the Left, 10, 88n10, 287–9
- death of, 77, 89n13
- *The Descent of Man, and Selection in Relation to Sex* (Darwin), 44, 72, 88–9, 260, *see also* Darwin and human evolution
- education of and influences on, 45–9, 60
- *The Expression of Emotions in Man and Animals*, 72, 260
- on extinction, 51, 81–3
- on the fossil record, 60–1, 69n68, 185, 244
- health of, 49n21
- human evolution, and 10, 13, 44, 46n11, 72, 88–9, 93, 186, 260n25, see *also Descent of Man*
- *Journal of Researches into the Geology and Natural History of the Various Countries Visited by HMS Beagle*, 47n14, 48
- London and, 43, 49, 54, 56n37, 57, 73n80, 82
- *Natural Selection*, 56
- *On the Origin of Species* (Darwin), 6, 8n8, 9n11, 11, 13n21, 13n22, 14, 29n30, 35n43, 43n1, 44–8, 51–83 *passim*, 85–90, 93, 96, 98, 102n48, 111–13, 122, 126, 131, 140, 144–7, 150, 156–9, 182, 184, 186, 193–4, 209, 218, 230, 244, 250, 282, 289–90, 299, 301
- religion and, 10, 13, 40, 43, 46n8, 63n53, 65–6, 70–2
- on species *and* speciation, 10, chapter 2 *passim*, 87–8, 93, 113, 131, 140, 142, 145, 150, 158–9, 195–6, 199, 212, 218–19, 229–30, 244, 280, 292
- on 'survival of the fittest', 59, 75, 77
- teleology, critique of, 9n11, 17n2, 64–70
- theory of descent of species by natural selection, chapter 2 *passim*
- tree of speciation and "tree" sketch, 51–2, 82, 233
- unpublished writings (1837–44), 45, 49–56, 58
- use of evidence from paleontology, 69, 219
- on variation and varieties, 51, 57–9, 61, 64

Darwin, Erasmus, 48, 58
A Darwinian Left (Singer), 289-2
Darwinian liberalism, 289-93
Darwin-inspired approaches to human morality, 250, 265n38
Darwinism, 10, 70-1, 73n83, 77, 100, 250, 283n81, 290, *see also* social Darwinism *and* Darwinian liberalism
Darwin Online Correspondence Project, 47-8n16
Davis, Mike, 89n12, 112n4
dead labor, 140, 273n54, 274
dedication myth, *Capital,* 100-1, 100n43, 284n83
deforestation, 239, 240
degrowth communism, 305-7, 309, 313
de Lamarck, Jean-Baptiste, 53, 54
The Descent of Man, and Selection in Relation to Sex (Darwin), 44, 72, 88-9, 260
Desmond, Adrian, 46n9, 46n11, 55n34, 78n93
determinism, 10, 35, 143-6, 219, 252, 256
development, concept of, 17n7, 19, 19n7, 20-3, 29, 36, 37, 41, 53n29, 66, 75, 91, 93, 102, 103, 124, 157, 177, 206n19, 221, 232, 254, 291, 308
dialectical materialism, 22, 92, 103, 112n5, 233, 262n30, 275, 283
dialectic of nature, 275-84
Dialectics of Nature (Engels), 276-7
division of labor, 137, 140n57, 209, 217, 243n87, 306

E
Earth, as translation of *die Erde,* 142n60, 303n33
earthiness and Marx's conceptual language, 38, 161n33, 172n59
ecological communism, 241, 295, 303, 305, 307
ecological Marxism, 4, 88n11, 107, 112, 130, 169n52, 200-1, 213n29, 238, 283, 300
ecological socialism, *see* ecological communism
ecology, Haeckel on, 80, 88n11
Economic and Philosophic Manuscripts (Marx), 21-31
Edmonstone, John, 46
"1844 Essay" (Darwin), 29, 54-5, 58
Einstein, Albert, 79n97
elite goods, 213
Empedocles of Akragas, 64
The End of History and the Last Man (Fukuyama), 315n61
Engels, Friedrich
  Anti-Dühring, 296-7
  critiques of, 283n80, 295-300
  Dialectics of Nature, 276-7
  The German Ideology (Marx and Engels), 23n18, 30-1, 36-40, 99n39, 108, 113n1, 188n98, 251n10, 266n41, 279, 310n48, 311n52
  letter to Kautsky, 282n79
  letters from Marx to, 13n21, 85n5, 103, 140n57, 185-7, 243n87
  letter to Marx on Darwin, 79n96
  letter to Zasulich, 207n19
  Manifesto of the Communist Party (Marx and Engels), 23n18, 85, 93n25, 133, 253, 300

Origin of the Family, Private Property, and the State, 104–5
The Part Played by Labour in the Transition from Ape to Man, 295–6
speech at Marx's graveside, 102–5, 187, 298
Enlightenment, 8, 15–16, 29n31, 37, 67, 198, 246, 248, 261n29, 296
Enquiry into the Nature of the Corn-Laws (Anderson), 157
Entwicklung, 93n23
environmental determinism, 219, see also determinism
epigenetics, 44, 88n9
eschatology, 249n8, 300
Essay on the Principle of Population (Malthus), 74, 76
The Essence of Christianity (Feuerbach), 22
eudaimonia, 33
Everyman's Library, 92n19
evolution
  Darwin's theory of, *see* Darwin, theory of descent of species by natural selection
  Darwin's wariness concerning the word, 53
  etymology and meanings of, 53–4
  evolutionists, 65, 70–1
  of Homo sapiens, 6, 10, 13, 41, 72, 76, 117, 128, 157, 161, 195, 198, 279
  Marx on, 93n23, 107, 161, 186, 218, 221
  theory of, by writers other than Darwin, 29n30, 41, 44, 56, 61n50, 63n55, 64–5, 68, 73n83, 75n85, 82n102, 264, 292
Marx-Darwin exchange of letters, 100–1
exchange-value, 118, 164, 165, 170n54, 174, 181, 215, 224
expanded accumulation, 162, 168, 177, 182, 208, 214, 291, 310
The Expression of Emotions in Man and Animals (Darwin), 70, 260
extinction, Darwin on, 51, 81–3
Extinction Rebellion, 6, 306

F
Feenberg, Andrew, 33n39
Ferri, Enrico, 282n79
fetishism, 123n20, 163–82, 250, 273
Feuerbach, Ludwig, 31–4
  The Essence of Christianity 22
Fick, Heinrick, 78, 78n94
First Principles (Spencer), 75n85
fitness, 61–2, 62n51
FitzRoy, Robert, 47
forces of production, 200n7, 209, 252, 254–5
form of society, *see* social form/formation
form of value, 94, 165n41, 167–70, 298
fossil fascism, 5n4
fossil fuels, burning of, 174n62, 179, 225, 286, 309
Foster, John Bellamy, 4, 12n19, 130n38, 276n61, 277n64, 281n76
Franklin, Benjamin, 125, 164
Fraser, Nancy, 118n10, 221n42, 318

Index 327

free labor, 117, 230–1
Freud, Sigmund, 11, 41n53, 103n49, 261–4, 273, 282, 288, 294, 299
Fridays for Future, 6
Fukuyama, Francis, *The End of History and the Last Man*, 315n61

G
Garfinkel, Alan, 68–9
Geist, 17–9, 22, 35, 243
gene flow, 63
generatio aequivoca, 41
genetics, 44, 62–3, 88n9, 146
genetic drift, 63
genetic mutation, 63
genetic variation, 63
Geras, Norman, *Marx and Human Nature*, 35, 259n23, 260n25
The German Ideology (Marx and Engels), 23n18, 30–1, 36–40, 99n39, 108, 113n1, 188n98, 251n10, 266n41, 279, 310n48, 311n52
global heating, 5, 6, 27, 227, 291
  See also climate change/crisis
Google's ngrams, 175n66
gorilla, 72–3, 233n74
Gould, Stephen Jay, 44n2, 47n15, 53n27, 60n47, 64, 284n83
Gramsci, Antonio, 7, 39n48, 184n85, 246n1, 255–6, 264, 279–83, 307n43
Gray, Asa, 67
Great Cultural Revolution, 312
Greenland ice sheet, 225n52
Grundrisse (Marx), 22, 56n36, 86n7, 92n21, 108, 111, 115–8, 127, 134–5, 140, 148–9, 156–7, 162, 184, 228n60, 230–4

H
Haeckel, Ernst, 80, 88n11
Harvey, David, 112n3, 141n58, 156n18, 300n30
Hegel, Georg Wilhelm Friedrich,
  background and achievements of, 17–23
  on colonialism, 215–6, 300
  critiques of, by Marx, 6, 12–3, 21–3, 28–9, 30–1, 36–42, 83, 96–9, 202, 232, 243n87, 298
  on "cunning of reason," 314n59
  on Greeks, 219n34
  influence on Marx of, chapter 1 *passim*, 193, 201, 302
  on market society, 298
  on nature, 18–23
  object of analysis for, 17, 36–7, 66, 134, 232
  The Phenomenology of Spirit, 18, 20n11, 243n87
  The Philosophy of History, 18–21, 23, 36–8, 197n4, 217n34
  Philosophy of Right, 18, 216n32
  Science of Logic, 18, 148n3
  on theodicy, 37–8
Heidegger, Martin, 226n57, 282n78, 283n81
Henslow, John Stevens, 46n12
Herschel, J., *Introduction to the Study of Natural Philosophy*, 47
Heyer, Paul, 99n40, 158, 297n21
HH×HN, 27–8, 31, 34, 36, 38–40, 95, 123, 127, 139, 143, 160,

169, 175, 178, 182, 197, 200–1, 213, 216, 226–7, 238, 249–50, 263
historical causality, Marx's four hypotheses on, 250–60
historical materialism, concept of, 7n7, 282, 288
Historical Materialism, annual conference, 89n14
historicist naturalism, 66–7, *see also* naturalism
history, *see* natural history, philosophy of history
  "antithesis of nature and history," 38–42, 240, 280
  "earthly" basis for (Marx), 30–1
  original (Hegel), 19
  philosophical (Hegel), 19–21, 36–7, 314n59
  reflective (Hegel), 19
History and Class Consciousness (Lukács), 276–81
hoarding, 136, 167, 173–4, 213
Hobbes, Thomas, 79n96, 140n57, 243n87, 250, 261
Hobsbawm, Eric, 70–1, 73, 76, 85, 134n46, 300n30
Hodgskin, Thomas, 158–9
Homo sapiens, 3, 6, 76, 160, 182, 198, 224, 240, 259
horse, in *Capital*, 267n42
human body, 31n36, 39, 67, 114, 128, 172n59
human consciousness
  Darwin on, 81, 88n9
  Gramsci on, 264
  Hegel on, 17, 20, 21, 37,
  human labor and, 272–4

  Marx on, 22, 28–9, 33, 41, 96, 104, 170, 188, 194, 254, 264–75, 302,
  metabolism and, 26, 28–9, 98, 265–75
human-human relations, *see* social relations (HH) *and* social form/formation
human labor, *see* living labor *and* labor power commodity *and* metabolizing of nature
human nature, concept of, 40–1, 259–64, 288, 290, 314–5
  in *Capital*, 127–32
  Gramsci on, 264
  and praxis, 31–6
  repressed by Left thinkers, 259n23, 259n24
human-human relations (HH), 26–8, 36, 38–40, 160, 259, *see also* HH×HN
human-nature relations (HN), 26–8, 31–40, 95, 122–5, 126n28, 127–32, 139, 143, 167–71, 175, 178, 182–3, 187, 197, 200–1, 213, 217, 226, 227, 237–8, 249–50, 261, 265, *see also* material exchange
human relationships with nature, *see* human-nature relations (HN)
human society, capitalist form of, *see* capital *and* social form/formation
Humboldt, Alexander von, *Personal Narrative*, 46–7
Hutton, James, *Theory of the Earth*, 81–2, 132n43,

I

ice melt, 224n50, 225
"Idea for a Universal History from a Cosmopolitan Point of View" (Kant), 65–6
idealism, of Hegel, 18–22, 104n50, 281
indigenous people, 108, 199, 205, 249, 288–9, 292
Introduction to the Study of Natural Philosophy (Herschel), 47
Invisible Committee, 288–9

J

Jameson, Fredric, 103n49, 112n3, 143–4, 243–4, 274n57, 280, 281n77, 283, 284n83
Jenyns, Leonard, 47n12
Jessop, Bob, 118–19n11
Johnson, Curtis, 50n24, 55n34, 59n46, 63–4, 81n101
Journal of Researches into the Geology and Natural History of the Various Countries Visited by HMS Beagle (Darwin), 47n14, 48
Just Stop Oil, 6

K

Kant, Immanuel, 18n4, 19n7, 65–6, 261n29, 287, 289, 310, 313–5
  "Idea for a Universal History from a Cosmopolitan Point of View," 19n7, 65–6
Kant Society, 279
Karatani, Kōjin, 9, 64, 65n60, 94, 103n49, 112n3, 122n17, 171n57, 174–5, 277, 283, 286n1, 298n25, 308n44, 310–15

Kingdom of Ends, 314
Krätke, Michael, 107n59, 189n99, 242n84

L

labor power commodity (C_{LP}), 108, 113–22, 127–9, 167n45, 173–4, 181, 200, 202, 204, 210–11, 213–4, 304
labor struggles, *see* class struggle
Lamarck, Jean-Baptiste de, 53, 54n30, 75n85
Lassalle, Ferdinand, 86, 98n38
law of identity or unity of opposites, 277, 277n62
law of the negation of the negation, 277, 277n62
law of transformation of quantity into quality, 277, 277n62
Lenin, Vladimir, 106, 234n74
  State and Revolution, 106n56
Leninism, 6, 16, 150n10, 183, 229, 309
letters, exchange of by Marx and Darwin, 100–1
liberalism and liberal ideology, 5, 10, 16, 49, 71, 81, 137–8, 186–7, 199, 220n38, 248–9, 260–1, 286, 289–93, 308, 312
Liebig, Justus von, 12, 13, 130n40, 238
life activity, 25, 28, 104, 268n43
life forms (*and* forms of life), 52, 58, 60–1, 64, 71, 81, 88n10, 244, *see also* species
living labor, 108, 115–16, 127–8, 134–5, 149, 158–9, 231, 234, 238, 265–75, 309
Locke, John, 106

London Pigeon Clubs, 57–8
Lukács, György, 276–81
   History and Class Consciousness, 276–81
Luxemburg, Rosa, 124n23, 215–6
luxury goods, 213, 222n44
Lyell, Charles, 48, 60, 77n91

M
MacGillivray, William, 45–6
machinery, 125n26, 126n28, 134–46, 156, 162, 175n67, 188n98, *see also* means of production, technology
Mackintosh, James, 45–6
Malm, Andreas, 5n4, 55n139, 112n4, 175n67, 220n40, 225n51, 272n50
Malthus, Thomas, 23, 72–9, 147, 149–50, 155–62, 183, 209, 226, 243n87
   Essay on the Principle of Population, 74, 76
Malthusian theory, 155–62
Manifesto of the Communist Party (Marx and Engels), 23n18, 85, 93n25, 133, 253, 300
Mann, Geoff, 5n4, 18n3, 148n3, 183n83, 223n46, 247n3, 290, 301, 304n35, 315n60
Maoism, 309, 312
market competition, 202–3, 208–12, 227
market society, concept of, 140, 195, 197, 204, 248–9, 298
Márkus, György, 30
Martins, Mauricio Vieira, 17n2, 46n8, 99n40, 292n12

Marx, Eleanor, 100
Marx, Karl, *see also Capital: A Critique of Political Economy* (Marx)
   base and superstructure metaphor, 103n49, 254, 255, 308, 312
   on bourgeoisie, 39, 133–4, 178n74, 253, 306
   on capital and capitalism, *see Capital* (Marx)
   on Christianity, 39–40
   compared with Freud, 261–4
   contributions to the Enlightenment tradition, 16
   Contribution to the Critique of Political Economy, 38n46, 66, 99n39, 111n2, 255n15
   "Critique of the Gotha Programme," 189
   on dead labor, 140, 273n54, 274
   death of, 89n13, 102–5
   Economic and Philosophic Manuscripts, 21–31
   on Feuerbach, 31–6
   on Fraas, 13n21, 123n19
   gendered language of, 21n13
   The German Ideology (Marx and Engels), 23n18, 30–1, 36–40, 99n39, 108, 113n1, 188n98, 251n10, 266n41, 279, 310n48, 311n52
   Grundrisse, 22, 56n36, 86n7, 92n21, 108, 111, 115–18, 127, 134–5, 140, 148–9, 156–7, 162, 184, 228n60, 230–4
   Hegel and, 13, 17–23, 31, 37–8, 96–9, 132, 156

on human labor, *see* living labor and labor power commodity
on labor, *see* living labor and labor power commodity
Manifesto of the Communist Party (Marx and Engels), 23n18, 85, 93n25, 133, 253, 300
on metabolism, *see* material exchange
on natural obstacles, 161n33
on natural selection, 126n28, 145, 160–1
negation of the negation, 98–9, 229, 277, 302–3
Paris writings, 15–31
"Preface to the *Critique of Political Economy*," 209, 254–5
standpoint in *Capital*, 88–101, 104–15, 127, 132, 138, 144, 173, 179, 182–4, 189, 194, 218, 249–50, 263, 275
on species *and* species-being, 22–6, 28, 31, 40n52, 93–4, 112, 125, 128–9, 132, 140n57, 142, 145, 157–60, 176, 197, 316
Stoffwechsel, *see* material exchange
on land and "the land," 11, 125, 130, 132, 178, 204, 207, 231, 256–7, 303n33
"Theses on Feuerbach," 31–6
on evolution of language, 161–2
on Wakefield, 161n32, 233–7
Marx and Human Nature (Geras), 35, 259n23, 260n25
Marxian natural history, 4, 7, 21n13, 23, 36, chapter 3 *passim*, 157, 160, 162, 167, 173, 179, 193–6, 220, 249–50, 259n24, 264, 275–6, 279, 282–4, 285, 287, 290, 303–4, 308, 315
mass consumption, *see* commodities, consumption of by proletarians
material exchange [Stoffwechsel], 12–3, 112, 122–32, 162, 163, 167–76, 265, 268, 269, 271, 274–5
materialism
Darwin's, 46, 65, 105n55, 141n58
dialectical, 22, 92, 103, 233, 262n30, 275, 278n65, 283
Feuerbach's, 31–4
historical, 7n7, 282, 288
new, 5n3, 272n50
praxis and, 33n38
Matthäi, Rudolph, 38n47
Maurizi, Marco, 265n40
Mayr, Ernst, 44n2, 55n34, 292
McCormick, Robert, 47n15
means of labor, 122–32, 135, 137, 146, 151, 169, 178, 180, 199, 203, 207, 211, 216, 231, 241
means of life, 25–6, 306
means of production, 117, 120–4, 135, 151, 153, 176, 180, 181, 203–5, 210–1, 234–5, 238–9, 241–2, 254–8, 274, 296–9, 303, 313
medieval Europe scenario, 180–1
Mendel's pea plant experiments, 44
Mercer, Kristin, 62n52, 63,55
merchants and merchant capital, 222–3, 234
Merrifield, Andy, 135n50, 142n59
metabolic exchange, *see* material exchange
metabolic break *and* metabolic rift, 130, 238–9, 240–1

metabolism [Stoffwechsel], *see* material exchange
metamorphoses of nature, Marx on, 242
Minima Moralia (Adorno), 139
modes of exchange, 9n12, 251, 312, 313n55
mode of production
  concept of, 66, 92, 124, 150, 168, 179, 204, 206, 207, 235, 252, 254, 257, 306
  relationship with social formation, 92, 228n60
modern population theory (Malthus), 149
modern synthesis, 13n21, 44, 73n83
monism, 5n3, 220n40
money, 113–7, 119–21, 123n20, 136, 163, 165, 168, 171n56, 173–7, 182, 202n10, 211, 224, 230, 272, 311, 313, *see also* hoarding
Monthly Review school, 4
Moore, James, 46n9, 46n11, 55n34, 78n93
Moore, Jason, 5n3, 201n9, 220n40
Morgan, Lewis Henry, *Ancient Society*, 104–5

N

nations *or* nationalism, 133, 154, 160, 163, 176n71, 178n74, 183, 188, 205, 241, 282, 292, 295, 310–4
natural disasters, 304
natural history
  as adjective (natural-historical), 85n5, 88n11, 91n16
  of capitalism, 108, chapter 6 *passim*
  Darwin and, 45–9
  described, 7–10
  etymology of, 8n8
  of humanity, *see* Darwin and human evolution
  Marxian, 4, 7, 21n13, 23, 36, chapter 3 *passim*, 157, 160, 162, 167, 173, 179, 193–6, 220, 249–50, 259n24, 264, 275–6, 279, 282–4, 285, 287, 290, 303–4, 308, 315
  material exchange, means of labor, and, 122–32
  as philosophy of history, 246–50
  of technology, 140–6
  value form, commodity fetishism, and, 163–81
Natural History Museum, 8n10, 84–9
naturalization, of capitalism, 115, 175, 195, 198, 248
natural law, 77, 140n57, 153, 262
natural selection
  defined by Darwin, 59, 70n69
  Engels on, 79n96, 102
  Marx on, 126n28, 145, 160–1
  theory of descent of species by natural selection, chapter 2 *passim*
  random chance and, 45, 52, 62–5, 67, 71
Natural Selection (Darwin), 56
nature
  "antithesis of nature and history," 38–42, 240, 280
  dialectic of, 275–83
  as Earth, *see* Earth
  as exchange-value, 123, 129, 170, 176
  Hegel on, 18–23
  metabolizing of, *see* material exchange

metamorphoses of, 242
rational planning *and* state regulation of, 133, 295–301, 306
relationships between humans and, 5n3, 7, 9, 27, 40, 95, 128, 143, 168, 174–5, 177, 182, 189, 200, 206, 217, 227, *see also* HH×HN
subsumption into history, 9, 135, 281
"true resurrection of," 27
naturalism, and Marxism, 27, 66, 241, 288, 316
Nazi Party, 11, 282
Neanderthal (*Homo neanderthalensis*), 72
negation-of-negation, 98–9, 229, 277, 302–3
New Introductory Lectures on Psychoanalysis (Freud), 11, 261–3
new materialism, 5n3, 272n50
Newton, Isaac, 70, 80, 89n13, 262
Nicomachean Ethics (Aristotle), 31n37
Nietzsche, Friedrich, 288

O
objective labor, exchange of living labor for, 134–5, 139, 169, 231n66
object of analysis, Marx's, 5, 22–4, 28–9, 31–2, 66, 91, 92n20, 93–5, 114, 151, 166, 173, 200, 203, 249
Occidentals, 288–1
omnia sunt communia (all wealth shall be common), 302
omnis vir enim sui (every man for himself), 302

On the Origin of Species by Means of Natural Selection, or the Preservation of Favoured Races in the Struggle for Life (Darwin), *see Origin of Species* (Darwin)
"On the Tendency of Varieties to Depart Indefinitely From the Original Type" (Wallace), 48n20
ontology, 33, 66, 220n40, 248, 264–5, 278–81, 308n45
original accumulation (so-called), 151, 199–200, 203–11, 216, 226–9, 234, 238–42
original history, 19
Origine et Transformations de l'Homme (Trémaux), 185–6
Origin of Species (Darwin), 6, 8n8, 9n11, 11, 13n21, 13n22, 14, 29n30, 35n43, 43n1, 44–8, 51–83 *passim*, 85–90, 93, 96, 98, 102n48, 111–3, 122, 126, 131, 140, 144–7, 150, 156–9, 182, 184, 186, 193–4, 209, 218, 230, 244, 250, 282, 289–90, 299, 301
Origin of the Family, Private Property, and the State (Engels), 105
overpopulation, *see* population, theories regarding human

P
Paul, Eden and Cedar, 92n19
Personal Narrative (Humboldt), 46–7
The Phenomenology of Spirit (Hegel), 18–20
Phillips, Leigh, 297n22, 307n42
philosophy of history, 4, 13, chapter 1 *passim*, 194, 246–59, 280

The Philosophy of History (Hegel, Lectures on), 18n5, 19–23, 36–7, 217n34
Philosophy of Right (Hegel), 18, 216n32
Physics (Aristotle), 64n59, 68
Physiocrats, 5n3, 142n60
pigeon, 57–8
planetary crisis, 5, 7n7, 74, 76, 82, 99, 112, 124, 138, 216, 238, 249, 263, 285–7, 290–1, 304, *see also* climate change/crisis
Plinian Society, 46
Pliocene, 224–6
Political Marxism, 2008, 208–12
population, biological, 54, 61n49, 62, 75
population, theories regarding human, 74, 79n96, 130, 137–8, 142, 146–62, 182, 297, 301
praxis
  absence of concept in *Capital*, 31, 94, 131–2
  ambiguity of, 33n38, 33n40
  human nature and, 31–6
  uses by Gramsci, 7n7, 256, 281
"Preface to the *Critique of Political Economy*" (Marx), 209, 254–5
price, 163, 165, 167n46, 168, 170n54, 207, 210, *see also* exchange-value
primitive accumulation, *see* accumulation, original (so-called)
producers, market competition between, 202–3, 208–12, 227
production
  forces of, 200n7, 209, 252, 254–5
  of use-values, 223, 268, 271, 275,

productivism, 297, 306, *see also* Prometheanism
proletarianization, 118n10, 152n11, 214, 271n45
proletarian-consumers, 118n10, 136, 140, 152, 203, 214–6, 222, 226, 308, 310
proletarians, consumption of commodities by, *see* commodities, consumption of by proletarians
Prometheanism, 132n44, 135, 297, 299n29
proto-psychoanalytic analysis, *Capital* as, 272–5
psychoanalysis, 11, 273

R
random chance, natural selection and, 45, 52, 62–5, 67, 71
reflective history, *see* philosophy of history
Reitter, Paul, 4n1, 12n20, 90–1, 127n30, 142n60, 187n95, 303n33
relational value, *see* value theory, *Capital*
relative surplus population, 152–5, 162, 183–4
religion, 15, 40, 41, 240, 247, 312
  Darwinism and, 46n8, 71, *see also* Darwin, on divine revelation
  Gramsci on, 256, 281
  Marx on, 37, 40, 41, 171n56, 176, *see also* fetishism
renewable electricity, 179
Ricardo, David, 23, 73–4, 94, 106, 115, 156n18, 164–5, 170n54, 215

right-wing ideology, 19, 11, 66, 73, 156, 176n71, 276, *see also* fascism
Robinson Crusoe scenario, 180–1
robotics, 138

S
Saitō, Kōhei, 4, 12–3, 112n4, 123n19, 127n30, 127n31, 130n39, 168n50, 189, 226n58, 228–9, 277, 100, 303, 305–7, 309
Sasaki Ryuji, 112n3, 127n30, 269–71
Sayers, Sean, 249–50, 259n23, 277n62
school strikes, 305, *see also* Thunberg, Greta
Science of Logic (Hegel), 18
Seccombe, Wally, 150n10
self-estrangement, 27–8, 37, 240–1, 316
serfdom, 180, 204, 231, 253
simple commodity production (C-M-C), 119, 121, 122n17, 222, 224
Singer, Peter, 289–90
  A Darwinian Left, 77n91, 289–90
Smith, Adam, 23, 47n16, 82n102, 94, 115, 140n57, 163–5, 169, 170n54, 197–8, 204, 206, 248
  The Wealth of Nations, 163
social Darwinism, 71–9, 95, 147, 250, 259n24, 282
social democracy, 283n80
social form/formation
  Althusser on, 24n19
  capital as a distinctive, 93–4, 98, 108, 113–22, 128,131, 146, 177, 181, 223, 273, 286, 289

competitive nature of, 178–9, 223
described, 92, 179–81, 228n60, 254
Marx on, 92n21, 93, 112, 140n57, 169, 170, 175, 180, 228n60, 275
toward a new, 128n33, 184, 258, 301–16
transition to capitalism as a, 202–16 *passim*, 219–21, 228, 241, 244
socialism, *see* communism
social reproduction, 117–8, 200, 213, 305, 308
social unsociality, Kant's conception of, 315n59
social welfare provisions, 74–5
socio-economic formations, *see* social form/formation
Der Sozialdemokrat (newspaper), 102
species *and* speciation, Darwin on, 10, chapter 2 *passim*, 87–8, 93, 113, 131, 140, 142, 145, 150, 158–9, 195–6, 199, 212, 218–9, 229–30, 244, 280, 292
species *and* species-being, Marx on, 22–6, 28, 31, 40n52, 93–4, 112, 125, 128–9, 132, 140n57, 142, 145, 157–60, 176, 197, 316
Spencer, Herbert, 59, 72–9, 282n79
  First Principles, 75n85
Spivak, Gayatri, 283
Stalin, Joseph, 277–8, 283
Stalinism, 259n24, 278, 283
state, the, 10, 75, 102, 106, 107, 113, 130n40, 133, 182, 202n10, 206, 216, 286, 295, 298, 299, 309–12
State and Revolution (Lenin), 106n56
Stedman Jones, Gareth, 14, 103–4
Stoffwechsel, *see* material exchange

strike, labor, 116, 302, 304–5, 309
The Structure of World History (Karatani), 174n64, 298n25, 308n44, 310–14
surplus population, 112, 147–55, 162, 183, 227, 258, 301
surplus value, 94, 116–7, 128, 142, 152n11, 162, 173, 176–7, 182, 200, 202–3, 208, 211–12, 214–15, 218, 221, 224
"survival of the fittest," 59, 75, 77

T
Taussig, Mick, 126n29
technology, 113–46 *passim*, 152, 162, 166, 178, 198, 209, 213, 252, 263, *see also* machinery, means of production
teleology, 9–10, 17–18, 31, 36–7, 45, 59, 64–6, 76, 81, 83, 84n2, 86, 97–9, 101, 103, 132, 184, 185n89, 189, 196–7, 199, 208, 218, 233, 243, 246, 248–52, 256, 283, 287, 292, 298, 306
Theodicaea, 36–8, 67
Theōria in Aristotle, 31n37, 33
Theory of the Earth (Hutton), 82, 132n43
"Theses on Feuerbach" (Marx), 31–6
Thesis VI (Marx), 34–6
Thomas, Paul, 259n23, 298–9
Thunberg, Greta, 305
Tōhoku earthquake and tsunami, 314
transcendent Being, 54–5
transgressive segregation, 58n41
Trémaux, Pierre, 185–7

U
unemployment, 15, 142, *see also* labor power commodity
uniformitarian school of geology, 60, 82
United Nations (UN)
 establishment of, 15
 reformation of, 313–4
 Universal Declaration of Human Rights (1948), 15–6, 154n15
United Workers' Party of Germany, 166–7, 306
urbanization and urban life, 57, 130, 160, 178, 239
use-value, 121, 128, 163, 167n46, 170, 172, 174, 181, 213, 224, 275, 268
USSR, 16, 283, 286, 300
utility, *see* use-value

V
valorization, 114–15, 154, 160–2, 227, 274, 301–2
value, *see* exchange-value, form of value, use-value, value theory
value theory
 in *Capital*, 152, 163–74, 176–7, 181–2, 221–4, 238, 272, 307, 309
 neo-physiocratic reformulation of, 5, 142n60
 Smithian-Ricardian labor theory of value, 106, 114, 163–5, 169
van Wyhe, John, 47n16
Varoufakis, Yanis, 197n2

W
wage, the, and wages, 78, 117, 119, 121, 122n17, 124, 128n33, 129,

136, 152–3, 164, 167, 173n61, 181, 203, 213–14, 223, 227, 230, 231n66, 233
Wainwright, Llama, 265n39
Wakefield, Edward Gibbon, 161n32, 233–7
Wallace, Alfred Russel, 48, 56, 58–9, 68
"On the Tendency of Varieties to Depart Indefinitely From the Original Type," 48–9n20
The Wealth of Nations (Smith), 163
Western gorilla, *see* gorilla
Williams, Howard, 17–18n2
Wills, Vanessa, 34n41, 259n23
Wolf, Martin, 260–1, 263
The Crisis of Democratic Capitalism, 260–1
Wood, Ellen Meiksins, 198–200, 208–9, 234, 248n4,

worker-consumers, *see* proletarian-consumers
world-history, 17n2, 37, 107–8, 188–9, 301, 305n36,
World Republic of Nations, 313–4
world revolution, 183, 286–7, 309–15
World War II, 89n13,24, 263
world war, threat of, 5, 286, 287, 315

Z
Zasulich, Vera, 189–90, 207n19, 301